WILLIAM HASTIE

WILLIAM HASTIE
Grace Under Pressure

GILBERT WARE

New York   Oxford
OXFORD UNIVERSITY PRESS
1984

Copyright © 1984 by Oxford University Press, Inc.

Library of Congress Cataloging in Publication Data
Ware, Gilbert.
William Hastie : grace under pressure.
Bibliography: p.   Includes index.
1. Hastie, William.   2. Afro-American judges—Biography.
Biography.   3. Civil rights workers—United States—
Biography.   I. Title.
KF373.H38W37   1984   347.73′2434 [B]   84-5657
ISBN 0-19-503298-5   347.3073334 [B]

Printing (last digit): 9 8 7 6 5 4 3 2

Printed in the United States of America

For
G. James Fleming
and
Benjamin Quarles

[William Hastie embodied] my highest hope for what education is all about: the development of trained and disciplined intelligence at the service of a mature sense of self, and a clear awareness of what is worth doing. We need such heroes for the simple reason that by them we may imagine what we may yet become.
John William Ward
Presidential Address
Amherst College Commencement
6 June 1976

Preface

She was a little girl, but she sang loudly. As the bus rolled along the streets of Manhattan, her anthem made another passenger think of William Henry Hastie. It had been his battle song.

The anthem was composed with black Americans in mind, but its universality was evident during the thirties and forties, when Hastie made history. The Great Depression, the New Deal, World War II, and the Red Scare highlighted the era. Cruelty, suffering, and turmoil were widespread, and when "Lift Every Voice and Sing" was sung one thought not only of the murder of blacks in America; one thought also of the colonization of natives in the Caribbean and of the massacre of Jews in Germany.

Hastie knew that the beastliness was cut from whole cloth, fascism. A graduate of Amherst College and Harvard Law School, he was a modest man of monumental achievement. As a lawyer in the Department of the Interior, he was the main architect of the Organic Act of 1936, which eased the Virgin Islands out of colonialism. As a federal judge there in 1937-39, he tutored those who chartered the course and led the way.

Before going to the Virgin Islands, Hastie had won lasting glory for his work with the New Negro Alliance in securing historic rights for pickets; with Howard University Law School in developing a cadre of black lawyers who became civil rights warriors; and with the NAACP in litigation that terminated the payment of racially discriminatory salaries to public school teachers; as well as in the first in a series of lawsuits that culminated in the ban on racial segregation in public schools in 1954.

Hastie resigned his judgeship in 1939 to return to the law school at Howard as its dean. He took leave in 1940 to serve as a civilian aide in the

War Department. Insofar as any person can be designated the catalyst for racial integration of the armed forces, he was that person. He viewed his opportunity and responsibility in the broadest sense, that is, to be fulfilled in order to hasten the end of fascism abroad as well as at home.

Hastie excelled as a lawyer of the last resort, one to whom others turned when all seemed lost. Resuming this role upon his return to Howard in 1943, he continued his running battle with the military services by trying to prevent the Army from carrying out its version of the Scottsboro case. But two other judicial encounters attest even more to his noteworthiness: One overturned the greatest obstacle to voting by blacks in the South, the white primary; the other invalidated a Virginia law that required segregation on interstate carriers.

It was a mark of Hastie's versatility that he also served as governor of the Virgin Islands and while in office played a role in securing the surprising triumph of the Democratic presidential candidate in 1948. The next year, he was nominated to the federal appellate bench. The Superpatriots—those practitioners of virulent anticommunism called McCarthyism—again tried to block his path. They disapproved of his constant civil rights advocacy, but it was fashionable to try to discredit militant blacks by accusing them of being Reds or Pinks—fashionable and effective, especially among black leaders, who did everything possible to avoid having the stigma of being communist added to that of being black. Charges of criminal conduct were not necessary; accusations of association with communists sufficed to ruin the lives of persons far more protected than Hastie.

Given the social and political circumstances of his day, Hastie enjoyed success that was uncommon among Americans and unprecedented among Afro-Americans. He was an assistant solicitor in the Department of the Interior, a United States District Court judge, a governor of the Virgin Islands, and a United States Court of Appeals judge. Highly acclaimed, he never forgot how little it meant for him to be "the first" among black Americans as long as they were "the last" among all Americans. It is in the nature of things, especially regarding blacks, to publicize "first" in its sequential meaning. The word, however, has another meaning, one that stresses quality: To be "the first" is to be the earliest, but to be *first* is to be best—superlative in whatever one does.

The essence of Hastie's story is not that he was "the first" among any Americans, but that he was *first* among all Americans in waging a bittersweet war against bigotry.

Philadelphia   G.W.
March 1984

Acknowledgments

I am indebted to many who eased my way in tracing William Hastie's path through history. Judge A. Leon Higginbotham, Jr., suggested this biography to me; Brenda Susan Spears brought it to my publisher's attention; Susan Rabiner guided it through the vicissitudes of publication; and Henry Krawitz admirably copy-edited it. I am grateful to them and, for her exquisite editorial touch, to Naomi Schneider.

Had it not been for fellowships from the Ford Foundation and the American Council of Learned Societies, this would still be a work in progress. At every stage progress was made because librarians and archivists are among an author's best friends. These friends of mine include Richard A. Binder and Helen M. Kildea, Drexel University Library; June A. V. Lindqvist, Enid Baa Public Library; Denise D. Gletten, Maricia B. Bracey, and Clifford L. Muse, Jr., Moorland-Spingarn Research Center; Dr. Genna Rae McNeil, Howard University Law School; Jeane B. Williams, University of Pennsylvania Law School; Michele F. Pacifico, David Keply, Richard C. Crawford, and Reneé Jeausau, National Archives; John C. Broderick and C. F. W. Coker, Library of Congress; Judith A. Schiff, David Schoonover, Edward A. Jajko, and George Miles, Yale University Library; Erika Chadbourn and Harry S. Martin III, Harvard Law School Library; Darlene Holdsworth and John Lancaster, Amherst College Library; Elizabeth R. Mason, Oral History Research Office, Columbia University; Benedict K. Zobrist and George H. Curtis, Harry S. Truman Library; R. J. Booker, Carolyn G. Jones, Margaret R. Gaiter, and Joseph Hopkins, Beck Cultural Exchange Center.

The William Henry Hastie Papers at Harvard Law School Library were indispensable to me, as were the related collection at the Beck Cultural Exchange Center and the Howard University Law School records housed at the Howard University Archives. Individual as well as institutional sources were beneficial to me, and I am indebted to Georgianna McGuire, Dutton Ferguson, Charles Ray, Mary G. Hundley, and Pauli Murray for giving me access to their private files. I am also grateful to persons who affirmed the value of oral history when they shared with me their recollections about Judge Hastie's life and times. That I, in turn, am able to share with others much of what I have learned from all sources is a mark of the skill that Kathryn S. Campbell, Thomas E. Fry, Karen Melton, Veronica McLean, and Lettitia Dolores brought to the task of typing the manuscript. As from them, I received constant assistance and encouragement from Basil Phillips at Johnson Publications and Drs. Harriet F. Berger, Thomas L. Canavan, and Phillip V. Cannistraro at Drexel University.

Karen Hastie Williams and William H. Hastie, Jr., entrusted their father's papers to me for as long as I desired and made their good offices always available to me. I want to express my gratitude for their consideration. But how does one express the ineffable?

Contents

1. Black Diamond, 3
2. Amherst College, 12
3. Old Ironsides, 21
4. Harvard Law School, 29
5. Gideon's Band, 35
6. Teachers' Pay, 55
7. New Negro Alliance, 66
8. A Founding Father, 81
9. Stimson's Stables, 95
10. Domestic Enemies, 110
11. Invincible Man, 124
12. Call Dupont 6100, 142
13. A Fighter for Us, 175
14. Caribbean Outpost, 192
15. Truman's Rescuer, 213
16. The Greatest Need, 225
17. Epilogue, 242
    Notes, 244
    Bibliographical Essay, 293
    Index, 297

WILLIAM HASTIE

CHAPTER ONE
Black Diamond

One day an African princess, the next an American slave. That misfortune befell William Hastie's ancestor. Her ship sailed from Madagascar in 1794, bound for France, where she was to enter a convent school. English privateers intercepted the ship, captured the princess, and sold her into slavery in Philadelphia, Pennsylvania. In 1835 a man named Samuel Childs took seven of her descendants—all children, all slaves—to Alabama. There, in Marion, another descendant was born in 1867. Her name was Roberta Childs, and she was Hastie's mother. She attended Lincoln Normal Institute, which blacks had built and given to the American Missionary Association (AMA). She was taught by AMA members who had come from New England to educate the freedmen and indoctrinate them with such Christian values as brotherhood, diligence, thrift, and strict rectitude. She would tell her son how Ku Klux Klansmen attempted to destroy her school but were driven off by her four older brothers and other blacks. Her stories impressed upon him the notion that if he steadfastly fought racism, he could win.[1]

After completing her education at Fisk University and Talladega College, Roberta Childs became a teacher in Chattanooga, Tennessee.[2] There she met a young man who had studied mathematics at Ohio Wesleyan Academy and pharmacy at Howard University but had been unable to find work as either an actuary or a pharmacist. He did manage, however, to obtain the first appointment ever given to a black as a clerk in the United States Pension Office. He bore the name of his father, William Henry Hastie, who once had been a small boy called Willie.

Willie and his big basket had been a familiar sight on the waterfront in Mobile, Alabama. The basket contained food that his mother, with their master's approval, prepared and Willie sold to riverboat passengers. His sales enabled her to buy freedom for herself and him; the price was one thousand dollars. She and her son moved to Ohio, where she presided over two generations of her family, including the Pension Office clerk whom Roberta Childs was to meet and marry in Chattanooga. That young man was the grandson whom she had raised after the deaths of Willie and his wife. She instilled in him the dignity, self-confidence, and fierce antiracism that she personified. And he, in turn, joined his own wife in passing those qualities on to their only child, a son born in Knoxville, Tennessee, on 17 November 1904, the third William Henry Hastie.

Although blacks and whites often lived in the same neighborhood in Knoxville, they did not socialize. However, since children were allowed to mingle, young Bill Hastie found it possible to have white friends, provided they observed the rules of peaceful coexistence. As he explained, "Most of our neighbors were white. Their children soon learned that to me 'nigger' was a fighting word. So we became friends and playmates in mutual respect."[3]

Young Hastie's parents were his models of unrelenting resistance to racism. Rather than riding the segregated streetcar, they bought a horse and buggy for transportation to town (after they moved to the suburbs, where they owned a small chicken farm) and to his school, and only if their private means of transportation was unavailable and the distance involved was too great did they take the streetcar. When Roberta Hastie did not take her son to school by horse and buggy, he skipped and trotted or bicycled there. Expressing the family creed, she declared, "They can't Jim Crow us!"[4]

Roberta Hastie is remembered for her flaming racial pride. "She was a *race* woman," Etta Childs Walker, her niece, said. Another relative, Louise Russell Hall, recalled her dynamism and determination, saying, "She was a live wire. Civil rights–minded. Very strong character."[5]

Hastie's mother did not talk much about her interests and activities, but, as those around her noted, racial fair play was a passion in her life. Her other consuming interest, according to everyone who knew her, was the welfare of her son. No one questions Mrs. Walker's assertion that "she gave her life for him."[6]

Hastie senior was less demonstrative than his wife, which explains, in part, why even less is known about him than about her as a molder of their son's character. But perhaps no more need be known than this: When William Hastie died at the age of seventy-one, in his possession was a black tin box containing three letters that he had received from his father when he was eight years old.

In one of these letters, written in Washington, D.C., on 3 April 1913, Hastie's father told his son about having visited two men who were to be important in Hastie's life. They were Cousin Plug (William L. Houston, a black attorney in Washington) and his son, Charles Hamilton Houston, then one of three blacks among four hundred students at Amherst College, later a civil rights lawyer, and (as it turned out) young Hastie's prototype. In his letter of 16 April 1913, Hastie senior congratulated his son for high marks in school. He was disappointed, however, to see that in one month the mark for deportment had slipped from his son's usual 98 to 92. Saying that he was confident that his son would stop laughing during classes, he invoked a high ideal: "After all, no deeper satisfaction can come to us than the feeling that we have fought against wrong—and won."

Young William's reported improvement the next month drew an inspired response from his father on May 16. "Always aim high, never aim low," his father counseled. "If you aspire to lofty things, you have accomplished much even though you have not reached the topmost round. What most concerns your mother and myself is not how perfect you have been but how much your heart yearns for what is good and noble. Even should you sometimes meet with failure, we could fold you to our bosoms, kiss away your tears of disappointment and point the way by which you will finally succeed."

Sometimes his parents folded young Bill across their knees—he was no angel. Once he was part of a group that raided a neighbor's tomato patch. That, however, was not the reason he received his worst spanking ever. "What really made my father mad was the excuse I gave. I claimed I hadn't stolen the tomatoes but had just helped the other children to eat them." At times it was his energy rather than his excuses that unsettled his father. Preparing to go to work, Hastie senior often discovered a jato unit in his home. Corrective action was required for those mornings. "So we trained William . . . to sit in his little rocking chair, perfectly quiet, and still, and relaxed," Roberta Hastie said. "I think that is how he learned the calm and composure that people talk about." It was part of their raising him "by the book," says Mrs. J. Herman Daves, his childhood friend. Parents and son adhered to the principle that Hastie's father had explained to him in his letter of 16 May 1913. "Two things . . . should grow stronger in you each day: 1st a desire for well-doing; 2nd a willingness to be guided by your parents."[7]

Hastie's parents guided him with loving discipline. For both of them, a family friend recalls, "everything was William." Another friend, Robert C. Weaver, says that this was especially true of Hastie's mother.

"She made him the center of her life without enveloping him. . . . She was a very dominant force without being a dominating force in his life." She was strict and he was obedient, Weaver adds, "but I think President [Lyndon B.] Johnson would have said they 'reasoned together.'" Beautiful and gracious, Roberta Hastie was also intelligent, principled, energetic, and dignified. About her drive, a friend recalls: "She had it to the 'nth' degree." After the birth of her son, she discontinued teaching in school. Her favorite pupil was in her home.[8]

"We once thought of Bill as a bookworm," says John J. Johnson, his friend. "He wore glasses and studied a lot." Other kids also studied after school, but none as thoroughly as Hastie. His mother allowed him about an hour for games such as marbles. "Then he went home to her tutoring," Johnson says. William's friends at first misunderstood his regimen. Later they understood. His parents were preparing and pushing him to excel. He would be especially guided by his mother's standards, those of a woman of courage, dignity, and commitment to intellectual pursuits, says Weaver, who thinks that her singular attributes were passed on to her son through personal example and with salutary results: "Because she instilled in him the idea that he should be there, Bill would always be in the top ranks." When her husband was transferred to Washington, D.C., in 1916, Roberta Hastie was overjoyed because Dunbar High School, an institution that had produced, and would continue to produce, many outstanding black graduates, was located there.[9]

The formal education of black children in the District of Columbia dated back to 1807.[10] In that year, when there were just two schools for white children, three former slaves—George Bell, Nicholas Franklin, and Moses Liverpool—built one for black children. Known as the Bell School, the one-story frame building was named in honor of George Bell. It had a single teacher, a Mr. Lowe, and functioned for a few years before it was converted into a dwelling. Meanwhile, in 1809 an Englishman named Henry Potter established another black school, and in 1810 an Englishwoman, Mrs. Mary Billings, opened one in Georgetown, which was then a city separate from Washington. In the following decade, Mrs. Anne Marie Hall and the Resolute Beneficial Society, a black self-help organization, each started a school to educate blacks. The society's was the first night school in Washington for blacks. Its first teacher, Mr. Pierpoint, taught for three years and was succeeded in 1821 by John Adams, the second black person (Mrs. Hall was the first) to teach in Washington.

In the years that followed, blacks continued to build schools and employ teachers while paying taxes that helped to provide education for white people's children but not for their own. They received help from whites—usually but not always. In the pre–Civil War period, for example, whites and blacks operated many schools, but in that same period white rioters drove an educator, John F. Cook, out of Washington in 1835. The next year, however, Cook returned and reopened his school, Union Seminary. This was almost thirty years before public education was made available to black Washingtonians, the first public school for them opening in 1864, two years after the abolition of slavery in the District of Columbia.

John F. Cook, his sons John and George, John Prout, Henry Smothers, and Myrtilla Miner are several of the many persons who were pathfinders in black education in Washington and Georgetown. Another was William Syphax, the black who became chairman of the Board of Trustees of the Colored Public Schools of Washington and Georgetown in 1868, three years before the two cities were merged into a Government of the District of Columbia. (Prior to 1901, there were separate boards of education and superintendents for black and white schools.) Led by Syphax, blacks opened a high school in 1870, nine years before the opening of the first public high school for whites.

Financed by philanthropists in memory of Myrtilla Miner, who had established the Seminary for Colored Girls in 1851, the black high school was first located in the Reverend Henry Highland Garnet's Fifteenth Street Colored Presbyterian Church. It began with fifty seats, four students, one room, and no playground. Within a year its enrollment increased to forty-five students, most of whom were actually completing their last two years of grammar school. Subsequently, its location changed to increasingly larger quarters in public school buildings, the school having been placed under the authority of the black Board of Education. It took on the name of its location when it moved to M Street in the northwest. About twenty-five years later, the school made its final move and was given a name to match the magnificence of its new facilities: The arrival of the Hasties in Washington coincided with the opening of the new Paul Lawrence Dunbar High School.

Hastie's education in Washington began not at Dunbar High School but at Garnet Elementary School. In his second year a public outcry forced the Board of Education, for the first time, to allow blacks to participate in a reading proficiency contest. Garnet Elementary School entered its best readers. In the contest, Annette Hawkins, a fourth grader, read at an eleventh-grade level. As for Hastie, an eighth grader, Hawkins says, "Heaven only knows how far he read." The next year, 1917, Hastie entered Dunbar High School, which a friend, W. Montague Cobb, dubbed "two galaxies with many stars." The first "galaxy" consisted of teachers who were exceptionally qualified by virtue of their training. "They didn't take 'Education' in those days," Weaver, a Dunbar graduate, explains. "They learned some subject matter." And some of them taught it in memorable ways.[11] More than half a century later, Montague Cobb recalls the day that Clyde McDuffy spread across the blackboard a newspaper that headlined a battle then raging during World War II. McDuffy, who graduated Phi Beta Kappa from Williams College in three years and taught Latin at Dunbar, then said to his class, "Now, let's see how Caesar waged war on this terrain." McDuffy was one of the most demanding teachers, and under Dunbar's policy college-bound students were assigned to classes taught by these faculty members. The other Latin teacher was an able man, Cobb says, "but if you signed up for the go-to-college program, you got McDuffy." And according to Mercer Cook, students got more than brilliant instruction and discipline. "Clyde was the kind of teacher who would bring you into his room after class and inspire you."[12]

Whether enrolled in the college preparatory program or not, Dunbar students profited from the attitudes as well as the abilities of their teachers. The teachers instructed them in the use of cultural, historical, and other institutions in Washington. They urged students to seek out the limitless educational treasures in the city, and Saturdays and Sundays were days devoted to "mining" these treasures. Students made such heavy use of the Library of Congress when preparing for debates that the library placed time limits on them to give others an opportunity to consult its collections. And well before Montague Cobb entered medical school, he had acquired a solid understanding of anatomy through visits to the National Museum and the District of Columbia Zoo. Dunbar alumni Hastie, Montague Cobb, and Charles R. Drew were to be unimpressed by the bands they heard later at Amherst College, but some time passed before they realized the reason: They were comparing those bands with the U.S. Marine Corps and U.S. Navy bands whose free concerts they had attended so often in Washington.[13]

Although teachers, as Ernest McKinney asserts, "poured everything they knew into our heads," their impact was not strictly limited to pedagogical matters. As Dr. Adelaide Cromwell Hill, an alumna, commented, "Dunbar High School was not designed to create Uncle Toms." There were constant reminders of what it meant to be black in America. How else could one explain the concentration of so much pedagogical talent at the school? Or the complete absence of it—as embodied in black teachers—at white schools? How else could one explain the exclusion of Dunbar teachers from white professional organizations and of Dunbar students from white schools, even from concerts? The answers to such questions were reasons that Dunbar teachers and students were not Uncle Toms. With the approval of their principal, Garnet Wilkinson, teachers recruited members for the National Urban League, a leading civil rights organization. One of them, E. B. Henderson, head of the Physical Education Department, regularly opposed Jim Crow by writing letters to the Washington *Post*. Another, Nevall H. Thomas, the history teacher who was "the big NAACP man" at Dunbar, led students in "nuisance raids" on downtown restaurants that barred black customers. As they staged the sit-ins, he would tell them: "You are honored members of our race for doing this."[14]

Hastie was about to enter his junior year at Dunbar when everyone there was reminded that racism took more brutal forms than exclusion from cafeterias, schools, and teachers' organizations. In the summer of 1919 a race riot occurred in Washington. Trouble started on Saturday, 19 July, when white marines, sailors, and soldiers stormed into the southwest part of the city and stomped blacks. On Sunday they were joined by civilians and dragged blacks out of street cars and beat them. Others they shot and left in the streets. In all, three hundred blacks were wounded. On Monday, at a cost of two lives and some injuries, blacks killed four whites and wounded many more.[15]

At Dunbar, teachers and the stars of the second "galaxy"—their students—chose brains rather than bricks or bullets as their ammunition in the war against racism. This choice was made in collaboration with the teachers' auxiliaries, who were parents in families (such as the Hasties and the Weavers) that owned books, held intellectual discussions, and, according to Weaver, had "no question" about what the children would achieve. "You would do well in school, go to college, and enter a profession."[16]

Parents of many Dunbar students pushed their children toward academic achievement. Dr. Weaver's mother read Tennyson and Longfellow to him and his brother, Mortimer, telling them repeatedly that "the way to offset color prejudice is to be awfully good at whatever you do." Bill Hastie's mother continued to tutor him. She and other parents made a point of visiting Dunbar to motivate their children. "If you did not do well," Weaver says, "you felt that you had let your parents down—and they let you know it. There was competition among parents as well as among their children." The emphasis on academic achievement was a manifestation of the middle-class status of many of the families of Dunbar students. The middle-class sensibility among these blacks was determined not merely by income but even more by a certain self-perception, said by Dr. Michael R. Winston, Director of the Moorland-Spingarn Research Center, to have been "the key functional element in whether they were anxious about education and whether they supported their children and all the things that made the Dunbar situation as unusual as it was."[17]

One of the things that made Dunbar unusual was the sense of urgency that many students, like their parents and teachers, felt about educational excellence. For all their seriousness, however, students were pranksters as well as academic types. Often one of them would hide another's cap or other possession. Only the victim would be unaware of the culprit's identity. According to Hastie, the game progressed as follows: "The victim, after fruitless efforts to find the cap, would begin in this fashion: 'Whoever took my cap is a _____!' Then would follow a long description of the culprit's ancestors, his immediate family, and his vices and habits. If the boy doing the talking was good at that sort of thing, the joke would be turned on the joker. He would have to stand by and pretend to be amused while all the others howled with glee. It was a good stunt."[18]

Many students were also athletes, coached by Willis Menard, a Williams College alumnus. Hastie starred in track with Montague Cobb, Lancess McKnight, and Charles Drew, who was the best all-around athlete. "We were always confident that we could beat the white high schools at sports, even though we could never compete with them," he would say later. There was a time when Hastie wanted most to be an athlete. "Everybody knows a good athlete," he remarked. "Nobody pays any attention to your scholastic work." He talked it over with someone who did—his mother. The light continued to burn in the small study at 608 Que Street, N.W.[19]

Hastie graduated from Dunbar in 1921. Forty of the graduating class of 174 were to go on to college on scholarships. Hastie's father did not live to see his son graduate as the valedictorian. He had died during Hastie's senior year, leaving to Roberta Hastie even more of the responsibility that she considered her life mission. Her accomplishment of that mission places her in the company of such outstanding mothers as Mary Hardy MacArthur, Sara Delano Roosevelt, and Martha Truman, whose sons—Douglas, Franklin, and Harry—influenced the course of history. Commenting on those women of vitality, culture, strength, and determination, David McCullough has written: "They were people to rise to the occasion . . . and to the occasion of a son most of all. Being mother was their calling, entirely serious business. Education—direction—began with them, and as early as possible."[20]

We hear little about such women and even less about the mothers of notable blacks. Many mothers, however, have not been merely incidental to their children's lives but have significantly affected their children's careers. McCullough concludes that we can often judge a man and his achievements by looking to his mother. About the mothers he describes, McCullough adds: "Perhaps as important as anything provided by . . . [them] was the sense of mission they imparted to their sons—to be useful, to be good, even to be great. . . ."[21]

An intensely private man, Hastie rarely talked about his mother. It was clear, however, that he carried the torch she passed on to him. Dunbar was many years behind him when his own son, a lawyer, won a court victory over employment discrimination against blacks and women. "My father was pleased about that," says William H. Hastie, Jr., "but the *true* accolade that he gave me was not his saying he was pleased. He said, 'This is something your grandmother would have been very proud of.' *His* mother."[22]

Long before Bill Hastie, Jr., received that accolade, his father had tried to explain to whites the reasons that blacks were angry with some of them all of the time, and with all of them some of the time. Emancipation was a century behind blacks when he spoke at the University of Rochester in 1967, but equality was still ahead. Or so it seemed to black militants, who many whites thought should be appreciative of opportunity already granted rather than aggressive about opportunity still withheld.

In regard to that militancy, many whites were bewildered or belligerent, and we cannot know what they made of Hastie's observation about its pervasiveness: "Every Negro can tell a different story of how he became a militant in the war against racism . . . but one way or another, most of us have acquired the spirit of battle. . . ." It was not only about himself that he revealed a great deal by saying, "With some of us, militancy against discrimination and racial indignity is a heritage from our forebears."[23]

Militancy and excellence were two sides of the same coin in the eyes of Hastie's parents. That conviction was the core of their legacy to him. Those combined qualities would be his trademark at Amherst College, an institution that reflected the racism of the larger society but nevertheless further prepared him to justify his parents' constant belief that he was rare, special, precious—a black diamond.

CHAPTER TWO
Amherst College

"I was free but there was no one to welcome me to the land of freedom," Harriet Tubman, the abolitionist, said about her arrival in the North. "I was a stranger in a strange land. . . ."[1]

New England was a strange land for black students in the twenties. The three who left Dunbar for Amherst in 1921 knew that. But when the Federal Express pulled into Union Station in Washington, D.C., they boarded it and headed north. They reached New Haven, Connecticut, at three o'clock the next morning, took another train to Springfield, Massachusetts, and a third to Northampton, where they switched to the electric streetcar for the ride to Amherst. Bill Hastie, Montague Cobb, and George Winston Harry had arrived in the North. Hastie and Cobb were to follow in the footsteps of Edward Jones, who, on 23 August 1826, had been the first black man to graduate from Amherst College.[2]

Hastie's attendance at Amherst was predestined. His parents wanted the very best college education for him and had saved money with that goal in mind. Amherst's alumni included Charles Hamilton Houston, the son of the Hasties' close friends William L. and Mary E. E. Houston of Washington. Hastie's father had called to his attention Charles Houston's attendance at Amherst. Then, too, Dunbar High School was the main springboard to Amherst for black students in the twenties.[3]

In its treatment of black students, Amherst typified New England colleges. "You were not made to feel either that you were or were not an integral part of the scholastic life in those colleges," says Robert C. Weaver, a Dunbar graduate who attended Harvard. "You weren't stared at, but you weren't accepted socially."[4]

Social acceptance of blacks at Amherst was out of the question. For one thing, the "crass racism" that Hastie detected was not practiced only by white students. Without naming the man, Hastie has recalled an Amherst president who hosted evening socials at his home for all seniors except those who were black, and he has discussed the exclusion of a black from the Amherst Glee Club because the college would never have asked its friends to extend that student the hospitality shown the club during nationwide tours. Hastie was not that student, says Mercer Cook. "Bill was a monotone," Cook adds, laughing. "The only thing he could do musically was to play on the piano a few bars from 'Charleston, Charleston.' I can see him now." Cook's own talents were another matter. With Montague Cobb and Charles R. Drew he sang at the junior prom in 1924. In fact, he wrote most of the music, including an uncopyrighted composition that was introduced there and, after having been plagiarized, became a "Hit Parade" song several years later. It was called "Sweetheart of All My Dreams."[5]

Cook was the son of musical notables. His mother, Abbie Mitchell, a soprano, was not allowed to sing in opera—she was black—but she became a star on the radio and stage. Cook's father, Will Mercer Cook, was said by the legendary Duke Ellington to have been the person from whom he learned all that he knew about composing music. In addition to composition, Cook senior's high reputation was based on his performance as a musician and bandleader. His preparation included the study of violin at Oberlin College and under Joseph Joachim in Berlin, and further study under the Bohemian composer Dvořák, in New York at the National Conservatory of Music.[6]

It was, of course, possible that the Cooks' son, Mercer, was far less talented than all of the whites in the glee club and the orchestra at Amherst. But it was not possible that a white would be excluded from the glee club or orchestra for the reason applied against Cook and another of Hastie's college classmates, Benjamin J. Davis, Jr. A talented violinist, Davis was for a while part of the singing ukelele group (with Cook, Cobb, and Drew) that performed "Sweetheart of All My Dreams" at the junior prom in 1924. Even while they danced to the group's music, white students were unaware of the blacks' resentment of racism at Amherst. Typical of white students in this respect, Oliver B. Merrill failed to see the racial pride beneath Hastie's cool exterior. As he recalls, Hastie did not appear at all as "a rebel."[7]

Emerging at that very moment, however, was the graceful rebel who was to become one of Amherst's most distinguished alumni. Who but a rebel would have protested against an article that appeared in *Amherst Writing*, a student publication, in Hastie's sophomore year? A white student, Winthrop Tilley, had written about "niggers" waiting on tables, "coons" playing cards, and porters singing. He and his companion had watched five "high-paid niggers, such as the chef and the butcher," gambling. Tilley's companion found this game unexciting but dropped dark hints about knives and bloodshed in other games. The two of them moved on toward the porters' quarters—the "glory holds"—drawn by the sound of singing. "The harmony was wonderfully soft and pleasing," Tilley wrote. "Jack cautioned me to keep out of sight. 'They don't like to have any white men hanging around here,' he told me. 'If they catch sight of any of us, they'll stop until we go away.'"[8]

Complaining about the publication of Tilley's article, Hastie wrote the following to President Alexander Meiklejohn: "Every publication bearing the name of Amherst College as well as every team or other organization bearing that name represents the college. I well remember your attitude, Mr. Meiklejohn, when during last spring a college organization was guilty of conduct which cast discredit upon the college. It is with the hope, sir, that you may deem this matter likewise worthy of consideration that I am writing you this letter."[9]

Meiklejohn was "of course very much troubled and grieved" by Hastie's complaint. But the president said that Hastie's argument was invalid and his reasoning mistaken. He asserted instead, "You say that Mr. Tilley's hero must be either Mr. Tilley himself or 'a man who tells a story of sufficient merit to be published in the Amherst Writing.' It does not follow, of course, from the publication of an autobiographical story, that the hero must be represented by a person who could secure the publication of his story in the literary magazine." Meiklejohn asserted that a contributor had literary license as a story's narrator. "I am not ready to say that Mr. Tilley's representation of a given type of young American as using the objectionable terms to which you refer is an accurate one. On the [other] hand, you will of course admit that there are men who do use these terms. That being the case, I do not believe you can attribute either to Mr. Tilley or to the Amherst Writing the intention or the implication of insult which you find in this story."[10]

Hastie blamed both *Amherst Writing* and Winthrop Tilley, one of its editors, for the racist piece. He would not have objected to Tilley's use of "such crude and vulgar language" in dialogue written with character portrayal in mind, but Tilley had used the epithets as narrator. That abuse of literary license offended Hastie but not Meiklejohn. In an unrelated instance, President Meiklejohn had written: "The purpose of education is to make young people ready for living life in this world of theirs." Inside as well as outside Amherst, the world of white students was pockmarked by prejudice against blacks and Jews. Inside Amherst this bigotry was evident in the practices of fraternities, which Professor Walter Gellhorn of Columbia University School of Law describes as "the central social organisms." Another alumnus, Stephen H. Millard, adds that even before classes began, fraternity rushing was conducted. To enable students to meet each other informally, and to acclimatize freshmen to "a good and friendly feeling for the college," invitations to visit went out from fraternities to all freshmen except those who were Jewish or black. "You just didn't get into a fraternity," Cook says. He remembers that the admission of one Jewish student during his years at Amherst became "quite a story" on campus. The nonadmission of all blacks, according to one of Hastie's white classmates, Oliver B. Merrill, was not, at bottom, racial rejection. "The two [races] just didn't mix in that way." Negroes just were not considered for membership in fraternities and did not press the issue.[11]

Although not devastated by this ostracism, Hastie and his friends were irked by it. By way of retaliation, they filled the walls of their rooms with photographs of the loveliest black girls in Washington and other cities. Cook recalls that this had the intended effect in one noteworthy instance. "Charlie Drew got a picture of a Washington girl who was particularly beautiful," he says, "and John Coolidge, son of Calvin Coolidge, who was then President, tried to steal it." Cook's clique had more than photographs for company, as *The 1925 Olio*, published by the junior class, seemed to underscore with its lighthearted comments about Cook and Hastie. It noted Cook's "habit of shaking the dust or snow of Amherst off his feet every Friday evening and not returning until Monday morning," and proclaimed "a crowning mystery" his ability to win Phi Beta Kappa honors on five-day weeks. And if Hastie had ever flunked anything, it added, no one had ever heard about it, "and we verily believe that if he did he'd just turn over and die of shame." How did he manage it? Through hard work, readers were told. And perhaps through occasional visits to "the little girl who is not only his best pal, but his severest critic. And it wouldn't do at all to let her have anything to criticise!"[12]

The *Olio* ribbed Hastie, Cook, Davis, and Cobb along with other fellows, but they were not allowed to be just regular fellows at Amherst. There must have been other coteries on campus, but none was more circumscribed in terms of opportunity than theirs. Merrill remembers that they "almost moved as one. They lived in the dormitories and ate together." They relaxed by playing pinochle for a quarter a game. Hastie played more for fun than for profit. He received a weekly allowance of ten dollars from his mother, and never expressed a desire for more. When losing, the other guys kept playing, all night if necessary, to recoup their losses—or trying to. But Hastie always set a deadline for himself, and whether he was winning or losing, when that hour struck he quit. "He loved the game," Cook says, "but he just had that willpower."¹³

Hastie thought that his black predecessors had promoted in New England colleges "a genuine belief in the Negro's worth," but that a seemingly inescapable necessity militated against black students' acceptance as equals by whites. Davis, Cobb, and Cook were allowed to sing in the choir but not in the glee club, the reason for this exclusion being the same as that for the exclusion of Davis (an accomplished violinist) from the orchestra: the inevitability of social contacts. But the ostracism of blacks at Amherst was also a result of discrimination reinforced by the menial jobs that many blacks, of necessity, performed. As Hastie put it, "Say what you will, a man does not choose as his boon companion an individual whom he never sees in dress clothes except when carrying a tray at his banquets."¹⁴

Sterling A. Brown, his friend from Dunbar, knew what Hastie meant. He quit waiting on tables at Williams College in his sophomore year. Later he explained why he chose unemployment. "I about starved, but I just couldn't stand waiting on those guys and then going from the lunchroom to the classroom—and I was beating them in the classroom—where they were snide, nasty."¹⁵

Blacks at Amherst wanted nothing more than a chance to compete with other students as equals. Bill Hastie urged blacks to persevere:

> You with brown skins will not make Chi Phi nor even get bids to fraternity dances, not now. But continue sending more and more representatives of your race against whom no negligibility can be found except the color of their skins, and unless human nature is basically rotten, recognition will come in time.¹⁶

Sports and scholarship became instruments in the battle against racism at Amherst. Hastie's ambition was to captain the school's track team. He seemed to have a chance for that honor. His build and stamina were assets. Cook commented that Hastie was "slim but not as puny as in high school." Newton F. McKeon added that he was "trained right down, lean, fit." Hastie not only looked like he could run but, in fact, he could. Oliver Merrill described him as "poetry in motion," and McKeon coined another metaphor for Hastie: "He always made me think of an arrow in flight." Lawrence K. Blair, a teammate who sometimes acquired shin splints chasing Hastie around the outdoor board track, remembered him as "a tough competitor." This toughness and determination manifested itself in a contest between Williams and Amherst in the spring of 1925. In the quarter mile Wendell Phillips took the lead for Williams. "I was doing fine till I got halfway round the track, when a guy named Hastie passed right by."[17]

In three years of varsity competition, Hastie ran in eleven meets, placing first in seven events, second in six, and third in three. His specialities were the 440-yard dash and the 880-yard run. His best times were 51.7 seconds in the 440; 2.07 minutes in the 880; 23.6 seconds in the 220; 2.18 minutes in the half mile; 55.2 seconds in the quarter mile; and 10 seconds flat in the 100. He also competed in the pentathlon, which consisted of the 100-yard dash, shot put, broad jump, 200-yard low hurdles, and one-mile run. He placed third in this event in his sophomore and junior years and second in his senior year, when Drew (who won each year) edged him by amassing 422.9 to Hastie's 422.2 points. Hastie's years of walking, skipping, and running the three miles to school and the three back home in Knoxville were paying dividends.[18]

According to Cook, there were three blacks on the track team during Hastie's years in college: "Monty Cobb in the distance races, Bill Hastie in the middle distance races, and Charlie Drew in everything." Drew was in an athletic class all by himself when he arrived in 1922, and there he remained. With Monty Cobb, Benjamin J. Davis, Jr., and George Gilmer, Drew played on the football team. He was a six-foot, two hundred-pound star whose exploits remain fresh in some memories. E. King Graves, for example, remembers Amherst's second game in Drew's sophomore year. Amherst was Princeton's annual sacrificial lamb. The team went by train to Trenton, New Jersey, and spent Friday night at the Stacy-Trent Hotel. Telephone calls from "rabid racists" prevented Drew from sleeping until the early hours of Saturday morning, when Coach D. O. "Tuss" McLaughry heard about and put an end to them. Drew said nothing to him or to any teammate.[19]

Hours later the team traveled by bus to the game. Princeton lost no time in sending its superstar, Jake Slagle, on a run around Drew's end. Recalling the game, E. King Graves thought the end run was a mistake. "[Charlie] brushed aside the interference and tackled Slagle by picking him up in his arms and, still running, carried him over to the Princeton bench and literally tossed him into the coach's lap. The coach, Bill Roper, made history. With the crowd roaring from Drew's tackle, Roper got up and shook Drew's hand, whereupon the house came down. When they went back on the field, Slagle had his arm around Charlie's shoulder."[20]

A meaner spirit prevailed during the track team's visit to Brown University in April 1925. Brown beat Amherst, the Purple and White, 80 to 55 in a dual meet. Drew led Amherst by placing first in the 120-yard hurdles, second in the running broad jump, and third in the shot put and 200-yard hurdles. Cobb placed second in the one-mile run and third in the two-mile run. Hastie was second in the 440-yard run and third in the 220-yard run.[21]

The twenty-five-man team had left Amherst by taxi that morning and planned to return that evening. After the track meet, some of their teammates were at the cars when the three blacks approached. Several turned away, embarrassed when the student manager cut short his conversation with a coach and walked over to Drew, Cobb, and Hastie. He hesitatingly said to them, "Fellows, I don't know just how to put this, really." They knew what would follow. "The hotel says it's against their policy to serve . . . colored people . . . and . . . there just isn't any place that can serve a group this large on such short notice. How about you fellows having dinner at the Commons here on campus?"

They did. Later, the team made the quiet trip back to Amherst.[22]

Hastie thought his record would be enough to assure him the team captaincy in his senior year. "It wasn't, though, and even being named president of the college Phi Beta Kappa couldn't compensate me," he acknowledged. His yearbook noted that he had done two things, "both very well": track and scholarship. Or as McKeon put it, "Bill was finely honed as a runner, even as he was as a mind."[23]

Amherst and Hastie were well matched in many ways. The curriculum and faculty fulfilled his desire for "a thoroughly rounded education." His lifelong appreciation of poetry and the written word probably was nurtured in Robert Frost's course in creative writing. He himself credited Stewart Lee Garrison with having inculcated in him an idea—invaluable to persons in "talking professions"—that was to distinguish his work: "the idea of saying simply and directly what one wants to say, and no more." Otto Matheny-Zorn, a controversial figure who taught stimulating courses (German and modern German philosophy) also helped Hastie to develop intellectually, as did President Meiklejohn, who once greeted prospective freshmen by stating: "When a man chooses to go to college, he declares that he wants to be different, that he is not satisfied to be what he is. If any one of you is satisfied with himself, he had better go back and keep still for fear something may happen to disturb his perfection. If those who stay are rightly dissatisfied with themselves, they will satisfy us."[24]

Hastie satisfied Amherst. His mettle showed early. Cook provides a pertinent anecdote. "We had eight o'clock chapel every morning then. So you'd eat your breakfast, and you'd run up what we called Indigestion Hill and get in chapel, sometimes with a sheepskin over your pajamas." The bell tolled fast, and students had to be seated by quarter past eight. There he and Hastie sat when Meiklejohn announced William Hastie the winner of the three-part (Latin, English, and Mathematics) Porter Prize Admission Contest. Cook recalls being "deliriously happy." He then looked at Hastie. "Bill's face was just stoic. No indication that he had won. I think that's why, in later years, some of his close friends called him Fish Eye," Cook adds. "They'd say you could never get Bill's reaction from the way he looked. Well, he claimed later that I was happy because I wanted to borrow some of that fifty-dollar prize."[25]

Hastie continued to excel.[26] He was one of the Kellogg Five as a sophomore. He won the Walker Prize in Mathematics and four honorable mentions, one for work in German, which he had wanted to study in high school but could not because it had been cut from the curriculum, a casualty (he said) of "the special type of thought control" practiced during the "War to Save the World for Democracy." (In preparation for his visits home, his mother took lessons in German.) Winner of awards in philosophy and Greek, Hastie was so far advanced in his major (mathematics) that in one class he was the only student.

As a junior, Hastie earned an A in each subject, including physics, in which he won the Porter Prize for first-year proficiency. He was elected to Phi Beta Kappa on the first drawing, one of three juniors with a grade average of 88 percent. He went on to win the Addison Brown Scholarship, which was awarded to a senior already on the scholarship list who had led all others in the first three years. He received the Woods Prize for "outstanding excellence in culture & faithfullness to duty as man and scholar." One of six who graduated magna cum laude, he was the valedictorian of his class.

The 1925 Olio gave this appraisal of Hastie: "The _____ prize of _____ dollars awarded annually for proficiency in (supply your own subject) has been awarded to William Henry Hastie of the Class of 1925. . . ." His success was no surprise to Hastie's friends, one of whom, Sterling Brown, was to say, "I looked on Bill as excellence."[27]

Hastie had been groomed for success from infancy. And according to Monty Cobb, at Dunbar High School the teachers "hitched our wagons to a star." Those teachers made students believe that they could ride the star, thus instilling everlasting confidence in them. "We just felt that we could achieve what anybody else could achieve," says Cook. Making the same point, Weaver notes, "We-knew-who-we-were." Sterling Brown describes Dunbar students' outlook by saying, "We knew that we were set apart. We never felt that the people who set us apart were better than we."[28]

Dunbar prepared Hastie and his friends for college. They even felt sometimes that the education they had received in high school was superior to that provided in college. For example, after hearing an expert lecture at Amherst about King Tut and the pyramids, Hastie and Cobb met outside the lecture hall.

"Are you thinking what I'm thinking?" Cobb asked.

"Yeah. Cat was better."

Nevall "Cat" Thomas had taught them history in high school. A typical Dunbar teacher, he groomed them spiritually as well as academically. This socialization stood them in good stead when they were caught in the crosscurrents of systemic racism and academic excellence in college. So admirably did Hastie rise to the challenge that he became for others a confidant, counselor, or, as Cook puts it, "just the rock of ages."[29]

Hastie found Amherst to be an imperfect but meritorious environment that, on balance, prepared him to hitch his wagon to ever brighter, higher stars. He would become one of its most faithful and distinguished alumni, the personification of the qualities cherished by the college. Amherst professed to blend humaneness and intellectuality, says Robert J. McKean, Jr., an ardent alumnus and a partner in a Wall Street law firm. "The more important of the two is humaneness," he adds. "That's what Amherst was trying to say. And Bill heard it."[30]

CHAPTER THREE
Old Ironsides

Hastie's academic record at Amherst entitled him to a fellowship for graduate work at Oxford University or the University of Paris. Intent on continuing his education, but not abroad, he turned down the fellowship and accepted a faculty appointment at a black high school in Bordentown, New Jersey, to earn money for graduate school.[1] In just two years he was prepared to enroll at Harvard Law School. But preparation was not merely a matter of time or money, and Hastie would have been the first to say that his experience at the high school contributed immeasurably to his preparation for law school and life.

The school would eventually enroll 450 students, occupy a 400-acre estate on a high bluff overlooking the Delaware River in Burlington County (with the campus proper on 35 acres), employ 50 vocational and academic teachers, own prize-winning livestock, and reach an overall value in excess of two million dollars. But it began with 8 or 10 students in 2 frame buildings in 1886, the year it was founded by the Reverend Walter A. Rice of the African Methodist Episcopal Church and 6 associates.[2]

At first privately funded, Reverend Rice's school began to receive state support in 1894, and for the next several years the state provided annual appropriations of two thousand dollars. In 1897 the state superintendent requested higher appropriations for the school, saying that it had pursued "a devious, difficult and well-nigh friendless way" to rise from humble beginnings and become a "star of hope" to blacks. James M. Gregory, who had succeeded Reverend Rice as principal that year, testified to the school's fine reputation and supported the superintendent's plea for financial aid. On the superintendent's recommendation, in 1900 the state assumed control of the school and placed it under the State

Department of Public Instruction. The next year authorization to buy the Parnall Estate on the outskirts of Bordentown was given, and according to one official publication "the work of developing a first-class school was begun in earnest."[3]

Officially named the New Jersey Manual Training and Industrial School for Colored Youth, the school was called the Bordentown Manual Training School (BMTS) or simply Bordentown. It was also affectionately known as Old Ironsides because the estate on which it was located had been the home of the commander of the man-of-war *Constitution*, or to many Old Ironsides.

The institution was unique among high schools in the East. It was a boarding school, and students wore uniforms that they purchased through the school. Organized into cadet corps for purposes of physical training and experience in discipline, self-government, and order, boys wore khaki uniforms similar to those of soldiers. Girls wore white shirtwaists and blue skirts. In addition to lending uniformity to the students' dress, the wearing of uniforms made savings on clothing possible.

Tuition was free, but students paid a registration fee and a monthly fee for board, medical attention, uniforms, and laundry. Male students paid fifteen dollars per month; female students paid one dollar less because they did their own laundry. To help keep expenses down, students worked one hour daily: girls in the laundry or dining room; boys in the automobile shop (often servicing state police cars), in the orchards, or on the farm.[4]

Some students at Old Ironsides were public high school graduates who wanted to learn a trade, while others were older students who wanted to devote themselves entirely to given trades for which they showed a particular aptitude. The latter admissions were unusual. The typical student at Bordentown received both vocational and academic instruction daily and graduated with a diploma that was equivalent to the one awarded by other high schools, except that it also certified the recipient's competence in a given trade.[5] In other words, whether desirous of going from high school into industry or to college or technical school, students were well advised to apply for admission to Old Ironsides.

When he hired Hastie, William R. Valentine, the principal at Bordentown, was in his tenth year in office. Born in Loudon County, Virginia, Valentine had attended public school in Montclair, New Jersey, and had graduated from Harvard in 1904. In Indianapolis, Indiana, he had served for four years as the principal of the Graded School (Colored) and for seven years as the supervising principal before accepting the position at Bordentown. Among his traits, one remained etched in the memory of a lady who had a chance conversation with him while visiting the school. Responding to his question about her marital status, she said that she was dating a lawyer. She knew that Valentine was thinking of her economic future, not prying into her private affairs, when he said, "A lawyer? A doctor would be better."[6]

Valentine selected teachers with care. He wanted only competent instructors like Frances O. Grant (Phi Beta Kappa, Radcliffe), Charles Ray (Bates), Solomon Fuller and Thomas C. Williams (Harvard), Lester B. Granger (Dartmouth), and Fred Work (original Fisk University Jubilee Singers). Many were alumni of New England colleges, most of them held master's degrees, and all of them spent evenings in wide-ranging discussions at the home of the Valentines. Teaching at Bordentown was comparable to graduate school and, in some ways (though they would not have called it this, and perhaps the designation is a bit exaggerated) a revival meeting, with time out for cards (bridge, not poker) and strawberry shortcake (Hastie's favorite dessert, which Mrs. Valentine baked especially for him). Had opportunities for blacks been better, many of these bright and ambitious teachers would not have been there.[7]

To benefit students and faculty, Valentine also brought to Old Ironsides men and women, both black and white, who were role models. Among them were Paul Robeson, James Weldon Johnson, Mary McLeod Bethune, Albert Einstein, and Eleanor Roosevelt. In addition, some blacks came to the school as travelers seeking food and shelter. (Hotels turned blacks away.) Still others came as political strategists. "They were champions trying to do something for the Negro in their own quiet way," says Vivian Anduze. All sorts of conventions were held at Bordentown, giving Valentine the pleasure of utilizing dining rooms and dormitories to help keep the school going. Working or studying, students were on campus twenty-four hours a day, every day, even in summer, barring emergencies. Most of the 350 students were residents of New Jersey before arriving at Old Ironsides, but many were from out of state. In fact, the school attracted so many students from out of state that in time enrollment was made contingent upon graduation from elementary school in New Jersey.[8] The requirement that teachers live on campus served the students' economic, intellectual, and social development. It had the same effects on the teachers.[9]

If one thinks metaphorically of Valentine's school, one envisages him as the captain guiding Old Ironsides through seas of black educational theory whipped to storm proportions by two men of far greater renown, Booker T. Washington and W. E. B. Du Bois. The Washington and Du Bois camps had forged an armistice at the Amenia Conference in 1916, the year after Washington's death, but the issue concerning the best education for blacks was still unresolved fourteen years later.[10] Differing views among prominent blacks gave rise to the controversy that Henry Lee Moon, a longtime NAACP warrior, considered the most momentous intraracial dispute that has ever occurred.[11]

Du Bois was primarily interested in the development of a Talented Tenth, the best of the race, whose task it would be to lead the black masses from bondage to freedom, from barbarity to civilization. He declared that the creation of the Talented Tenth was the foremost problem in the education of blacks.[12] In a commentary published in 1903, Du Bois described four decades of education in the South, beginning with the end of the Civil War. The first decade was characterized by the uneven and unsteady establishment of schools by the Army, missions, and Freedmen's Bureau; the second by an effort to build a public school system; the third by the creation of manual training schools at a time when the industrial revolution, racism, and economic desperation of whites made education for blacks a very practical matter; and the fourth by the full blossoming of the institution that had begun to gain notice in the preceding decade, namely, the industrial training school.[13]

But should industrial training be accepted as the totality of black education, especially when the legacy of slavery and the currency of imperialism were being melded with racial bigotry that held men of color at a disadvantage? Du Bois thought not, for education in all its beauty was not the exclusive right of white men. And as for industrial training, which had been standard in almost all schools, Du Bois said that it was given a new dignity which increased its relevance to industry and "an emphasis which reminded black folk that before the Temple of Knowledge swing the Gates of Toil."[14]

The leading black gatekeeper was Booker T. Washington, who urged blacks to seek their salvation through industrial (i.e., vocational) education and to remain in the South. Washington believed that blacks had to convince the white South that the former's education was in the latter's best interest. Rather than demanding classical education for blacks, which white Southerners opposed, he advocated industrial education, which they approved, particularly after Washington's assurance, given in his speech at the Atlanta Exposition in 1895, that social equality definitely was not among the goals he advocated. Both in the North and South, Washington gained white support for his philosophy and for his college, Tuskegee Institute, in Alabama. He did not deprecate the education that Du Bois advocated, but he de-emphasized it in the belief that it was impractical because it offended whites and, to make matters worse, offered scant opportunity for blacks to gain the economic foothold required for advancement. Washington, the former slave educated at a center of industrial training (Hampton Institute in Virginia), was chastized by Du Bois, the intellectual educated at Fisk, Harvard, and in Berlin.[15]

Although Du Bois disputed much of Washington's political thought, he did admit the importance of some of Washington's teachings.[16] For example, he placed industrial training second only to the establishment of black colleges, which were an absolute necessity in the production of teachers, including those in industrial education. And while Du Bois insisted that "the object of all true education is not to make men carpenters, it is to make carpenters men," he also assigned equal importance to the two means of making carpenters men. One was to provide "liberally trained teachers and leaders" to instruct the carpenter-men and their families about the meaning of life. This, said Du Bois, made black colleges and college-trained blacks necessary. The second means of making carpenters men was to ensure that they developed the mental and manual ability characteristic of efficient workmen. In Du Bois's judgment, this necessitated "a good system of common schools, well-taught, conveniently located, and properly equipped. . . ."[17]

Old Ironsides was such a school. It was not devoted exclusively to the training of carpenters, to use Du Bois's term. However, after graduation the overwhelming majority of its students went to work rather than to college or to technical school. This would have pleased its founders. The Reverend Rice had been determined to emulate Booker T. Washington, and, indeed, another nickname given to the school was "Tuskegee of the North." But he, his associates, and his successors maximized educational practicality and minimized the dysfunctionalism of Washington's and Du Bois's philosophies, which in their exclusive rather than complementary emphases were "detrimental to the best educational interests of the American Negro."[18]

When Hastie went to Bordentown, high school education in New Jersey was characterized by rising costs, an emphasis on vocational training in an increasingly urbanized and industrialized state, a de-emphasis on fads and frills, marked public interest, and widespread concern about the quality of instruction. It was said that the schools were being "democratized," that is, opened to an ever-growing and varied student population. Years earlier (1881) the state had enacted a law that prohibited the exclusion of children, between the ages of five and eighteen, from public schools on grounds of nationality, religion, or color. And in 1884, in Burlington County (where Bordentown is located), a black won a court order against the exclusion of his children from white schools in the city of Burlington, which operated three schools for whites and one for blacks.[19]

But public schools ran the gamut from integration to segregation in New Jersey when Hastie became one of the 231 black teachers hired between 1919 and 1930. In that period the number of schools for blacks increased by 26 percent, from fifty-two to sixty-six. Most students and teachers were primarily concerned about vocational training and placement after graduation. Old Ironsides trained its students, placed its graduates, and through its extension department helped them to find opportunities not only for employment but also for additional training. The New Jersey Conference of Social Work (whose Interracial Committee was chaired by Valentine) praised the school as "one of the first agencies of importance in racial adjustment," and noted that it alleviated many of the social ills in New Jersey while promoting responsible citizenship. According to the conference, "The Manual Training and Industrial School for Colored Youth at Bordentown serves the Negro population in more ways than its name implies."[20]

The poet Sterling A. Brown appreciates that description of the school. "Bordentown was special," he says, remembering a group of blacks whom history, as recorded, has obscured. These blacks were not students, teachers, or administrators at Old Ironsides. Nevertheless, they were bound to the school as star tennis players. Some were called Biscuits from the moment that one of them strolled onto the courts in Baltimore, an overdone biscuit from breakfast strung around his neck. "They were a bunch of hell-raisers," adds Brown, amending the latter statement by adding, "*We* were." The amendment was in order, for Brown was the Biscuits' leader, the Grand Dough.[21]

The Biscuits and the Muffins, the women's auxiliaries, led by Jackie Jamison, were part of the American Tennis Association (ATA), the black counterpart of the United States Lawn Tennis Association. In the year that Hastie began to teach there, the ATA held its ninth annual tournament at Bordentown before fifteen hundred fans. The first tournament had been held in Baltimore, and others were held elsewhere. "But Bordentown was our Forest Hills," says Brown.[22]

The tournament at Bordentown was something of a reunion for Hastie. Growing up in Washington, he had seen some of the star players in action behind the Twelfth Street YMCA. They played on courts owned by the James Walker Tennis Club, which Brown and other founders named in honor of the major who had commanded the black battalion that fought alongside French troops in World War I. Like many other youngsters, Hastie had become a Biscuit admirer and, in time, a Biscuit. He was not in the same class as Talley Holmes, Ted Thompson, Lulu Ballard, Ora Washington, L. C. Downing, C. O. "Mother" Seams, or James Trotman—all champions—but he was not lacking in enthusiasm for the game.[23]

"Bill Hastie loved tennis," Brown says, "but he wanted to do everything by the book. He wasn't going to look awkward to get a shot. If he had to look awkward, to hell with the shot. I'll never forget Bill's windup on his forehand. It was a thing of beauty. He might have missed the damn ball, but he did it the way it was supposed to be done. And he did the same thing with his backhand. He was a stylist."[24]

Hastie's style off the court is well remembered by former colleagues on the faculty at Bordentown. "He was class all the way," Frances O. Grant recalled. "He was noble."[25]

Hastie was to return to Old Ironsides in a different capacity, namely, that of a role model for students. It was during such a visit that Helen Reid, who was then Valentine's daughter-in-law, first met him. She and her husband were giving a party in their cottage at the end of the campus. "We were having a *good* time," she says. Someone knocked, and she opened the door. There stood Hastie. He appeared to be so reserved, but during the party she also saw his sociable side. "This was in the days of rug cutting," she says, "and everybody around the room was swinging everybody else, and his coattails were flying." Hastie looked so proper, and in fact was, "but he had that other side to him. He *liked* a good time. He liked life. He really did." Reid thought of him as "a well brought up Negro youth of his time, who had this other little side of him where he could get out there and go with the best of them."[26]

Both sides of Hastie's personality made him appealing to his colleagues at Bordentown. The fact that he only planned to stay a short while created stress for some of them. He was young and attractive in a setting where designs on matrimony were not unknown. But, says the friend who often gave him a lift to Philadelphia to see the girl he was sweet on, marriage just was not in his plans. "He was pleasant company, a witty conversationalist, and a friendly person liked by faculty and students," another acquaintance says. They did not want to see him go. "But we knew that Ironsides was only one step on his way to reach higher goals."[27]

After two years at Old Ironsides, Hastie drew closer to those higher goals by entering Harvard Law School. It was 1927, and blacks who aspired to become lawyers had to pass through a veritable mine field, a portion of which was legal education. The day that Harvard Law School would begin to display active concern about the racial unfairness that typified legal education was almost forty years away.[28]

Much closer to 1927 was the year that a professor at the school, while addressing a class attended by one of Hastie's black friends, remarked, "I have said over and over again that if you colored fellows are not absolutely tops, you are making a mistake going into law. Law is the one field where you can't be mediocre and get any real success."[29]

Hastie would not be mediocre at Harvard Law School. Indeed, Professor Felix Frankfurter, who taught him constitutional law, was to say that he was "not only the best colored man we have ever had but he is as good as all but three or four outstanding white men that have been here during the last twenty years."[30] Nevertheless, Hastie was offended by the remarks reported to him by his friend. Harvard produced incompetent as well as competent white lawyers. Furthermore, as Hastie viewed success and failure, neither was causally related to color. But what if his friend's professor had been encouraging rather than disparaging black law students? The professor's remarks would have been no less offensive to Hastie, who told his friend, "This notion that Negroes have got to be better than other people is about as disgusting as the notion that Negroes are inferior. As a matter of fact, I very much fear that they are rationalizations of the same thing."[31]

That "same thing" awaited Hastie at Harvard.

CHAPTER 4
Harvard Law School

One could be forgiven (or understood) for wondering why any black would have wanted to become a lawyer in 1927, when Hastie enrolled at Harvard Law School, or in 1930, when he graduated. Black lawyers were scarce—only 1,230 among 160,000 attorneys according to a 1930 census report, and even that figure was said by Charles Hamilton Houston to have been exaggerated. Houston deplored the distribution as well as the scarcity of black lawyers, observing, for example, that there was only one per nine million blacks in the South.[1]

Not many blacks had run the gauntlet reserved for those who wished to emulate Macon B. Allen, the first to pass the bar in Maine eighty-four years before Hastie enrolled at Harvard. The hurdles included financial difficulties that caused interruptions in legal studies and delays in career beginnings. Additional discouragement resulted from their lack of standing in the eyes of public officials, established lawyers (black and white),[2] the black community, and the white American Bar Association, whose exclusion of them led to the establishment of the black National Bar Association in 1926.[3] Opposition to black participation in the legal profession parallelled opposition to participation in government, said Houston, the leading black lawyer of the period. "Yet it is where the pressure is greatest and racial antagonisms most acute," he added, "that the services of the Negro lawyer as a social engineer are needed."[4]

For Hastie and many other black lawyers, Houston's creed about the law was, in a sense, a law unto itself. Houston contended that white lawyers could handle ordinary legal business but not civil rights cases because they themselves benefited from the very exploitation that civil rights advocates contested. That advocacy, he insisted, represented black lawyers' "social justification."⁵ As with Houston, it was the main reason that Hastie enrolled at Harvard.

Harvard was not an "equal opportunity" law school. While the first black formally trained in law in the United States, George Lewis Ruffin, received his LL.B. degree in 1869, during its first 150 years the law school graduated fewer than fifty blacks. Credible indeed would be the comment made generations later by the university's president, Derek A. Bok, about the law school's provision of legal education to blacks. "If our tradition is long," he said, "it is also very thin."⁶

Hastie was one of nine blacks on whom Harvard conferred LL.B. degrees between 1920 and 1930. The first of these graduates, Charles Houston, received his degree in 1922, eight years before Hastie completed his studies at Harvard. Both Houston (in 1923) and Hastie (in 1933) went on to get their doctorates in juridical science. They alone among the nine blacks won that distinction. Both had gained even higher distinction through membership on the editorial board of the *Harvard Law Review*, which Houston had achieved in his third year and Hastie in his second.⁷

Hastie was therefore the second black to become an editor of that prestigious student journal. Benjamin J. Davis, Jr., who had gone from Amherst to Harvard while Hastie went to Bordentown, had expected as much of his friend. He told Louis L. Redding, a fellow black student at the law school, "Bill has a mind that can see around corners." The accuracy of Davis's assessment became apparent almost immediately upon Hastie's arrival at Harvard. William J. Brennan, who was destined for the Supreme Court of the United States, did not know Hastie at the time. "But, of course, everyone knew *of* him. He was not only one of the ranking students but outstanding as a student."⁸

Hastie's reputation rested in part on performances such as that which secretly delighted another student in Professor Samuel Williston's class. This student, who was passing for white, enjoyed telling a cousin about the routine. Williston would put a question to the hundred or more students in his class, entertain responses, and eventually say, "Mr. Hastie, give them the answer."⁹

Williston knew that he could count on Hastie. But the latter knew, in turn, that a student could not count on recitation alone to carry the day. A white classmate explains: "You stood or fell by the four-hour annual examination, and the professor did not know whose book he was marking." That pressure and anonymity were part of the rigorous academic program at the law school. "It was pretty rough," adds this interviewee, a future *Harvard Law Review* officer, Wall Street lawyer, and diplomat. "I was out on the golf course when my first-year grades arrived. My mother rushed out and gave them to me. When I saw them, I burst into tears. I was so relieved."[10]

Hastie had been warned that not more than one third of his class would survive the first year. Dean Roscoe Pound regularly cited that statistic and the faculty enjoyed repeating it. Students who survived the first year of law school delighted in adding, "The cowards never started; the weak died on the way." Then, as later, the "HLS achievement ethic" held worthless or disgraceful anything less than superb. At first, it seemed to Hastie that he would not measure up, at least not in Dean Pound's criminal course. To alleviate students' anxieties—their not knowing how they were faring in their courses—the law school scheduled midyear examinations. These were marked but no grades were recorded. Hastie scored a low mark on Pound's midyear examination. Disappointed, he intensified his program of study. "I did not cross the Charles from January till June," he explained. "I even had a hometown girlfriend attending school across the river at Boston University. But, I did not cross the Charles from January till June."[11]

In June an average of 75 percent (out of a possible 80 percent) entitled an examinee to an A grade. Hastie's average was 76. One of 5 blacks in his class, he ranked fourteenth among 690 students. As a member of the top 2 percent of his class, he became a *Harvard Law Review* editor and entered the charmed circle of the favorites whose careers Professor Felix Frankfurter aided whenever possible.[12]

Giving his views about the editorship twenty years after he received the honor, Hastie wrote, "I suppose one never entirely gets rid of the awe . . . felt as a first-year law student in contemplating the giants pointed out . . . as members and officers of the law review boards."[13]

Hastie was a fine editor and a respected contributor of "Case Notes" to the *Harvard Law Review*. In total points earned for published editorial contributions, he outstripped his fellow editors, and according to a contemporary, "His work was done with such promptness and such finish, requiring a minimum of editing, as to make his presence on the board a constant pleasure to those who worked with him. There were only a very few others on the board during those two years of whom it would be possible to say that. . . ."[14]

Hastie was respected yet reserved. "For a man of his outstanding and recognized talents, Bill was pretty shy . . . retiring," a colleague recalls. "Let's put it this way: You had to make the advances, though when you did they were very cordially received." Maynard J. Toll, the journal's treasurer, says by way of further explanation, "I think . . . the main reason was that Bill was always quite a private person, and that he was careful not to 'thrust himself' on any of his friends."[15]

Toll remembers Hastie as an "extremely cultured and courteous" fellow who, though not aggressive, readily participated in intellectual discussions. One classmate's appraisal of Hastie was: "He is not only agreeable without being unctuous, but the sort of person who makes friends and, in spite of the real drawback of his color, makes an effective leader of men. . . ." A faculty member spoke for many of his colleagues when he said later about Hastie, "The opinion here is without reservations strongly in his favor . . . He was modest and attractive in his social contacts; and the faculty are equally enthusiastic over his mental quality . . . It is rare to find such a unanimity of opinion."[16]

Alger Hiss might well have summed up the sentiment about Hastie held by students and faculty. Recalling the discussions by *Harvard Law Review* editors in Gannett House, Hiss says that he had been impressed by Hastie's "charm of manner and keen intelligence . . . I was drawn to Bill, and despite his gentle personal dignity—almost aloofness—I regarded him as a student friend."[17]

It has been said that Harvard Law School denied equal opportunity to blacks largely because many of its students were sons of Southern aristocrats.[18] But even if Harvard and its sons of aristocrats did not grant equal opportunity, they did grant some opportunity, and in doing so they unwittingly planted the seeds of destruction—at least in a legal sense—of the racist order. Two of the three principal lawyers who brought about Jim Crow's downfall, Charles Houston and William Hastie, were trained at Harvard, and at Howard they, in turn, trained the third, Thurgood Marshall.[19] Hastie could not have known, while at Harvard, that history would take that turn, which he considered one for the better. But he did know that Harvard could—and, in his case, would—provide or sharpen the skills required for battle. That realization accounted in part for his stellar performance at Harvard, which bedazzled many people, and for his enduring appreciation of Harvard, which bewildered others.

Among the bewildered is Derrick A. Bell, Jr., who in 1969 was to become the first black given a tenure-track appointment at Harvard Law School. Bell had been present when Harvard added Hastie's portrait to the gallery of greats in the law school. Bell, taken aback by Hastie's "most moving talk" about his years at Harvard, recalls, "I sat there in wonder and admiration that a black man who attended Harvard during its pre-liberal days could speak with such warmth and affection about what must have been a very hostile place. Perhaps, though, Hastie's intellectual gifts were so overpowering that they drew people to him despite the racial factors. . . ."[20]

But Hastie had not gone to Harvard to collect grudges. One of his purposes was to prepare for the battles that were to come. Another was to help pave the way for other blacks by making himself, as a friend has said, "a demonstration project of what he thought would be true of others."[21] Ironically, some of these beneficiaries were to become his detractors. These critics resented him because his record was cited by school administrators—but never by Hastie—as the standard by which black aspirants to the faculty would be judged. Consequently, these students developed "a love-hate response" to him, in effect blaming him for their bombardment with "'Judge Hastie' stories . . . about incredible discipline and superiority of intellect over normal men."[22]

But there were other "Judge Hastie" stories understandably left untold by white administrators and apparently unknown to black students. Judge Charles E. Wyzanski, Jr., could have told the students about the editor from South Carolina who, when Hastie began his first report on cases under consideration, stalked out of the room. Before the end of the year, however, the South Carolinian and Hastie were dining in Harvard Square. The other fellow had learned, says Wyzanski, that "he would be lucky if he were the equal of Bill." And William T. Coleman, Jr., a black whose record at Harvard (journal editor; LL.B., magna cum laude, 1946) made him an object of the same resentment as that directed against Hastie by black students, could have told them about another incident in Gannett House, one that Hastie did not allow to pass without comment.[23]

When the editors met to choose the next year's president, one said that selection of a particular candidate would place a Jew in the top post.

"If that's a relevant factor," Hastie said, "I'm going to leave the room."

"He shut off that conversation," says Coleman.[24]

"You can't imagine the amount of racism and anti-Semitism in Cambridge and Harvard," a president of the *Harvard Law Review* says. "No Negro or Jew was admitted to membership in the law clubs. At the Harvard Club of Boston, no Negroes were admitted *to the building*." The rule stipulating that blacks were not to be admitted to the club posed a problem in 1928 when the *Harvard Law Review* planned to hold its annual dinner there. One account holds that an editor suggested that Hastie not be invited, since there was always the chance that some doorman might turn him away. Another account claims that the editor merely said that the possibility of such a reaction by a doorman created some awkwardness. Yet Hastie attended the dinner without incident.[25]

Hastie shared the conviction that Houston, his mentor, held, namely, that no one in this world made allowances or accepted excuses "even for understandable shortcomings."[26] He was willing to work hard. "I'm afraid I am a bad person to advise you on shortcuts to legal knowledge," he was to tell a friend. "It just happens that I don't believe any of them are any good. I know what your problem of time is. But during the first year in particular I think there is no substitute for the laborious work of reading and analyzing cases."[27]

While doing this tedious work and reaching the zenith of academic achievement, Hastie managed to enjoy a lighter side to life. Most students lived in rooming houses near the law school but dined elsewhere. Hastie, John Preston Davis (a law student), and Ralph J. Bunche (a graduate student in political science) lived at the home of the Clarks at 417 Broadway. Two other blacks, Robert C. Weaver, an undergraduate in economics, and Louis L. Redding, a law student, lived next door. Hastie paid two dollars a week for a room, towels, and two changes of linen. Only on Sundays did Ma Clark furnish breakfast, which featured thick biscuits—and her husband's complaint: "Me and God both don't approve of poker!" Nevertheless, the fellows played the game every night, with dawn their time limit on Saturday. Mr. Clark was known to rush into the room shouting "Stop it! Stop that evil game!"—only to find that it had long since given way to discussion of politics, race relations, and other serious issues.[28]

Poker, bull sessions, sports, and meals (the forty or so blacks at the university often dined together) made Harvard more than a place of study to Hastie. For him Harvard symbolized camaraderie as well as scholarship. He valued both. It was easy to miss that point, as Charles R. Drew, his friend from high school and college, noted. "Bill's calm manner makes some people think he's cool," said Drew, who knew better. While at Bordentown, Hastie had helped him to meet the costs of medical school at McGill University in Montreal. Drew adds, "A cool man doesn't do things like that."[29]

Nor did a cool man take to the battle for equal justice as naturally as Hastie did. It was not that he did at first. As he said, ever since his childhood, when his mother told him stories about the construction of the Panama Canal, he had been "firmly convinced that I was going to build great canals and dams and stand the whole universe on its head with new and startling things." In fact, when he graduated from Amherst he had planned to become an engineer.[30] And, indeed, Hastie did become an engineer—not a civil engineer but a "civilizing engineer."

CHAPTER FIVE
Gideon's Band

In the year that Hastie graduated from Harvard Law School (1930) the National Association for the Advancement of Colored People (NAACP) formulated a strategy for all-out war against racism in America. Concentrating on civil and political rights, the NAACP neglected economic rights. The organization, said Hastie, was "a great watchdog" that guarded blacks' rights, but it could not have foreseen a phenomenon that was to have a tremendous impact on their economic well-being. The strategists of 1930 could not have known that the federal government would soon be exerting "enormous influence in the entire life of the nation."[1]

Nevertheless, the New Deal made the strategists' oversight grist for the mill of NAACP dissidents, who demanded that the organization assign top priority to economic programs. Moreover, Abram L. Harris, a Howard University economics professor, led others in arguing that blacks should join forces with those whites who were most receptive to the idea of a black-white alliance, namely, those in the labor movement. Accordingly, the NAACP's primary function should be to lead the way in forging an alliance of black and white workers. NAACP leaders resisted this demand. In August 1933 dissidents and leaders met at the estate of Joel E. Spingarn in Troutbeck, New York, to try to reach an accord.[2]

The Amenia Conference dramatized the growing insistence on economic considerations in civil rights and left the NAACP with no choice but to take some relevant action, principally through the Joint Committee on National Recovery, whose work became an invaluable basis for NAACP analysis of New Deal legislation, whose research remains insightful about the New Deal's effect on blacks, and whose functioning depended on NAACP support, which, in turn, was contingent upon Hastie's recommendation to Walter White. On White's instructions, Roy Wilkins, the NAACP's Assistant Secretary, wrote to Hastie immediately after the Amenia Conference. He told Hastie that representatives of black churches, lodges, and other organizations would meet in Washington to consider the proposal that John P. Davis and Robert C. Weaver be authorized to continue monitoring the National Recovery Administration (NRA), but under the auspices of those organizations rather than the Negro Industrial League. White wanted to know whether Hastie, with "full authority to speak for the organization in the determination of any plan of action," would represent the NAACP at that meeting.[3]

Hastie was a good choice as the NAACP's representative, in part because of his friendship with Weaver and Davis, who were active in the Negro Industrial League, which concentrated on the status of blacks under the New Deal. To be more precise, Weaver and Davis *were* the Negro Industrial League. Both were Harvard educated (Weaver having received a doctorate in economics and Davis a degree in law), in their twenties, and smart and bold enough to take it upon themselves to testify at the early hearings on NRA codes, advocating fair treatment of black industrial workers and specifying codes that militated against it. White knew that he could not disregard an invitation from Weaver and Davis to attend the meeting at which they would propose to black organizations an alliance that would be known as the Joint Committee on National Recovery.[4]

Hastie attended the meeting as White requested. Within a month the NAACP was one of fifteen (later twenty-four) organizations whose resources were used to launch the Joint Committee on its mission as the paladin of black workers, consumers, businessmen, and farmers under the New Deal. Hastie became a member of the Advisory Committee to the Executive Secretary and White became a member of both the organization's Budget Committee and the advisory board of its Committee on Consumers. With Davis as its executive secretary and Weaver (until he became Secretary of the Interior Harold L. Ickes's adviser on the economic situation of blacks) as its technical adviser, the Joint Committee paid special attention to the NRA, which to black newspapers stood for "Negroes Ruined Again." According to Davis, in its first two years the Joint Committee testified at more than one hundred hearings on NRA codes, conducted field studies "sometimes at the risk of the life of the investigators," and prevented the NRA from imposing differential wages on black workers despite the efforts of some of the most powerful industrialists in the nation. It participated in hearings on cotton contracts that the Agricultural Adjustment Administration

(AAA) held in 1934 and 1935, and it brought farmers to Washington to file complaints with the AAA and the Farm Credit Administration.[5]

Neither the Joint Committee nor any other organization representing blacks had enough power to prevent their mistreatment in the recovery programs. In Hastie's opinion, "The Negro, who shared least in the prosperity of the twenties, is getting more than his share of the suffering of the thirties." When asked years later whether the President had been the black's friend, Hastie said that this had little to do with friendship, for where racial justice was concerned, Roosevelt went only as far as political expediency allowed, though he "genuinely hoped" that the goal would be reached at some future day. But the black's present pain overshadowed the President's future hope, and two years after Franklin Delano Roosevelt took office, John Davis felt that he could say "with reasonable certainty" what the New Deal meant for blacks. It meant that the number of blacks on relief rolls increased from 2.1 million in 1930 to 3.5 million in 1935. It meant that thousands of other black families in rural areas were not reached by relief agencies. Severe black unemployment despite all the recovery programs meant failure, said Davis. "It points as well to the fact that we are fast becoming a race of paupers." Davis and the Joint Committee's technical adviser, Rose Marcus Coe, laid at NRA's feet much of the blame for that pauperism. No mention of race was made in NRA codes; no mention was required to guarantee unfair treatment of blacks. For the NRA permitted exemption from minimum wage and maximum hour standards for those occupations in which black workers were concentrated, sanctioned the establishment of lower minimum wages and higher maximum hours in regions whose work forces were mainly black, and condoned lower wages for blacks than for whites in the same industry. As for the AAA, Davis said that it had used "cruder methods in enforcing poverty on the Negro farm population. . . ." These included acreage reduction, eviction of tenants, extensive foreclosure on properties owned by blacks, and referral of blacks' complaints to their local AAA boards.[6]

As Davis pushed ahead with the Joint Committee, Walter White was turning the NAACP against him. No one disputed Davis's exceptional ability in lobbying among congressmen, philanthropists, and black leaders. Giving him high marks for what the Joint Committee accomplished "under terrific handicaps as to finance, human endurance, facilities for getting the facts, etc.," White said that evaluation of Davis's work must take into account the evils prevented as well as the good achieved. But Davis's personality was not to White's liking, and Davis had been a member of the International Labor Defense, which had bedeviled the NAACP in the Scottsboro Boys case. To White and some

NAACP board members Davis was, as Roy Wilkins put it, "a little too pink for us." Suspicious of the Joint Committee's sources of funds, White had planted a close friend on its staff as its volunteer treasurer and his spy. No funds came to the Joint Committee from the Far Left, but by September 1935 White argued that in the interest of self-protection the NAACP ought to "make a clean break as John may turn the Joint Committee over to the Communists, at least tacitly, if not openly." Unmentioned by White, there was something else that Davis might have done: He might have outdone White in obtaining scarce philanthropic backing; indeed, he might have done White out of his job.[7]

The Joint Committee was doomed because Davis incurred White's disapproval.[8] In White's opinion Davis's demerits included his support from the young dissidents who had carried over from the Amenia Conference their demand for an NAACP economic program anchored in black-white labor solidarity. In 1934 and 1935 the NAACP propagated solidarity and could not do more as long as white workers and their unions, especially the American Federation of Labor (AFL), were Negrophobic. Interracial cooperation became possible with the emergence of leftist unions that originated in opposition to the AFL in the South. The Communist National Textile Workers, the Sharecropper Union, and the United Citrus Workers all fell short of becoming major forces, but they promoted interracialism. So did the Southern Tenant Farmers Union, which fought for both racial and economic fair play. But it was not until A. Philip Randolph forced the Pullman Company to recognize his Brotherhood of Sleeping Car Porters (BSCP) that the AFL gave even slight thought to its own discriminatory practices. The BSCP, NAACP, and Joint Committee tried to persuade the AFL to drop its color barriers, but their pleas were disregarded until another challenge to the AFL arose in the form of the Committee for Industrial Organization, which in 1938 was renamed the Congress of Industrial Organizations (CIO). The CIO reflected the racial fairness of leaders such as John L. Lewis of the United Mine Workers (UMW) and David Dubinsky of the International Ladies Garment Workers Union (ILGWU), who were cofounders of the CIO, and Philip Murray, Lewis's second in command in the UMW and CIO. Some labor leaders remained bigots, but that did not discourage Hastie. "We are a working people," he said. "Although there may be bad leaders in labor, and ministers in the church, whose character we do not admire, we cannot, for that reason alone, stay out of the fold of either." Humanitarianism played a part in the action of many racial progressives in labor, but so did pragmatism. Outside of the union, black workers (as potential strikebreakers) were a source of concern; inside they were a source of strength.

Whatever the motives on either side, the fact that labor and blacks were reaching an accord added weight to the arguments for an NAACP policy of biracial cooperation among workers. In July 1934 the association appointed a Committee on Future Plan and Program, chaired by Abram Harris, to take up this issue and another, namely, a change in organizational structure that would make the association less elitist. The Harris committee's report was drawn up by its radical members, including Harris, after the more conservative ones refused to attend the meeting called for that purpose. Among other things, it urged the NAACP to assume the work of the Joint Committee, to hire Davis to direct an NAACP economic program, and to place that program under the direction of an Advisory Committee on Economic Activities.[9]

White questioned the feasibility of financing the proposed economic program, saying to Hastie that he heartily approved almost all of the proposal but pointing out that Harris, "like many people who have excellent ideas but little practical experience in financing programs," had not addressed that problem. But Hastie thought the NAACP could find the necessary funds if it really wanted to adopt the program, without which it stood no chance of gaining and holding the average man's interest. The problem, as he saw it, was not financial but personal, manufactured by leaders to conceal their reservations about the proposed program.[10]

Hastie, who believed that the NAACP should "emphasize and embody . . . the making of common cause by all working people,"[11] was involved in the controversy as a member of one of several advisory committees to the NAACP Board of Directors. In addition, he chaired a special NAACP committee to combat racism by labor unions. But just when the economic program suggested by Harris's committee was allegedly being implemented, White and Wilkins complained that the NAACP imperiled its own operations through its financial aid to the Joint Committee. As Raymond Wolters has noted, however, "[White] may well have decided that it was best to let the infant joint committee die a natural death because of lack of funds."[12]

For White the last straw might have been the outcome of the National Conference on the Economic Status of the Negro, which the Joint Committee and the Social Science Division of Howard University sponsored in 1935. Taking as their theme "The Position of the Negro in Our National Economic Crisis," and representing eighteen states and the District of Columbia, 250 delegates assembled on 18, 19, and 20 May in Frederick Douglas Memorial Hall at Howard University.[13]

Hastie attended the conference, at which the New Deal was uniformly criticized. His attendance aroused the interest of two investigators from the Department of the Interior, where he was an assistant solicitor. Two months after the conference, they asked him to prepare a written statement about it. He wrote that at the gathering sharecroppers, farmers, industrial workers, and people on relief had given personal accounts of the impact of New Deal programs on blacks. Current and planned recovery agencies were discussed by government officials and other commentators the next day, when the audience had an opportunity to participate. That evening speakers discussed the philosophical and programmatic underpinnings of the economic betterment of black life.[14]

The investigators were particularly interested in Hastie's recollection of any remarks about "revolution" or "bloodshed." Hastie contended that perhaps two members of the audience had said economic betterment for blacks could only be achieved through revolution. He recalled no remarks about bloodshed but remembered that some speakers "recounted instances of their own blood being shed because they had protested against local administration of particular federal recovery measures." Hastie reminded the investigators that the conference had been publicized in advance, that the public and their department (he believed) had been invited to attend, and that "highly reputable" organizations constituted the Joint Committee.

"The conference was noteworthy both for factual presentation and for critical discussion of the economic problems and issues which are the serious concern of American life today," Hastie added. "It is difficult to imagine a subject matter or a type of discussion with which the Division of Social Science of an American University could more appropriately concern itself."

The conferees' announced purposes were to evaluate the New Deal's impact on blacks and to discuss strategy for reducing its harmfulness. No strategy resulted from the conference itself, but at a post-conference meeting—a rump session, one might say—John Preston Davis and others concluded that a new organization was needed to provide the leadership that no other, particularly the NAACP, was providing. As Davis put it, "*Perhaps a 'National Negro Congress' of delegates from thousands of Negro organizations (and white organizations willing to recognize their unity of interest) will furnish a vehicle for channeling public opinion of black America.*"[15]

Such an organization, named the National Negro Congress (NNC), met for the first time in Chicago in February 1936. Attracting 817 representatives of 585 organizations from the District of Columbia and twenty-eight states, it elected John Davis executive secretary and A. Philip Randolph president.[16] That same year, rather than waiting for hours for a train to Washington, Hastie stayed overnight in Philadelphia to attend the regional meeting of the congress the next day. He was impressed both by the enthusiasm of the three hundred delegates and by the fact that the NAACP's own effectiveness in Philadelphia would depend on many of them. He wanted the NAACP to support the NNC, which he thought would give blacks their best opportunity in twenty years to improve their lot. He believed that the NNC would be too important for the NAACP to shun. "It seems to me that the Association should go in or stay out because it anticipates that this is or is not going to be a representative gathering," he told Houston. "If the gathering promises to be significant and representative, as I believe it will be, then it seems to me that mistrust of the inspiring personality is not a sufficient reason for remaining on the outside." To White he wrote: "If the Congress turns out well, it may well prove embarrassing for the Association to explain its aloofness. If the gathering should be prostituted, it is always possible to withdraw." Hastie was not the only NAACP stalwart who hoped that the NAACP would join the NNC. Yet it did not.[17]

At first the NNC seemed destined to succeed, and a number of black leaders supported the organization. But it soon encountered problems because it could not be all things to all black groups. It could not prevent the defection of black conservatives who disapproved of its militancy; it could not raise funds among white philanthropists or among black masses; and it either could not or would not ward off the communists, whose acquisition of control discredited the organization. As a member of the NNC presiding committee elected at the Chicago convention in 1936, Hastie had considered the convention's most important accomplishment to have been attendance by delegates ranging "from the most conservative to the most rabid radical, all of whom came together on a common ground to which all were committed." But by the time he spoke at its convention in Washington in 1940, the NNC was passing not so much into history as into oblivion.[18]

Hastie's support of NNC would come back to haunt him when anticommunist fanaticism swept the nation. He gave that support because he was an institutionalist. He did not think that any single emphasis—civil, political, or economic—should be adopted to the exclusion of the others in fighting Jim Crow, which, after all, was hydra-headed. And far from fearing new leaders, he welcomed them. Roosevelt's program was, as the community newspaper on which Hastie served as columnist and associate editor called it, the "New Deal but old deck."[19]

Hastie was skeptical about the role that established black leaders might be assigned. "The chances are 100 to 1 that the same reliable 'yes' men will be called in to shuffle the New Deal for us," he wrote, "and that they will say, 'Oh, Master, thou hast arranged the cards better in thy infinite wisdom than thine humble servant could ever hope to do. Deal on, and if Sam gets no trumps he will call it the luck of the game and keep on playing and paying as he always has.'"[20]

Hastie had no designs on leadership but was destined for it. While nurturing other organizations, he worked with the NAACP—"Gideon's band," he called it, a collection of blacks and whites who never gave in to apparent hopelessness and constant frustration in their resistance to racism.[21] Blacks had no influence in the White House or state house, the Congress or state legislature, the labor or business centers of power, or anywhere else except, perhaps, in the courts. The economic, political, and social well-being of blacks would depend upon the NAACP's success in laying an adequate legal foundation for equality. And soon after Hastie became a member of the bar, he became a member of the band and made his mark in the NAACP's first case against school segregation.

At the time (1933) Jim Crow had a death grip on equality. As Hastie explained, in the South blacks were not allowed to vote in the Democratic party's primaries, in which victory was tantamount to election, inasmuch as the Democrats dominated Southern politics. Schools were segregated in half of the states, and black schools were grossly neglected in financial and other respects. Hotels, motels, restaurants, parks, theaters, and many other public places were closed to blacks. Trains and other public carriers were segregated. Interracial marriages were prohibited. In Oklahoma blacks violated the law if they used public telephones. Interracial boxing matches were banned in Texas. In Tennessee places of worship were segregated. Segregated housing was commonplace throughout the nation. Adding the final touches to this grim picture, Hastie asserted that whites advocated or acquiesced in the legally segregated social order, while blacks accepted it as their essentially unalterable fate.[22]

In his commentary on the era Hastie wrote, "During the darkest days of World War I the French commander, General Foch, informed his government that his lines could no longer hold and, therefore, he had no choice but to attack. The plight of the American Negro was worse. His subordination in the legal order was so great that he had no lines to hold."[23]

The major legal obstacle facing blacks was the separate-but-equal doctrine that the Supreme Court had promulgated in *Plessy* v. *Ferguson*. On 7 June 1892, Homer Adolph Plessy seated himself in a whites-only car on an East Louisiana Railway in New Orleans. Plessy, a "seven-eighths Caucasian" (the distinction in color being a matter of law, not of vision, for his complexion was exceedingly lacking in pigmentation), had no sooner seated himself than the conductor, almost certainly by prearrangement with the railroad company, ordered him to move to the blacks-only car. He disobeyed the order and was arrested. At his trial in the Criminal District Court for the Parish of New Orleans, Plessy contended that the segregation law was void because it violated the Fourteenth Amendment of the Constitution of the United States. Although he was convicted, on appeal the Louisiana Supreme Court approved the petition that enabled him to put the issue before the Supreme Court of the United States.[24]

In 1896, the Supreme Court ruled in Plessy v. Ferguson that the Fourteenth Amendment had been intended to put blacks on an equal legal footing with whites, "but in the nature of things it could not have been intended to abolish distinctions based upon color, or to enforce social, as distinguished from political equality, or a commingling of the two races upon terms unsatisfactory to either." The all-important question was whether the segregation law was "a reasonable regulation." With its affirmative answer, the Court assured the triumph of segregationists in all facets of American life, seemingly for all time.[25]

In 1929 hopes for an improved legal order were lifted by a pledge of $100,000 from the American Fund for Public Service (AFPS) to the National Association for the Advancement of Colored People (NAACP). Commonly known as the Garland Fund, the AFPS had been established in 1922 by Charles Garland, a twenty-one-year-old undergraduate at Harvard. Unable to square receipt of his share of his father's estate with his belief that no one should inherit a fortune that he had not helped to create, Garland chose the life of a farmer over that of a billionaire and gave $800,000 to establish the AFPS as a supporter of liberal causes.[26]

A number of liberals of various callings administered the Garland Fund. Three of them—attorney Morris L. Ernst, literary critic Lewis S. Gannett, and NAACP General Secretary James Weldon Johnson—formed the special committee that recommended the $100,000 grant to the NAACP. The committee suggested that the grant be used for a legal campaign against those states in which entirely separate but definitely unequal education was provided for blacks. Roger N. Baldwin, founder of the American Civil Liberties Union (ACLU), and some of the other directors of the Garland Fund preferred a campaign to unify black and white workers, but the committee's report won out, the $100,000 was allocated, and $8,000 was given to the NAACP to map out a strategy.

When asked to suggest a chief strategist for the legal campaign, Felix Frankfurter advised the NAACP to consult Vice Dean Charles Houston of the law school at Howard University.[27]

Houston recommended Nathan R. Margold, his fellow editor at the *Harvard Law Review*. A small, brilliant, and intense man, Margold had been a private attorney, a federal prosecutor, and an instructor at Harvard Law School for a year. He had done considerable public-service legal work, and in the year that the NAACP selected him to chart its legal campaign (1930) he had represented the Pueblo Indians in land-title litigation. Financed by the Garland Fund, the campaign would be the first systematic NAACP effort to win redress for blacks in the courts. The committee that recommended the grant had advocated a strategy of filing taxpayers' suits to compel states to provide equal and not merely separate public educational facilities for blacks. The idea was to make segregation a luxury that these states could not afford. If the states lost and appealed cases to higher courts, the results were likely to be decisions of even greater benefit to blacks.[28]

In his report on the legal order as it affected blacks, Margold rejected the committee's proposal of a district-by-district, suit-by-suit, and year-by-year attack on public school segregation in the South, an attack made in hopes that courts would issue writs of *mandamus*—that is, orders requiring public officials to perform their legally prescribed duties. Margold proposed that the NAACP, without challenging the constitutionality of segregation itself, attack instead the states' practice of failing to make the facilities for blacks actually equal to those provided for whites. He maintained that the Supreme Court's decision in *Plessy* v. *Ferguson* required such equality as much as it permitted separation. If the NAACP presented "the proper case" to it, the Supreme Court would rule in its favor—or so Margold reasoned.[29]

According to Hastie, Margold's report became the central battle plan for the NAACP's legal campaign. The rub of the matter, though, was to file "the proper case." That in itself was no easy challenge, and it was but one of many difficulties confronting the NAACP. For example, the organization received only an additional $10,000 from the Garland Fund, which was not immune to the stock market crash. In addition, Walter Francis White succeeded James Weldon Johnson as secretary of the NAACP, fanning the fires of internecine conflict. Furthermore, the South was fiscally unable to provide genuinely equal education for blacks. The list of problems seemed inexhaustible: Few blacks were willing to become plaintiffs in a law suit; most knew little, if anything, about their rights; few black lawyers were available for litigation; and after Margold left the NAACP to become Solicitor in the Department of the Interior, the search for his successor to lead the legal campaign was acrimonious and protracted because Roger Baldwin of the Garland Fund insisted that the new commander be white. For two years Baldwin prevented the selection of the man whom Walter White wanted for the post, Charles Hamilton Houston. But in the summer of 1935 Houston left Howard and became special counsel to the NAACP, and the systematic legal assault on segregation got under way.[30]

As Hastie was to explain, the assault was not easily launched, much less sustained. "At the time there were not ten Negro lawyers, competent and willing to handle substantial civil rights litigation, engaged in the practice in the South," he wrote. "And it was almost impossible to find a white lawyer who dared to represent black clients in controversial litigation, even if he were so disposed. . . ."[31]

Even if disposed to take on civil rights cases, white lawyers were not preferred by Houston. He wanted black lawyers. Hastie's firsthand account was that Houston had taken the reins at Howard University in 1929 in order to recruit, train, and indoctrinate black lawyers. Houston considered it unrealistic to expect white lawyers to help destroy a system from which they themselves benefited. He thought that the inescapable duty of black lawyers was to serve their people. One of his students at Howard, Oliver W. Hill, recalls: "He kept hammering at us all those years that, as lawyers, we had to be social engineers or else we were parasites."[32]

Houston was certain that black lawyers would serve the civil rights cause.[33] Those who were to do so in the thirties and forties would implement the strategy devised by Houston, Hastie, and Thurgood Marshall, their former law student at Howard, who was to become a legendary figure in the struggle for equal rights. These three leaders selected, prepared, and argued cases both to influence judges and to educate the public about racial injustice.[34] Public education remained the core of NAACP legal strategy, but Margold's recommendation of carrying the separate-but-equal issue before the Supreme Court was set aside. Hastie, Houston, Marshall and other NAACP lawyers thought the Supreme Court, if presented with that issue, would reaffirm rather than reverse *Plessy*. As recently as 1927, the Court had declared unanimously that states could constitutionally segregate schools. That was the last word about *Plessy*, and the lawyers thought it unwise to seek the Court's reconsideration of the separate-but-equal doctrine. The toppling of Jim Crow, Hastie would explain, was not an immediate likelihood, but segregated education "was vulnerable to attacks that would erode it and might lead to its destruction sooner than most people thought possible. . . ."[35]

To promote that erosion, NAACP lawyers chose as their first target segregated education on the graduate level in the South. One reason for this choice was that there were no separate, much less equal, facilities for such training of blacks there. Given the costs of providing that training for the relatively few blacks who would be applicants, state officials might relent and allow them to enroll at white institutions. The choice of battlefields should take into account the possibility that judges could be convinced that blacks were entitled to graduate and professional training when they qualified for it, rather than having to wait until separate institutions were provided for them. If judges were persuaded to that view, said Hastie, the "only meaningful remedy" would be to order the admission of blacks to "white" institutions until "black" ones were provided. Judges who did not want to decide the constitutionality of the *Plessy* doctrine could stop short of doing so, and yet "the state-erected wall of segregation" would be breached and blacks would be placed in "a strategically advantageous position" from which to improve their access to training.[36]

Commenting on the decision to attack segregation in graduate rather than primary or secondary education, Thurgood Marshall said, "Those racial supremacy boys somehow think that little kids of six or seven are going to get funny ideas about sex and marriage just from going to school together, but for some equally funny reason youngsters in law school aren't supposed to feel that way." Commenting on that reasoning, Marshall has said, "We didn't get it, but we decided that if that was what the South believed, then the best thing for the moment was to go along."[37]

Hastie and Houston could not have known that two of their former students at Howard, Conrad O. Pearson and Cecil A. McCoy, would put this grand strategy to the test during the period after Margold had left the NAACP but before Houston took his place. Pearson and McCoy took that initiative in March 1933 in Durham, North Carolina. For months they had searched in vain for a plaintiff to file a desegregation suit against the University of North Carolina. They trudged through the rain and mud, searching for one. None could be found. Finally their search led them to twenty-four-year-old Thomas R. Hocutt, an assistant headwaiter at the Washington Duke Hotel. Hocutt had always wanted to be a pharmacist, and had worked in a drugstore in a nearby town. But neither his aspirations nor his work experience served him well; the only public school of pharmacy in the state at the University of North Carolina excluded blacks.[38]

On 13 March Pearson, McCoy, and Louis E. Austin, publisher of the *Carolina Times* in Durham, accompanied Thomas R. Hocutt to the University of North Carolina. They requested his enrollment in the school of pharmacy and his assignment to a dormitory room. In doing so they angered not only white leaders, who opposed racial progress, but also black leaders, who feared racial violence. That fear was well founded, for in 1933 North Carolina was fourteenth among the forty-four jurisdictions (including "Alaska and Places Unknown") in which 4,951 persons were lynched between 1822 and 1927. The victims were 1,438 whites and 3,513 blacks. In North Carolina 80 out of the 100 victims were blacks.[39]

Because of his fear of lynchers, C. C. Spaulding, president of North Carolina Mutual Life Insurance Company, tried to discourage Pearson and McCoy when they requested his support. Spaulding, however, puzzled them. Having castigated them for taking Hocutt to the university, he commended them to Walter White when they sought the NAACP's assistance. He might have been trying to set the stage for compromise with state officials in hopes of obtaining out-of-state tuition grants for blacks seeking graduate education. If so, he neglected to inform Pearson, McCoy, or White.[40]

Considered by Pearson even more of a Judas than Spaulding was Dr. James E. Shepard, president of North Carolina College in Durham. Shepard and other black educators were alarmed by the attempt to integrate the university. They dreaded being caught between young black radicals, on the one side, and reactionary state officials, on the other. In addition to out-of-state tuition grants for blacks who qualified for (but were denied) admission to the in-state university, Shepard favored the establishment of departments of law, medicine, and pharmacy at his own college. He argued that blacks preferred racially separate education, provided it was indeed equal, a condition that had been made even more unlikely by the state legislature's reduction of the budgets of black colleges. In a discussion with Nathan C. Newbold, Director of the State Division of Negro Education, Shepard promised that he would try to calm the radicals.[41]

Shepard also made a promise to Pearson and McCoy. Even as he undercut them by refusing to release Hocutt's college transcript, he said that he would respect the confidentiality of their plans. But they awoke one morning to find that Shepard's public relations man had released to a Greensboro newspaper details of their plans, including their intention to file suit on Hocutt's behalf. To counter that disclosure, and to win favor with the white press, Pearson and McCoy offered to tell the editor of a newspaper in Greensboro all that was going on. The editor refused to hear them out and instead countered with, "I wish all of you were back in Africa."[42]

Shepard's treachery and the editor's bigotry were part of the discouragement that Pearson and McCoy experienced as Hocutt's advocates. They had been heartened by Spaulding's help in their bid for NAACP assistance but were dismayed by behavior that elicited from Walter White the colloquial complaint that when the kitchen got hot Spaulding got out. The local NAACP branch had chosen Pearson to chair its legal committee and McCoy to be the branch secretary. Encouraged by their election, they were soon disheartened, McCoy told White, by the "group of spineless persons . . . scuttling to cover at the first stage of the action."[43] Still, the lawyers remained resolute and earned a reward from White.

The reward was, in McCoy's words, "the good news" from White, who had told them to expect a "competent person" to join them the day before they were to go to court. The person whom White hoped to dispatch to Durham was Houston, his unofficial legal adviser. Houston would rather have had another time, place, and team for the historic confrontation. The economy was ailing, North Carolina was not a border state, and Pearson and McCoy were inexperienced attorneys. But the Tar Heel State was not the worst possible battlefield, plaintiffs were hard to come by, and although Houston could not free himself from his duties at Howard and his obligations as NAACP counsel in a murder trial, expert legal guidance was available to Pearson and McCoy. Hocutt's case was the first opportunity to initiate the revised NAACP legal strategy of challenging Jim Crow. Unwilling to allow the opportunity to slip by, Houston recommended Hastie for the assignment in Durham.[44]

Houston's recommendation was the second vote of confidence in Hastie to be registered with White. The first had been given by Felix Frankfurter, who in 1930 had advised White "to keep your eye on . . . one of the finest students who has ever studied at Harvard during my time."[45]

When Houston commended him to White, Hastie was back at Harvard, studying for a doctorate in juridical science. After having taught at Howard, which he said was "the one institution to which a colored man can at present look for an opportunity to teach law," and having practiced law in the firm Houston and Houston (father and son), Hastie had been accepted for graduate work at Yale and Harvard. He elected to return to Harvard on a fellowship, informing President Mordecai W. Johnson of Howard University that he hoped to return "a year hence with definitely improved equipment for the teaching of law."[46]

By telephone White talked to Hastie and Houston in Boston before Hastie departed for Durham, eager to help Pearson and McCoy in a lawsuit whose special interest to him we gather from introductory notes he was to scribble for a speech at a school of pharmacy many years later: "Friendship with pharmacy through father." His father had been trained as a pharmacist but had been unable to find work in that capacity.[47]

Hastie was also drawn to Durham by the friendship that bound faculty and students at Howard University. The appointment of Mordecai Johnson at that institution had provided more than a strong black man as president; it had also infused the entire university with what Hastie described as "a remarkably buoyant atmosphere." The bonds of comradeship developed at the law school withstood the separation that followed graduation in many, even most, instances, and certainly in the one that reunited Hastie and Pearson. Recalling their days as professor and student at the law school, Pearson says, "Bill had a reputation of being a very, very able man, very fair, affable."[48]

Immediately upon arriving in Durham, Hastie conferred with Pearson and McCoy. As Pearson recalls, "Bill saw several things he thought would have been better if we had done them . . . but we didn't have time to change anything." Judge M. W. Barnhill was to hear the case the next day. When Hastie, Pearson, and McCoy arrived at Superior Court, they found Victor S. Bryant, a prominent local attorney, several ranking state officials, including Attorney General Dennis G. Brummit, and representatives of the University of North Carolina. "They wanted to talk to us. So we went in to talk," says Pearson. Outside, some five hundred blacks chanted "Don't give in! Don't give in! Don't give in!" Hastie, Pearson, McCoy—and Thomas Hocutt—were not about to give in. Unable to reach an agreement that would have admitted Hocutt to the university, they told their rivals, "Gentlemen, we will have to try the case."[49]

The case was an event. Other courts recessed to enable lawyers to attend it. Duke University Law School professors and students also attended, as did their counterparts at the University of North Carolina. All were interested in the case, as well as in Hastie's performance. To Hastie their presence was indicative of a level of white communal hostility lower than he had anticipated. *Capacity Crowd . . . Town Agog*, he telegraphed Walter White. With blacks' closing of ranks behind Hocutt's team also in mind, he added: *Incalculable Good Done Whatever the Outcome.*[50]

Hastie might have been even more optimistic about the outcome had the trial been taking place in another state. North Carolina, says Richard Kluger, "might have enjoyed relatively peaceful race relations, but no one was putting up statues of Harriet Beecher Stowe in front of its county courthouses."[51] Certainly not the courthouse in Durham. In front of it stood a huge statue that featured a soldier and bore the inscription "In Memory of the Boys Who Wore the Gray."

Despite the symbolism of the statue in Durham, blacks "packed the courthouse like a sardine box," according to Conrad Pearson. Having held a mass rally on Sunday night in high spirit, they came to the courthouse on Tuesday. In court, State Attorney General Dennis G. Brummitt declared, "I think there is a deep motive behind this suit and I think that motive is that this 'Nigra' wants to associate with white people." Whites remained quiet, but blacks laughed loudly and derisively.[52]

Blacks laughed knowingly because they understood (as even whites must have understood) that whites themselves were all in favor of interracial association so long as it was on their terms. But there was greater irony in the situation, for during the period that Thomas Hocutt was refused admission to one university in North Carolina, Peele Hackney was given the chance to enroll at another. Paroled in 1930 after having served thirteen months for highway robbery, Hackney received a full gubernatorial pardon that made him eligible for admission to Duke University. Hackney was a felon, but Hocutt was a black.[53]

In court, the state and university parried Hocutt's thrust with three principal contentions. The first was that his failure to furnish a copy of his college transcript constituted noncompliance with the university's regulations; hence its refusal to admit him. Attorney General Brummitt also argued that the university did not have the authority to disobey a state law that required segregation in education, a requirement that the Supreme Court, if presented with it, would sanction. The third contention by the state and university was that the court would be taking improper action if it were to grant the relief that Hocutt had requested, namely, a writ of *mandamus*.[54]

The case revolved around the first and third of these contentions. Invariably its outcome—a loss for Hocutt—is explained in terms of the first, namely, his inability to provide a college transcript. In ruling against him, Judge Barnhill did cite Hocutt's failure to meet the requirement of "necessary evidence of scholastic qualifications." But Barnhill's decision did not turn on that point; rather, it turned on the state's third contention, which claimed that the court, if it were to honor Hocutt's request for a writ of *mandamus*, would be overstepping its authority. In other words, his court did not have the authority, in this instance, to compel a public official to discharge his legal duty. Although Hocutt wanted the court to order the university's registrar to enroll him, Barnhill held that he could not; at most he could have ordered the university to act on Hocutt's application in good faith and without regard to his African heritage.[55]

Barnhill did not want his opinion to rest solely on the technical point that improper relief (i.e., *mandamus*) had been sought. Nor did he want it to be devoid of facts that would figure in an appeal, should one be made. He therefore set forth the facts of the case as he had determined them. For example, he ruled that the university had not, in good faith and without regard to Hocutt's African heritage, considered his application in order to ascertain whether he had complied with its rules and regulations governing admission. But Hocutt had not sought, and Barnhill did not make, a ruling as to whether the university was obliged to admit for professional training a "person of African descent" who met its admissions requirements and for whom such training was not provided in the state-supported black colleges.[56]

But what if Hocutt, instead of requesting a writ of *mandamus*, had asked the court for a declaratory judgment? The writ involved Barnhill's ordering action by university officials, but a declaratory judgment would have involved nothing more than his specifying the rights of the parties involved in the case, or expressing his opinion on a legal question, without ordering anyone to do anything. "I am kicking myself all over the place for not having thought of that possibility long ago," Hastie told Walter White. Notifying Pearson and McCoy that the "perennial schoolboy is back in Washington about to undertake again the serious task of earning a living," Hastie asked them to consider the possibility of seeking a declaratory judgment. He thought "a judge like Barnhill," in a proceeding concerning denial of admission to the university because of the applicant's color, would have ruled in favor of Hocutt.[57]

Maybe Barnhill would have rendered a declaratory judgment favorable to Hocutt—and maybe he would not have. At the time of Hastie's arrival in Durham, Attorney General Brummitt and university officials wanted to strike a compromise with Hocutt. They argued that Hocutt's withdrawal of his lawsuit would improve the likelihood of North Carolina's paying the tuition of blacks who attended out-of-state schools. Hocutt's lawyers demanded assurance of such financial assistance to blacks. Brummitt and his allies could not give that assurance. Hastie suggested that they ask the court for a postponement in order to have time to persuade the state legislature to furnish the guarantee. When Brummitt refused to make the request, Hastie made it. Barnhill denied the postponement.[58]

Barnhill thought the lawsuit was a mistake. Although he had told Pearson and McCoy that, they were willing to have him hear the case because he had a reputation for being a decent man—as white Southerners went. The decency became evident as Brummitt argued that the goal sought by Hocutt was not education but racial intermingling; Barnhill upheld Hastie's quiet objections. But the Southerner in Barnhill was also evident as time and again he tried to derail Hastie. Hastie, Pearson, and McCoy drafted an order that they hoped Barnhill would issue to the university's registrar. The draft order stated that the court, "being willing that justice should be done," required reconsideration of the application without regard to Hocutt's color or race. The draft order would also have had the university inform Hocutt of any deficiency in the application and allow him reasonable opportunity to correct it before deciding whether to accept or reject his application. Barnhill rejected the draft.[59]

Hastie won the admiration of lawyers and other observers, including Barnhill, but he would rather have won the case for Hocutt. Having lost, Hastie and his allies considered the advisability of appealing Barnhill's decision. Hastie admired Hocutt's courage and thought that replacing him with another applicant "would seem rather harsh." Actually, Hastie would have preferred a scholastically stronger plaintiff. "Not only is it necessary that the person chosen be qualified beyond possibility of doubt, but it is also desirable that he be of *outstanding* scholarship," he told Dr. J. M. Tinsley, president of the NAACP in Richmond. The plaintiff should also be courageous, poised, quick of mind, neat, personable, and unmistakably a black.

"At first blush some of these qualifications may seem unimportant, but it is to be remembered that the applicant chosen may have to take the witness stand as the central figure in a much-publicized trial," Hastie continued in his letter to Tinsley. "The eyes of the whole state will be upon him . . . eager to find fault. It is important that he measure up. From another approach, it is a valuable object lesson which shows the whites in the community that there are Negroes in the community who measure up in every respect to collegiate standards."[60]

Hastie's performance served as an object lesson with every regard to professional standards. His artistry had attracted lawyers and laymen throughout the state, and Barnhill had said that *Hocutt* was among "the most brilliantly argued" cases he had heard in twenty-two years as a judge.[61] Walter White, informing Hastie that the state senate had killed a bill that provided out-of-state grants for blacks, also told him that all that President Frank P. Graham and faculty members at the university could talk about was "the way you showed up the Attorney General." From other whites as well as from blacks, "paeans of praise" for Hastie were all that White heard. He asked Hastie to reserve two days (in October 1933) to accompany him to Raleigh to organize a state conference of NAACP branches to "wage a relentless fight" not only to make professional training at the University of North Carolina available to Negroes but also to eliminate racial disparities in teachers' salaries. When White promised to try to bring Hastie with him, "[Blacks] whooped with joy and said we would have a tremendous turnout." Pearson and friends said they would "stock up" in anticipation of Hastie's return.[62]

White's letter drew a good-humored response from Hastie. "Did you say the gentlemen are 'stocking up' in anticipation of my return? What do they call 'stocked up'? Wherever I went on my last trip I found a gallon or more of that colorless fluid, so like water to the eye, but to the olfactory and gustatory senses like nothing else that man ever brewed."[63]

At the time, an appeal of Barnhill's decision was under consideration. McCoy favored that course of action, but Hastie opposed it, saying that an appeal probably would be "good money wasted." Barnhill had said that he did not want the ultimate outcome of the case to turn on a technicality, namely, Hocutt's having requested improper relief (writ of *mandamus*). But in retrospect Pearson says that the crafty judge knew full well that no appellate court, including the Supreme Court, would hear an appeal of a decision that, like Barnhill's, had been based on a procedural matter. Seeing no possibility of success on appeal, Hastie thought that any further action should take the form of a new lawsuit. In the end, however, neither an appeal nor another suit was filed.[64]

Hocutt, the man and the case, faded into obscurity, but both merit a place in history. Sixteen years after the decision in Durham, McCoy would tell Hastie how thrilled he had been "to observe the beautiful working of a fine and brilliant mind and the clear-sighted and judicious counsel your deliberations afforded us." He had not forgotten the "very exciting cause in which three of us were pioneers struggling for a claim of right since vindicated."[65]

Hastie, too, remembered Hocutt's lawsuit. It had "seemed no more than a ten-day wonder"—segregation was so entrenched—"but it started something. It was a first step toward eliminating the legal and moral contradiction of racism in the scheme of education for life in a democratic society."[66]

In the years to come, Charles T. Duncan, dean of the law school at Howard University, would claim that Hastie and other black lawyers had made possible the enjoyment of freedom's blessings by women, civil servants, military personnel, homosexuals, senior citizens, and virtually all other persons seeking equal justice under the law. Duncan saw this as a novel and unpredictable legal development and said that in all probability it originated with *Brown*.[67]

In *Brown* v. *Board of Education*, the Supreme Court overturned the separate-but-equal doctrine that had been enshrined in *Plessy* v. *Ferguson* for more than half a century.[68] As a lawyer, Thurgood Marshall led the forces that won the legally conclusive victory in the struggle over public school segregation. He had become a member of the Court when, assessing the significance of the lessons learned in *Hocutt*, he said that in it "the groundwork for the future was laid." The future, he added, was *Brown*.[69]

CHAPTER SIX
Teachers' Pay

Even as Hastie and his allies debated whether to appeal the *Hocutt* decision, one of them, Walter White, was laying the groundwork for another battle in the Tar Heel State. North Carolina was one of the states in which public schools were segregated and teachers in black schools were paid less than those in white schools solely because of their race. The NAACP intended to put an end to the widespread salary discrimination against black teachers, and, as White informed Hastie, the organization wanted him to lead the campaign, which would begin in North Carolina.[1] But why this campaign? Why North Carolina? And why Hastie?

The reasons for the campaign were fully appreciated by Hastie. He understood, for example, the individual hardships that discriminatory salary schedules created for black teachers.[2] And he understood the adverse effects—and not solely the economic ones—that black communities suffered as a result of losses in the incomes of their teachers. It was reported in 1940 that nationwide the losses amounted to at least ten million dollars annually.[3]

Hastie was fully cognizant of an additional reason for the campaign, one articulated by his colleague Thurgood Marshall, who saw the battle in terms of eradicating the general notion—no matter what the job—that blacks should not be paid salaries equal to those received by whites.[4] There was, in addition, merit in the notion that government should not racially discriminate in paying its employees. Although teachers were not the only victims of racial discrimination at the hands of government, they were, as Richard Kluger, the author of *Simple Justice* notes, important in their communities.[5] When Hastie referred to the fight for equal education as "the decisive campaign," he included this issue about teachers' salaries.⁶

As Hastie explained, the campaign not only provided relief for teachers but also showed other blacks the support available to them. According to him, it electrified the community. "Fear and frustration were replaced by boldness and hopeful eagerness to be identified with the struggle. It may reasonably be doubted whether impetus essential to radical change in the status of the Negro would have developed without this initial stimulation and mobilization."⁷

Durham was one of the communities stimulated, a development that encouraged the NAACP to initiate the campaign in North Carolina. In this regard, Walter White tried to capitalize on the popularity that Hastie had gained there.⁸ It was not, however, only those considerations that motivated White, who acted with one eye on blacks and the other on Reds—the communists. He had been informed that the International Labor Defense (ILD), the legal arm of the Communist party, might take up the salary-equalization cause and prevent the NAACP from obtaining "concrete results."⁹ As part of the Communist party's determined (but notably futile) effort to win blacks to their side, the ILD had defended nine blacks who stood trial in Scottsboro, Alabama, in 1931 on charges of having raped two white women on a freight train. The case seemed to be made to order for the NAACP, which entered it belatedly but left it eventually, forced out after the defendants, their parents, and their guardians chose the ILD as defense counsel.¹⁰

White agreed with his informant that the perfect antidote to the ILD in North Carolina was Hastie's involvement on behalf of the NAACP. Answering the call to arms again in North Carolina, Hastie said that, given "the Alabama mixup over the Scottsboro case," he understood White's position on the ILD, but that he himself had no "great concern" about the possibility of its intervention. In fact, he regretted the timing of White's announcement of the campaign; word about the NAACP involvement would not deter the communists if they had already decided to act, but it might well deter the teachers from taking the initiative themselves—they would gladly allow the NAACP to fight their battle—and *that* was Hastie's great concern. Hastie thought it best for the NAACP to be seen as an auxiliary to the teachers rather than as the instigator of the fight. He also believed that the NAACP should help the teachers only if they helped themselves. As he put it, "I feel very strongly that an educated professional group must be required to pay its own way if the National Association is to lend aid." Suggesting that this willingness on the teachers' part be ascertained and that the pertinent legislation and board of education rulings in North Carolina be obtained, Hastie declared that "it's full steam ahead, and to hell with the torpedoes."[11]

Soon thereafter Hastie, by then a member of the NAACP's National Legal Committee,[12] returned to North Carolina as the NAACP's representative. The comment by the Greensboro Daily News that "there is insurgency in the air and there is a moving toward the courts" led him to say to White, "If there is enough 'insurgency' we may be able to stimulate the 'moving toward the courts.'" But his misgivings about insurgency on the part of the crucial group (black teachers) was evident in his remark that if they prized their self-interest enough to finance the campaign, it could mark the opening of an important chapter in Southern history. Preparing for battle, he asked Conrad O. Pearson and Cecil A. McCoy, his co-counsel in Hocutt, to obtain all legislation and orders concerning school appropriations in 1933.[13]

Upon his arrival in Durham, Hastie discovered that blacks and whites alike were displeased with this legislation of 1933. He informed Walter White that the state legislature, succumbing to pressure from "certain large interests [that] pay heavy taxes," had cut appropriations for the public schools from twenty-five million dollars to sixteen million in that year. Centralized control and uniform practices had been instituted, but not so rigidly as to rule out discrimination. For example, state officials were required to set, separately for blacks and whites, the total costs of salaries for teachers, principals, and superintendents in the counties, but county boards of education were authorized to pay teachers less than the maximum allowed by the salary schedule.[14]

Hastie told White that he had called on the Director of the State Division of Negro Education, Nathan C. Newbold, who was white. "[He] was in every respect courteous . . . and actually said 'Mr.' without any effort," Hastie added. "He did not refer to his black mammy, but he did mention his Quaker 'pappy.'" Hastie thought Newbold probably performed his duties—which, said Hastie, were to keep blacks happy while trying to improve their schools—as well as any white man could. Newbold admitted there was salary discrimination and said that he was trying to reduce the differential it produced. He was willing to ask State Attorney General Dennis G. Brummitt, Hastie's adversary in Hocutt, whether the differential was unconstitutional, but the authority to make that request rested with the State Superintendent of Public Instruction, A. T. Allen. Newbold escorted Hastie, McCoy, and Pearson to see Allen, who refused to make the request of Brummitt. Allen admitted the theoretical possibility of unconstitutionality but stressed the impracticality of eliminating the differential at that time. Actually, Allen, Newbold, and Brummitt all doubted the constitutionality of the

1933 legislation. Brummitt preferred local rather than state control over education, probably because anyone who contested discrimination in salaries and conditions of employment would have had to take on several hundred jurisdictions rather than a single one.[15] Concerning Allen and Newbold, Hastie told White, "Needless to say, both gentlemen pointed with pride and emotion to the great progress the colored schools have made under their administration."[16]

Welcomed by some blacks, Hastie was resented by others, including the local NAACP branch president, J. N. Mills, who explained to White why state officials had not received a greater number of telegrams protesting salary discrimination. Mills said that Hastie had not pursued "the proper method," for he had gone to Raleigh and created a controversy with Newbold and Allen without having made "the proper contacts with the leading influences in Durham beforehand."[17]

Displeased because Hastie, Pearson, and McCoy had not informed local NAACP officers about their intentions, Mills said that this was the reason that "their steps were not altogether endorsed." He approved the local teachers' preference that the NAACP follow rather than lead their association.[18] Hastie shared that view, claiming the teachers' association was "the ideological organization to lead the fight." Unfortunately, it was controlled by reactionaries who were almost certain to be apologists for the state, which they would portray as making the best it could of bad circumstances. "We hope that such a move will be blocked by a stream of telegrams from all over the state."[19]

Developments up to that point provided greater reason for despair than for hope. For example, a delegation of blacks had petitioned the State School Commission, which (with Superintendent Allen) drew up salary schedules. These blacks requested the elimination of salary differentials and the replacement of Newbold by a black. For their efforts, however, the delegation's members—C. C. Spaulding, J. T. Taylor, and W. G. Pearson—were held in contempt by young blacks in Durham who viewed this petition as ineffective, self-serving, and even subversive. Hastie, who saw its maneuver as a futile and regrettable attempt by "old heads" to project themselves to their community as aggressive leaders, told White, "It is probably enough to state that the Negro spokesman on this occasion prefaced his petition with a statement that he was born and bred in North Carolina, and hoped to die there, and be buried in the soil of that state."[20]

Hastie felt that White had been unnecessarily worried about what the ILD might do in North Carolina. In his opinion there was greater reason to worry about what the North Carolina Negro Teachers' Association might not—indeed, probably would not—do. Trying to thrust the association into the battle, Hastie joined Pearson and McCoy in visiting several cities and sending telegrams to other communities. Meanwhile, White sent wires to local NAACP chapters. The editors of black newspapers in Durham and Raleigh promised to help, and Hastie believed that both would. He held no such belief about some black leaders, but felt nevertheless that the principal responsibility for the campaign belonged to North Carolinians—if not the association then the local parent-teacher associations (PTAs) or the state NAACP.[21] But would they, as he hoped, cooperate with each other?[22]

It was then September 1933, and Hastie had been in Durham for just four days, long enough to become pessimistic about the initiation of the protest he and White preferred. Still, he could tell White that some of the "moving toward the courts" was evident.[23] The national NAACP team in this campaign was the same as in the *Hocutt* case several months earlier, with the twenty-eight-year-old Hastie cast as leader and supported by Pearson and McCoy. Hastie was, therefore, principal among the first three NAACP lawyers who fought educational racism in court.[24] This made him a chief strategist in the campaign that led to the equalization of teachers' salaries.

For a while Hastie pondered the advisability of North Carolina teachers refusing to sign the discriminatory contracts offered to them. But as he told Walter White, those who did so might lose their right to complain afterward. He did not expect refusal by a significant number of teachers in any community, but he hoped that all of those in several communities would sign under protest, notifying the local administrative boards that they reserved "any rights they may have in the premises"—including, presumably, the right to challenge their salaries later.[25] This strategy contained the novel feature of seeking a middle ground—that is, conditional acceptance—in contractual terms, which had always involved only the alternatives of acceptance or rejection by the parties.[26]

Hastie doubted the legality of the proposed protest but thought it might have "some effect."[27] In the end, this point in strategy would figure importantly in the most decisive litigation, namely, that in Virginia eight years later.[28] His doubts notwithstanding, Hastie drafted the appropriate petition to be signed by teachers protesting discriminatory salary terms in their contracts.[29] The petition, which Hastie sent to Pearson for use by teachers at any school, notified school officials that the signers requested a new contract free of such discrimination; pending receipt of it, they would perform their duties for the salary offered in undetermined partial compensation. "I have been giving considerable thought to the matter of a teacher refusing altogether to sign his contract," Hastie explained. "If he does not sign at all, he is automatically out of the system. I believe the best procedure is for the group and also the individual teachers to submit such a protest as I am outlining. The teacher should hold his contract until the school board answers. If it refuses to increase his salary, he can then sign, submitting this contract and formal letter of protest reserving his rights in the premises. My personal thought is that such a procedure is more advantageous than an absolute refusal to sign."[30]

Hastie was idealistic enough to prefer that the teachers' association take the lead in the campaign but practical enough to suggest to White—who agreed—that the NAACP branches arrange meetings at which the petition would be made available. Telling Pearson and McCoy this, White passed on Hastie's suggestion that the meetings be held "on the night (the teachers) get their contracts, when they will be hottest over the discrimination."[31] Hastie hoped that at least one or two of the teachers would refuse to sign a contract for less pay than similarly qualified white teachers were to receive. "In case the discriminations are not wiped out," he told White, "these persons will bring suit. They will probably lose their jobs, so we are trying to get persons who do not plan to teach next year. We also hope to raise money to compensate them for any loss they may suffer this year. We are also trying to develop other plans for suits by persons who are not teachers."[32]

This was innovative strategy, devised for litigation that would itself be, as Thurgood Marshall pointed out, unprecedented. The campaign gave black North Carolinians their second chance (after *Hocutt*) to lead the way in toppling racial barriers, for had it been successful, one could have said about litigation there precisely what Marshall was to say about that which followed in Maryland several years later. "This case not only means that the salaries of teachers will be equalized and is a definite step in our campaign to equalize educational opportunities but also has an effect on our entire program," he explained.[33]

But black teachers in North Carolina would have nothing to do with such a destiny. As Hastie was to write, "Indeed, numbers of black school officials who enjoyed a vested interest in the segregated system were openly or covertly hostile [to litigation] . . . In North Carolina, where the black teachers had a well-organized professional association, pressure and persuasion by black school administrators effectively prevented teacher support of contemplated salary litigation, though thereafter North Carolina teachers were among the beneficiaries of successful litigation undertaken by Maryland and Virginia teachers."[34]

Stymied in North Carolina by what White called "the activities of those with weak knees," Hastie and the NAACP were on firmer ground in Maryland. For example, when William B. Gibbs, Jr., the acting principal of a black elementary school in Montgomery County, decided to sue for equal pay, he asked the NAACP to take the case. The next year (1937) an out-of-court settlement provided for the equalization of salaries.[35] Following Gibbs's example in 1937, Elizabeth Brown, an elementary school teacher in Calvert County, also forced another out-of-court settlement under which salary differentials were to be cut by one third that year and eliminated in 1939. Although Governor Harry W. Nice publicly endorsed equal salaries, the legislature raised white teachers' salaries, enlarging the differential and provoking a third lawsuit, one that would be settled in court. The plaintiff, Walter Mills, was represented by Hastie and other NAACP lawyers.[36]

Before taking the case of Walter Mills, an elementary school principal in Anne Arundel County, NAACP lawyers had been attacking on a county-by-county basis, an approach that involved prolonged and expensive litigation in state and local courts. Because the counties (and Baltimore City) received state educational funds, and the Fourteenth Amendment prohibits states to deny equal protection of the laws to persons subject to their jurisdiction, Hastie and his colleagues thought they might obtain from the United States District Court a favorable statewide ruling that would make the county-by-county campaign unnecessary. Such a ruling would spare the NAACP and plaintiffs as much as five years' effort in Maryland and establish a precedent that would be advantageous to them in other states. So Mills sued the state—and lost. One reason for his defeat was that the county board of education had not been made party to the suit along with the state (and in Maryland the counties and Baltimore City made and enforced policies governing teachers' qualifications and compensation). The judge added, however, that Mills could amend his complaint.[37]

Amended to apply only to the county board of education and superintendent, the complaint was the subject of Judge W. Calvin Chestnut's opinion on 22 November 1939. In nine months since his first decision, the state legislature had increased the minimum salaries paid to white teachers but not those paid to black teachers. In Anne Arundel County, an elementary school principal who had two to four assistants and had qualifications comparable to those of Walter Mills received a minimum salary of $1,550 if he were white and $995 if he were black. No black teacher earned as much as any white teacher.[38]

"The crucial question in the case," said Judge Chestnut, "is whether the very substantial differential between the salaries of white and colored teachers in Anne Arundel County is due to discrimination on account of race or color. I find as a fact from the testimony that it is. . . ." He ordered the cessation of that discrimination.[39]

After Chestnut's decision in *Mills*, other cases pending in Maryland state courts were dismissed, the parties stipulating to an equalization of salaries in the next school year, 1940–41. Sixteen months after the decision the state legislature equalized salaries. For his part in these developments Hastie had been placed on the 1939 honor roll of a national magazine, *The Nation*. The same recognition was accorded to his co-counsel in *Mills*, Thurgood Marshall, Leon A. Ransom, W. A. C. Hughes, Jr. (a Baltimore lawyer), and to Judge Chestnut. Congratulated by a friend for "the swell victory" in the case, Hastie described it as "something of real importance."[40]

Hailed for his participation in the Maryland campaign, Hastie won further commendation for his contribution to the one that followed in Virginia. Walter White's assessment of him a month after *Mills* was this: "Bill Hastie has been devoting his nights, days, Saturdays and Sundays to the Maryland teachers' salary case and since that decision was handed down in our favor, has been trying to catch up with his work at the Law School and to prepare for the teachers' salary case in Virginia in which we will have to go to trial in February. . . ."[41]

The rippling effects of *Mills* had reached Virginia, and black teachers there donated one thousand dollars to the NAACP and vowed to help it raise five thousand. But Hastie and other NAACP lawyers needed more than money. They needed litigants. In October 1938 Aline Elizabeth Black, a high school teacher, had filed suit against the school board in Norfolk. The board, however, made the effective countermove that Hastie had feared its counterparts in North Carolina would have made in 1933. Aline Black lost in the city court, and before her lawyers could file an appeal the school board refused to renew her contract, thus disqualifying her as a party to a lawsuit.[42]

Black's disqualification thwarted the NAACP in Virginia even before Walter Mills's first case in Maryland. When Hastie and other NAACP lawyers returned to Virginia, they again encountered the problem of finding one or more persons both eligible and willing to file suit against discriminatory teachers' salaries. Eventually Melvin O. Alston said that he would file suit provided the NAACP guaranteed he would receive one thousand dollars if he lost his job.[43] Before the second *Mills* case was decided, Alston had said he was willing to be a plaintiff, and black teachers in Virginia had raised more than thirty-five hundred dollars to finance litigation,[44] but none had been initiated at the time. Alston had not actually committed himself, and NAACP lawyers were, as always, shorthanded. Hastie joined the NAACP team in time for both the second *Mills* case in Maryland and for the planning and prosecution of the case in Virginia, which was to prove even more important in the overall campaign.

In one respect, the campaign in Virginia, which sputtered along as that in Maryland neared its climax, must have been an uncomfortable reminder of the abortive effort in North Carolina six years earlier. The NAACP made a practice of soliciting the cooperation of local black attorneys, but those in Norfolk were seriously at odds with the local teachers' association and its president, Alston. The latter's lawyer had been offended by the oversight that resulted in the omission of his own name from the petition that was sent to the local board of education after Alston's vacillation ended. Alston finally agreed to file suit against the board of education in Norfolk but then notified Marshall that the local black teachers' association refused to join him in the suit. He explained that some teachers were afraid of possible consequences for "their vital welfare" should litigation be conducted in the organization's name, and he was "somewhat averse to going to court" without its backing. Asked by Marshall for his views about this, Hastie said that he hoped the organization's refusal to authorize litigation "does not mean cold feet. I am sure the presence of the association as a party plaintiff will help."[45]

Complaining that "we should be able to concentrate our time, money and efforts on fighting the common enemy," Marshall decided to go to Norfolk. After conferring with Hastie and others in Washington, he attended a meeting of the teachers in Norfolk. There he calmed the waters and the Norfolk Teachers' Association voted unanimously to join the suit as a plaintiff.[46] The vote cheered him, but when the efforts of Hastie and his colleagues—Marshall, Ransom, and, in Virginia, Oliver W. Hill and J. Thomas Hewin, Jr.—took the case to the United States District Court for the Eastern District of Virginia on 2 November 1939, they lost. The court held that by signing his contract Alston had waived any right to contest its terms.[47] The innovative strategy explained by Hastie to Walter White in 1933 had finally been put to the test—and had been found wanting. But that only meant that Hastie and his colleagues would further put the adverse court decision to the test in the United States Court of Appeals for the Fourth Circuit, Judge John J. Parker's court, which might well be hostile toward the NAACP. As a Republican gubernatorial candidate in North Carolina in 1920, Parker had endorsed a state constitutional amendment providing for a poll tax, literacy test, and grandfather clause. He was reported to have said that "wise men," black and white, disapproved the black's participation in politics as "a source of evil and danger to both races." In 1930

President Herbert C. Hoover nominated Parker, then on the United States Court of Appeals for the Fourth Circuit, for a seat on the Supreme Court. The NAACP protested. And while Hoover disregarded the protest, the ensuing battle ended with the Senate's refusal to confirm Parker.[48]

The NAACP carried Alston's case to Parker's court in June 1940. When Hastie, Hill, and Marshall met at the law school at Howard University one Sunday evening to plan their strategy, they knew that the only way to protect their clients against retaliation by the Norfolk school board was to win a favorable verdict prior to the time that contracts were mailed to teachers. Since the court's session would end before that time, they would have to ask it to sit in special session.[49]

Hastie asked who would make that unusual request of the court. Hill said that he would. But he had second thoughts as he sat in court while Parker chastized the state's attorney for being late in filing a petition. If Parker treats a white lawyer like that, Hill thought . . . And then he was standing before the court, making his request and hoping that one of its members would ask, as Judge Morris Soper of Baltimore finally did, what was so important about Alston's case that the court should hear it in special session. Hill's answer interested the court. Instead of holding a special session, the court decided to hear the case in June, as scheduled, and instructed Hill to notify it if the school board retaliated against the teachers in the meantime. The school board, in turn, decided to withhold all contracts until the court acted on the case.[50]

In Washington Hastie held himself in readiness for the upcoming fight in Judge Parker's court. He managed a two-day trip to Amherst for his fifteenth class reunion and attended the NAACP convention in Philadelphia, but otherwise he kept his calendar clear for mid-June in anticipation of leaving "on short notice" for Asheville. Arising in Parker's court, the Alston suit would be particularly sensitive. This was made clear by Walter White's disclosure that he discussed with Hastie and the other NAACP lawyers the question of whether he should be in court when they argued the appeal. He had been urged to be present. "But because of the Senate fight over the confirmation of Judge Parker, it seemed to [us] that my appearance in the court would have been distinctly bad taste. . . ."[51]

In appellate court Hastie was in his element. With Marshall he laid out the case for a declaratory judgment to the effect that the payment of racially discriminatory salaries constituted state action that denied due process and equal protection of the law in violation of the Fourteenth Amendment. In addition, Hastie and Marshall sought an injunction restraining the Norfolk board of education from making distinctions based on race or color in fixing public school teachers' salaries. The first of three questions before the court was whether the school board's fixing salary schedules was state action prohibited by the Fourteenth Amendment. The court answered affirmatively, citing the decision that Hastie and his colleagues had won in *Mills*. The court also cited *Mills* in its affirmative answer to the second question, namely, whether the plaintiffs had rights that were infringed by the payment of racially discriminatory salaries.[52]

The third question before Parker's court was whether the black teachers were precluded from seeking relief from the discriminatory terms of contracts they had signed. The court, noting that the plaintiffs were not seeking relief from the terms of current contracts, said that they were entitled to the right to seek employment at nondiscriminatory salaries in the future. The board of education contended that only teachers under contract had a legal interest in salary schedules and that they could not sue once they signed their contracts. In his opinion for the court, Parker declared, "If this were sound [argument], there would be no practical means of redress for teachers subjected to the unconstitutional discrimination. But it is not sound. . . ."[53]

Parker's court handed Hastie and Marshall the victory, and Parker himself complimented them for their performance. When the Supreme Court refused to hear an appeal of the decision, the death knell for discriminatory teachers' salaries was sounded in Virginia and elsewhere. Hastie's optimism about overcoming Jim Crow in education moved up a notch or so. "Against such resistance," he had said, "there will be arrayed the aggressiveness and militancy of more Negroes throughout the south than those who still hear the guns of Shiloh and see the spectre of the reconstruction care to think about."[54]

But every town was as much Shiloh as it had been when Hastie had first enlisted in the bittersweet war against racism in 1933. From *Hocutt* to *Mills* to *Alston*, he had lived in Washington, D.C., where national as well as local power brokers still heard the guns and saw the specter—and where he still practiced the militancy that had bound him to the New Negro Alliance since its birth during the Great Depression.

CHAPTER SEVEN
New Negro Alliance

If asked how they were faring in 1933, blacks in the nation's capital might well have replied, in the words of a popular joke, "White folks in the lead." Hastie would have agreed. The Great Depression, he believed, had put blacks "in dire straits" in that city. To make matters worse, the city government discriminated against blacks in its employment policies. For example, only thirty-nine of Washington's 1,379 policemen were black. Unemployment among blacks was 40 percent, and when they went to the District Employment Center (an agency of the U.S. Employment Service), they found job listings for whites that included all available opportunities, but blacks had almost no choice other than listings for domestics and laborers. The picture, said Hastie, was not pretty. He did not expect it to change much by dint of efforts made by either of the two groups of black Washingtonians that he discerned. "The first [relatively advantaged] group," he said, "had little concern for the second group, whose insecurity was made more perilous by lack of cohesion and lack of comprehension of any tactics of struggle. . . ."[1]

Cries for relief went unanswered by civil rights organizations, and on 10 June 1932 the Washington *Tribune*, a black newspaper, declared the need for "a representative militant organization" in the city. Those in existence, it added, "have no sting." As a lawyer in the Department of the Interior, Hastie belonged to one of two groups (government employees and teachers) dominant in black organizations but delinquent in shouldering responsibility for community welfare. Hastie, however, worked not only with some of those organizations but also with the one that emerged in 1933 to fill the leadership void, namely, the New Negro Alliance. The Alliance originated with John Aubrey Davis, Belford V. Lawson, Jr., and M. Franklin Thorne, young blacks who were spurred into action in the summer of 1933 when they witnessed an act of indecency in the heart of their community. At the Hamburger Grill on U Street, the manager fired his three black employees and hired three whites. In retaliation Davis and his friends picketed the Hamburger Grill, causing the three employees to be quickly rehired at higher wages and fewer hours. With this action the New Negro Alliance was born.[2]

To Davis the firing of the black employees was a price paid for the "addle-brained, timid leadership" in his community. He and other young militants understood that in social protest what one attempts often is more important than what one achieves, what one dares more important than what one does. But when he began to organize the Alliance he was discouraged by most of the prominent blacks. He recalls, for example, that Ralph J. Bunche, head of the political science department at Howard University, suggested, "Why don't you go study some political science, John?" The attitude of Bunche's close friend Hastie was quite different. As Davis commented, "He told me not to worry about any 'ivory tower stuff.' He said that I was on the right track."[3]

The track led the Alliance into collision with retail stores in black neighborhoods. Demanding that blacks be hired in proportion to their patronage, the Alliance conducted surveys to determine the volume of patronage and then confronted store officials with its demands. If store officials took "no definite action" within a reasonable period, the Alliance began door-to-door distribution of materials that explained the reasons for its next action, picketing. At the same time, it obtained pledges to boycott the store unless and until its demands were met.[4]

The Alliance tried to rally black Washingtonians by contending that they would be supporting themselves, their children, and their community by helping to raise self-reliance, reduce dependence on charity, assure better goods and services in stores, increase the sense of neighborliness and cooperation, set a fine example for the young, and stimulate black business. Some blacks were unpersuaded. In Bunche's opinion the Alliance failed to understand that race had less to do with black unemployment than had an economy incapable of providing enough jobs. He said the organization's "narrowly racial policy" would split the ranks of organized labor and thereby harm black workers. In rebuttal, the Alliance contended Bunche took the position that the black's problem, because it was basically economic, would be solved when communism displaced capitalism in America, and that meanwhile blacks should join forces with white workers "and march on to victory." But the Alliance held that communism would not change human nature, and felt "intelligently controlled racialism" to be the solution to the black's problems. "We must fight as a race for everything that makes for a better country and a better world. We are dreaming idiots and trusting fools to do anything less. . . ."⁵

Hastie, like his friend Bunche, thought it important for black and white workers to unite. However, he saw no contradiction between this principle and another that held it necessary for blacks to use the power of the purse in realization of the fact that for them "a special racial disadvantage was superimposed upon the common difficulties of workers generally." If this were an ideal world, Hastie said in effect, there would be no risk of retaliation by whites. As things stood, however, blacks would simply have to run that risk.⁶

Hastie and other Alliance leaders were more realistic than Bunche about interracial solidarity. As H. Naylor Fitzhugh has said, "We didn't see an awful lot of white people who were interested in getting in the struggle with us, except those who had communist leanings, and we even thought that they were more interested in advancing their own cause than ours." Quoting Bunche in saying that current efforts to improve the black's lot had "policies of immediate relief and petty opportunism" as their goals, Hastie added: "But the impact of events makes policies of immediate relief necessary and desirable and their haphazard occurrence makes opportunism inevitable. . . ."⁷

While Bunche denigrated the Alliance's leaders, Hastie encouraged them. In fact, Hastie did more than that. He joined them and played multiple roles in the organization. At the outset, when Davis was the organization's administrator and Fitzhugh its assistant administrator, Hastie was (with Lawson) its legal counsel. And while the Alliance published its weekly newspaper, the *New Negro Opinion* (from 1933 to 1934), Hastie was the assistant editor and a columnist. During the Alliance's entire history (1933 to 1941), he served in a variety of capacities: Program and Policy Committee member (1933), Assistant Administrator (1935), Administrator (1936), and Treasurer (1940). During the two years (1938 and 1939) that he was in the Virgin Islands, his mother served as Executive Council member in 1938 and Deputy-at-Large in 1939. He was co-counsel for the Alliance in all but two of the years (1940 and 1941) that he was in Washington.⁸

Although the Alliance and Hastie as a member were primarily concerned about narrowly defined employment opportunity, both were simultaneously concerned about broadly defined equal opportunity.⁹ Or, as Hastie put it, the Alliance was "equally interested in this matter of civil rights." That he himself was similarly interested was suggested by the source of the latter statement, namely, a flyer soliciting support for a civil rights bill that the Alliance supported in an attempt to open public accommodations in Washington to blacks.

The idea for the bill originated at the Conference on Civil Liberties under the New Deal, which had been held in Washington in December 1934. Its most ardent advocate was Roger N. Baldwin, Director of the American Civil Liberties Union. Walter White was at first cool to the idea, both because Charles Houston (on whom Baldwin kept pressing it) could not spare the time for it, and because agitation for it might have reduced the chances of an antilynching law being obtained. Then, in a letter to Hastie, White indicated a change of mind. He said that it was probably too late (May 1935) to get Congress to act on a civil rights bill, but if one were introduced and made the subject of congressional hearings, its reintroduction in the next session would be facilitated. Baldwin had asked that an NAACP lawyer draft the bill, and White made that request of Hastie.

Hastie prepared and sent the bill to Baldwin. He did not think that Congress would act on it that year, but perhaps it would the next. Working in the Department of the Interior (where civil liberties for the Virgin Islands were under consideration), Hastie might have had those possessions in mind when he asked whether Baldwin thought it advisable to seek civil rights legislation for federal territories as well as for the District of Columbia alone. Baldwin, whose opinion was that racism in Washington was "scandalous," thought it inadvisable to seek broader coverage. A case could be made for legislation covering Washington but not the territories. As requested by Baldwin, Charles Houston submitted comments on Hastie's draft along with a draft of his own (which he sent to Hastie).

Had the civil rights bill been enacted into law, offenders who denied blacks "full and equal privilege" in public accommodations in Washington would have been liable to damages or a fine (each from one to five hundred dollars), imprisonment (thirty to ninety days), or both fine and imprisonment. Seeking support for the bill, the Alliance distributed copies of it, suggesting action that individuals and organizations could take. The flyer announcing a mass meeting scheduled for 22 November 1935 posed the question: "Will you help make Washington a better place to live?" But not enough of the right people (viz. congressmen) gave enough of the right help to bring the bill to the floor of the House of Representatives.

The civil rights bill would undoubtedly have been the topic of one or more of Hastie's columns in the Alliance's newspaper if it were still being published. Having won respect and popularity as a lawyer with the NAACP, a teacher at Howard University Law School, an assistant solicitor in the Department of the Interior, and a partner in the law firm of Charles H. Houston and his father, William L. Houston, he was valuable to the Alliance as a publicist who could inspire, defend, inform, and legitimize the organization. The column also gave him a forum from which he could address such broad matter of civil rights as the Scottsboro case.

The Scottsboro case involved an alleged rape of two white women by a group of black itinerants.[10] The crime was said to have occurred in a freight car filled with crushed rock, on a train bound from Chattanooga, where nine whites had hopped aboard, to Huntsville, Alabama. In Stevenson, Alabama, twenty or thirty blacks had climbed aboard. At first the two women, Victoria Price and Ruby Bates, cotton mill workers who moonlighted as prostitutes, denied repeatedly during questioning that they had been raped by the blacks. Two doctors confirmed their story, but then Victoria Price changed it and accused the blacks of having raped her and Bates six times during the thirty-eight-mile trip from Stevenson to Paint Rock, Alabama, where the arrests were made.

In the aftermath of the arrests, black churches in Chattanooga retained a white lawyer and established a defense fund for the accused, some of whom were from that city. As the NAACP hesitated in providing the legal aid requested of it, the International Labor Defense (the legal arm of the Communist party) began maneuvers that were to result in the defendants' selection of it rather than the NAACP as legal counsel. According to Hastie, after the blacks were convicted and sentenced (death for eight of them, life imprisonment for the fourteen-year-old), "the struggle between communist and noncommunist groups for control of this litigation seriously jeopardized the ultimate success of the effort to free the defendants." This Hastie deplored. Four years after sentences were imposed, he wrote in his column that there was "no recompense that Alabama or any other agency" could make for the time that the Scottsboro Boys had spent on death row and for the time they had yet to spend there. But Alabama, he added, was like Pharaoh, who, despite plagues reflecting "divine displeasure," kept both a hard heart and his captives.

If the Scottsboro Boys were to be saved, Hastie wrote in a second column, all available help had to be mustered. "I am not hopeful that a plague of appellate decisions, or a plague of communists, of a plague of adverse public opinion will soften the heart of Pharaoh," he wrote, "and it does not seem probable that Jehovah will smite the firstborn of the households of Egypt today. The penalty of economic, social, and moral disintegration that Egypt is paying for its oppression is, however, as dreadful, if not as spectacular, a penalty as Pharaoh paid."

In other columns Hastie crossed swords with W. E. B. Du Bois, editor of *The Crisis*, who had created dissension within the NAACP and the black community by arguing that since blacks could not escape segregation, rather than moaning about it as deprivation of opportunity and dignity, they should seize it as a golden opportunity to marshal their resources both to display negritude and to thwart the apostles of negrophobia.[11] Du Bois propounded his thesis in an editorial, contending that "the thinking colored people of the United States must stop being stampeded by the word segregation," and adding that segregation and discrimination "do not necessarily go together, and there should never be an opposition to segregation pure and simple unless that segregation does involve discrimination. . . ." To this assertion Hastie replied, in his column of 25 January 1934, "In theory there can be segregation without discrimination, separation without unequal treatment. But any Negro who uses this theoretical possibility as a justification for segregation is either dumb, or mentally dishonest, or else he has, like Esau, chosen a mess of pottage . . ."

Du Bois insisted that he was being pragmatic. Hastie, however, accused him of going beyond acceptance of segregation to a "startling conclusion that a fight against [it] is 'tilting against windmills.'" Du Bois approvingly noted that some blacks had favored segregated army officers' training camps, theaters, schools, and other institutions. "So what?" Hastie asked in his column of 3 February 1934. "Of course there are Negroes who will take insult lying down—will deliberately subject themselves to avoidable humiliation. If we were all united, grimly determined to fight for our rights as free men and women, there would be no great problem." The Alliance embodied that spirit. So had *The Crisis*, a prime function of which had been "to put backbone into some of the spineless ones, to awaken the sluggards from complacent acceptance, to hearten those in the fight. . . . What has happened?"

Du Bois was less direct in his rebuttal to Hastie, writing not about him personally but about the "excellent young gentlemen from Washington" who, having gotten the New Negro Alliance under way, had proceeded "to read me out of the congregation of the righteous because I dare even discuss segregation. . . ." He characterized them as ideologues who believed that the only solution to racial problems was total integration, something that would not occur in their lifetime. But he approved of what they were doing (if not of what they were saying), for they were fighting segregation with segregation (otherwise their demands would not be limited to jobs for blacks to serve other blacks in black neighborhoods). They were transmuting segregation into power that would bring an end to racial segregation someday. They did not have to use his terminology. "If the Negro Alliance wishes to say that it is not fighting segregation with segregation, it can call the thing that it is doing Transubstantiation or Willipuswallipus. Whatever they call it, that is what we both mean."

That was not what Hastie meant. In a letter to *The Crisis* that was reprinted in his column of 7 April 1934, he labeled erroneous Du Bois's statement that the Alliance's activities were (as the former called them) "efforts toward segregation." Hastie's column of 18 August 1934, which appeared after Du Bois had resigned as editor of *The Crisis*, suggested that his differences with Du Bois had much to do with the devotion of half a year of editorial and other space to the running battle between Du Bois and his critics. "Undoubtedly, the spirit, the energy and the genius of Du Bois have been the major factors in the development of the magazine," Hastie said, but he was pleased about the new editorial policy that fostered the airing of insights by gifted young analysts on a variety of issues. As for the dispute about segregation, it seemed that there really had been nothing to argue about.

Hastie's dispute with Du Bois was one of two controversies in which he spoke for the Alliance outside the courts. The second, involving the local board of education, featured him as the administrator of the Alliance.[12] This controversy illustrates the typical Hastie approach, namely, simultaneous representation of more than one organization, usually including, as in this instance, the NAACP. It began when the board approved recommendations that it received from Dr. Garnet C. Wilkinson, Hastie's former high school principal, who had become First Assistant Superintendent of Divisions 10 to 13 (black schools). On Wilkinson's recommendation, the board removed *Opportunity* magazine (published by the Urban League) from the approved reading list for senior high schools but kept it on the approved list for Miner Teachers College, and removed *The Crisis* (published by the NAACP) from both lists.

The board had ordered an evaluation of magazines and periodicals to assure the continued use of only those "that have educational value, that are factual, and that are free from objectionable material." Grounds for discontinuation were given as advocacy or regular publication of "any phase or aspect of any financial, economic, social, or political movement or condition in our country or in the world . . . the adoption of which in this country would be subversive to the traditional policies and institutions of the United States." Those instructions in hand, Wilkinson, the first black to hold his position, prepared his memorandum.

The board cut *The Crisis* from the approved reading list because, as Wilkinson noted in his memorandum, it conveyed "militant propaganda" unsuitable for educational purposes. Wilkinson cited four illustrative articles: one contained factual information about the status of women in Russia; another assessed the Communist party; the third contrasted the positions on pacifism taken by communists, Quakers, Ghandi's followers, and other groups; and the fourth interspersed an account of a lynching among the lines of the pledge of allegiance.

Just as the lines of poetry were blended with those of the pledge of allegiance, so the conflict within the local NAACP branch was mixed with that which Wilkinson's recommendations had ignited. Wilkinson, the branch treasurer (and for a while a member of the Alliance's Advisory Council) was allied with A. S. Pinkett, the branch secretary, whose hostility toward the national NAACP had created a rift in the branch on many issues, including the question of whether to back the national office in opposing the board of education in this controversy. Pinkett's adversaries included Hastie's mother but not Hastie (despite his close ties to Walter White and other national NAACP officers). Perhaps Hastie's nonalignment in the bickering helps to account for the adoption of his motion that the branch urge the board to restore the magazines to the approved lists and to grant the NAACP and allied organizations a hearing. Hastie was selected to chair the NAACP Special Committee on *The Crisis* and *Opportunity*.

In a letter to the board, Hastie argued that public schools had been too long established "on the false and vicious premise that Negro children, and white children as well, are going to live in a community where all are treated as white persons. . . ." Contending that the schools had discredited themselves by not preparing blacks to cope with racial discrimination and not informing whites about its heavy societal costs, Hastie said that to restrict blacks to separate schools and then deprive them of the two magazines best designed to help them deal with problems confronting their group was "an intolerable injustice." Continuation of the ban on *The Crisis* and *Opportunity* could only be interpreted as "a deliberate attempt to keep Negroes 'in their place' and satisfied with being there."

On behalf of the Alliance Hastie had praised the board for allowing instruction about communism to be given in the schools. Without such instruction "no intelligent criticism" of the Soviet system was possible. It was ironic, however, that the board condoned information about communism in Russia but condemned information about democracy in America, which was the subject matter of *The Crisis* and *Opportunity*. Hastie argued that "both magazines should be made as accessible as possible to the high school and college students in whose lives the matters discussed are and will continue to be so large a part. Any restriction upon the use of these publications impedes by so much the dissemination of vital truth and knowledge to the young people who most need to be thus instructed."

The dispute with the board of education occurred during a hiatus in the Alliance's campaign against job discrimination in Washington. Valuable to the Alliance as its administrator, Hastie was indispensable to the organization as its attorney. Making this assertion, H. Naylor Fitzhugh said, "What we were doing was not illegal, but it was very bold. It was reassuring to know we had some people on our legal staff who could go out and defend us if we got into trouble with the authorities."[13] Trouble of two kinds was a distinct possibility. One kind was harassment by the police doing the bidding of employers. That kind of trouble was the reason for the court appearance made by Hastie and other Alliance lawyers in the first of several lawsuits that were to mark the organization's campaigns leading from the Hamburger Grill in 1933 to the Supreme Court in 1937. The trouble began with the arrest of Dutton Ferguson and James Ward. The two were arrested while picketing an A & P store three weeks after the Alliance opened negotiations with the Great Atlantic & Pacific Tea Company (A & P), urging that black clerks be employed in the store at Ninth and S streets, N.W. The company balked, so the Alliance began picketing on 26 September 1933. On 29 September Ferguson and Ward were arrested and charged with violation of a local ordinance that prohibited the carrying of a sign over public space. In court Hastie, Belford Lawson, and Edward P. Lovett argued that the ordinance, which had long gone unenforced, applied only to commercial advertising, and that the extension of its coverage to peaceful picketing would deprive persons of their constitutional rights. After more than a month's consideration of briefs submitted by the prosecution and defense, the court sided with the Alliance.[14]

The Alliance resumed negotiations with A & P. After the latter had hired five black clerks but reneged on its promise to increase that number, the Alliance picketed and boycotted thirteen stores. Within seven hours thirteen more black clerks were hired. The total hired soon reached thirty. The Alliance then urged blacks to shop at A & P rather than at the rival chain, Sanitary (later Safeway) Grocery Company.[15]

That victory made a lasting impression on Dutton Ferguson. So did the experience of picketing, and he later recorded thoughts that had occurred to him at the time:

> Hammed (or is it hemmed?) in between one of these Sandwich signs again . . . gee, another customer going into the store; they must think that this sign advertises a dance or something . . . boy! It's a cop . . . hope he knows this is within the law . . .

look at all of these kids coming down the street . . . rather noisy . . . gee, they have quieted down . . . they get the point . . . beginning to get tired now . . . but these people are staying out of the store . . . can't stop now . . . this is our common fight . . . it must be done.[16]

The Alliance next carried the "common fight" to the High Ice Cream Company's store at Seventh and S streets, N.W., where picketing began on 28 November 1933. Fitzhugh reconstructs events: "I heard this squad car come screeching down Seventh Street, and I said, 'There's going to be a little excitement down here.' You know, break the monotony. And they wheeled into the curb, rushed into the store and talked to the manager. You would have thought there was a fire or burglary going on." The police then came out of the store. They hustled Fitzhugh into the squad car and took him to the 2nd Precinct. In light of the outcome of the Ferguson-Ward trial, however, Fitzhugh was released at once.[17]

The encounter with A & P had been a skirmish, and that with High's would be a battle. These encounters were part of a war, and its outcome would have constitutional meaning transcending the struggle for employment opportunities for blacks in neighborhood stores in Washington.[18] The Alliance's campaign for jobs did more than pit blacks against whites; it pitted labor against business. And that was why the second kind of trouble that the Alliance encountered involved the courts (rather than the police) doing the bidding of employers.[19]

The ultimate weapon of employers in labor disputes was the injunction, which curbed strikes, picketing, and boycotts. Judges tended to be such kindred spirits of employers that labor complained about "government by injunction." The Norris-La Guardia Act of 1932 was intended to provide relief to labor. Among other things, it exempted from federal judges' contempt powers a wide spectrum of labor disputes.[20] The Alliance hoped that its dispute with Kaufman's Department Store fell within that range.

On 15 December 1933 the Alliance began picketing Kaufman's. Five days later Justice F. Dickerson Letts of the Supreme Court of the District of Columbia granted Kaufman's request for a temporary restraining order (i.e., injunction) prohibiting picketing at the store for the next ten days. On 29 December Letts extended the injunction for ten days. No pickets would prick the conscience of shoppers or pinch the purse of management at Kaufman's during the Christmas season. Nor any other season, if Letts had his way. On 5 January 1934 he listened to Kaufman's attorney, then listened, the *New Negro Opinion* said, "reluctantly and with countless interruptions" as Hastie, Lawson, and Thelma D. Ackiss presented the Alliance's argument. The instant that they concluded, Letts signed a preliminary injunction that had already been prepared by Kaufman's lawyers.[21]

Hastie and other Alliance lawyers set out to appeal Letts's decision.[22] In what amounted to a preview of contentions to come, Kaufman's had claimed that the Alliance threatened "to use any, all and every means," including picketing, boycott, and publicity, to discredit and ruin Kaufman's business; that the Alliance had demanded the store fire white workers and hire black replacements; that disorderly pickets had molested patrons and interfered with their doing business at the store; and that the Alliance had created a situation that had been, and would continue to be, "dangerous to the life and health of persons on the highways" and to Kaufman's employees, property, and business. Kaufman's main arguments were that no employer-employee relationship was involved and that there was no labor dispute as defined by any statute. But the store's most revealing statement might have been that the Alliance's demands "go to the extent of requiring the plaintiff to place behind its sales counters colored and white employees working side by side, without regard to the natural and inevitable result of such a situation."[23]

Hastie and his co-counsel argued that the Alliance's controversy with Kaufman's was not subject to the court's authority to issue injunctions. The reason, they said, was that it was a labor dispute within the meaning of the Norris–La Guardia Act of 1932, which placed such disputes outside the jurisdiction of courts to issue restraining orders or injunctions. Therefore, said Hastie and his colleagues, the court could not prohibit the picketing of Kaufman's. According to them, even if the court refused to consider patrons as potential employees, the Norris–La Guardia Act proscribed judicial interference with the Alliance's boycott and picketing. The reason they gave was that an employer-employee relationship between the Alliance and Kaufman's need not exist in order for the controversy to be defined as a labor dispute within the Norris–La Guardia Act's meaning.[24]

The injunction obtained by Kaufman's prevented picketing at that location but not at other stores. Accordingly, the Alliance resumed its picketing of the High Ice Cream Company's store on 30 July 1934. Immediately High also obtained a restraining order. After the store and the Alliance failed to reach an agreement—High wanted the picketing to stop but would not consent to hire a black as the Alliance requested—they went before Justice Letts. High's arguments were essentially the ones that Kaufman's had made. Letts had also heard the Alliance's principal contentions before.[25]

The Alliance took heart in Letts's refusal to act at once on the petition for an injunction, as he had in Kaufman's case. But on 17 September 1934, at his home and with Hastie and co-counsel absent, Letts granted the preliminary injunction that prohibited picketing, boycott, or "announcing, advertising, or in any manner calling attention to the contention that plaintiff is unfair to colored people or to the colored race."[26] On 29 September the *New Negro Opinion* editorialized: "We can conceive of no broader or more vicious attempt to by law keep the Negro 'in his place' . . . If we feel that a firm is unfair to the colored people, we must not say so, we must not protest, we must silently take it and like it."

The Alliance had to take it (the injunction) but not lying down. While it could not picket Kaufman's or High's stores, it could still picket Sanitary's stores. The Alliance and Sanitary had clashed once before, in April 1934. They were to clash again in April 1935, just weeks before the United States Court of Appeals for the District of Columbia dismissed the Alliance's appeal of the injunction that Kaufman's had obtained. One year later, in April 1936, the third Alliance-Sanitary controversy occurred. With legal precedent in its favor, Sanitary obtained an injunction on 12 May 1936, the court holding that there was no labor dispute involved, only a protest by blacks to obtain jobs for blacks.[27]

With Hastie as its administrator and counsel, the Alliance filed an appeal in the United States District Court for the District of Columbia. In a rerun of the Kaufman's and High cases, Sanitary had complained that in this, a nonlabor dispute, the Alliance was bent on destroying the store. Giving the complaint a new twist, Sanitary accused the Alliance of having arranged or permitted the publication of articles about the dispute in a black newspaper, the Washington *Tribune*. Sanitary's bill of complaints reproduced the articles—selectively, that is. It omitted from one article the paragraph reporting the refusal of Sanitary's president to say anything about personnel plans for the new store, plans that the firm's personnel manager said did not include the hiring of blacks. Nor did Sanitary reproduce the paragraph containing the Alliance's allegations that the company had fired several black clerks and planned to dismiss all black employees. The store reprinted in full the article that reported Hastie's warning to consumers not to mistake "errand boys in white aprons" for clerks.[28]

Arguing that neither injury to Sanitary nor unlawful action by the Alliance had been established, Hastie said that a showing of both was the only proper basis for the injunction. Hastie contended that by forbidding the Alliance to refuse, or induce others to refuse, to shop at

Sanitary stores, the court violated established principles of equity and denied constitutional rights of personal liberty. His principal argument, however, was that the court had erred by refusing to treat the matter as a labor dispute under the Norris–La Guardia Act. He characterized as irrelevant the fact that the defendants were not Sanitary employees; after all, attempts by other "outsiders" to change additional employment practices without complaints from employees had been sanctioned by the courts. The only difference between those cases and the one at issue was that the latter concerned unfairness toward black labor rather than toward organized labor in general. And that, said Hastie, was not a valid legal reason for denying in this case the protection granted in the other cases.[29]

On 26 July 1937 the appellate court stood foursquare behind Sanitary. A majority of the justices held that because there was no employer-employee relationship involved, the controversy was a racial and not a labor dispute and was not covered by the Norris–La Guardia Act. The court upheld the injunction. Another appeal, another defeat. The trial and appellate courts were the double barrels of a shotgun aimed at the heart of black protest in Washington. Therefore, Hastie and the Alliance turned to the United States Supreme Court, which consented to hear the appeal during its October term in 1937.[30]

The facts of the Sanitary case were novel, but the problems they presented were common to all the Alliance's cases. The lower courts were free to reject Hastie's arguments provided the facts of the case constituted a nonlabor (or racial, as the courts had held) dispute. It was only in labor disputes that their injunctive powers were limited by the Norris–La Guardia Act to exceptional instances. But determining whether the facts constituted a labor dispute was neither a simple nor settled problem. Judicial answers to the question of what relationship between parties constituted a labor dispute had varied in prior cases.[31]

Hastie had argued the Alliance's case in the appellate court in April 1937. He would not be available to appear before the Supreme Court in October, but John Aubrey Davis says that the contributions he had made were the keystone in the case that other Alliance lawyers (i.e., Belford V. Lawson and Thurmond L. Dodson) would present. The case was part legal and part sociological, for Hastie and his colleagues believed that the Norris–La Guardia Act was specifically intended to cover controversies involving issues other than wages, hours, and working conditions, and those involving parties other than employees and employers.[32]

The Alliance felt that in these instances socioeconomic differences should be weighed; though subtle, they were powerful factors. For what were blacks to do? They did not sit on any board of directors whose corporation was doing business in the black community. Hence, in its brief the Alliance reasoned that "when the court enjoined the petitioners from peaceful picketing, it did more than prevent temporary damage to respondent; it gave him greater and unfair advantage over the petitioners before the petitioners could utilize the weapons at their disposal."[33]

The Supreme Court concurred with the Alliance's reasoning. The Norris–La Guardia Act stated: "A case shall be held to involve or to grow out of a labor dispute when the case involves persons who are engaged in the same industry, trade, craft, or occupation; or have direct or indirect interests therein; . . . or when the case involves any conflicting or competing interests in a 'labor dispute' (as hereinafter defined) of 'persons participating or interested' therein (as hereinafter defined)." The legislation characterized an individual or organization as one with an interest in a labor dispute "if the relief is sought against him or it and if he or it . . . has a direct or indirect interest therein. . . ." It defined a labor dispute as "any controversy concerning terms or conditions of employment . . . regardless of whether or not the disputants stand in the proximate relation of employer and employee." The Court was of the opinion that the Alliance and individual petitioners were persons interested in a dispute that was covered by the Norris–La Guardia Act.[34]

Reversing the lower courts, the Supreme Court declared that the Norris–La Guardia Act was not limited to controversies involving, on the one hand, employers and, on the other hand, employees, labor unions seeking to represent them, or persons seeking employment. The Court held that unless the controversy's "racial" features made Norris–La Guardia inapplicable, it fell squarely within the scope of that legislation. Those features, said the Court, had no such effect.

> In the first place, the Act does not concern itself with the background or the motives of the dispute. In the second place, the desire for fair and equitable conditions of employment on the part of persons of any race, color, or persuasion, and the removal of discrimination against them by reason of their race or religious beliefs is quite as important to those concerned as fairness and equity in terms and conditions of employement can be to trade or craft unions or any form of labor organization or association. Race discrimination by an employer may reasonably be deemed more unfair and less excusable than discrimination against workers on the ground of union affiliaton.[35]

This victory was momentous. As John Aubrey Davis asserted, "It was the first case in the United States to establish the right to picket by nonlabor union people. The case is one on which everybody else has been able to build."[36]

H. Naylor Fitzhugh added, "I think one of the things we were doing was what blacks have done so often in this country, and that is to set a tone that other people copy. Over the years and in various fields, I have seen a number of examples of how we have set the pace."[37]

That was to be expected. Although it is sometimes viewed as though it were divisible, freedom really is not. It outwardly appears to be selectively expandable or constrictable; in fact, it is not. Because the Alliance achieved something that no other organization—labor, civil rights, civil liberties—had achieved or even attempted, the Supreme Court expanded the concept of freedom on 28 March 1938.

Hastie was not in Washington that day. About him, the Alliance's bimonthly bulletin had said: "By his acts do ye know him."[38] The net effect of his acts was to make the name Hastie a byword for heroism. And what was true on the mainland would be true in the Virgin Islands.

CHAPTER EIGHT

A Founding Father

Hastie's transfer to the Virgin Islands was arranged by Secretary of the Interior Harold L. Ickes, in whose department Hastie had been employed while volunteering his services to the New Negro Alliance and the NAACP. He had been teaching at Howard and practicing law with William L. Houston and his son, Charles Hamilton Houston, in 1933 when Secretary Ickes, casting about for a black lawyer capable of being an assistant solicitor, heard about him. About the same time Clark Foreman, Ickes's adviser on the economic status of blacks, requested and received from Hastie a list of black lawyers in connection with "a rather important change" that Foreman anticipated at Interior. Called in himself for an interview, Hastie—according to Ickes—made "an excellent impression" on the secretary. Furthermore, Nathan R. Margold, the solicitor, remembered Hastie's brilliance at Harvard during the year (1927–28) that Margold had been an instructor there. Felix Frankfurter also put in a good word about Hastie. On 16 November 1933 Ickes appointed Hastie as assistant solicitor "on the basis solely of what [Margold] thought were the needs of his legal office," said Hastie.[1]

By his own account, Hastie was "by coloration something of an exotic" at Interior. A colleague recalls two white secretaries having refused assignment to Hastie because of his color. Rather than firing them, as Charles Fahy, the first assistant solicitor, suggested, Margold hired a black secretary for Hastie. She sat not in the secretarial pool but in Hastie's office.[2] Color consciousness at Interior also created problems for Robert C. Weaver, Ickes's adviser on race relations. He remembers the day he and his fellow "exotic" decided to lunch not in the cafeteria for messengers but in the cafeteria for other employees. Weaver explains their rationale: "First, we weren't messengers and, second, we knew damn well it was just a subterfuge for a segregated lunchroom." They flipped a coin to determine who would pay for lunch. Weaver lost. They went into the main lunchroom.

The cashier was flustered. "Do you work here?"

"Yes," said Weaver, reaching for his wallet.

"Well, would you mind giving me your names?"

"No. This is William Hastie and I am Robert Weaver. Now, would you mind giving me your name?"

In ten seconds she remembered her name. She gave it to Weaver. He wrote it down, said "Thank you," and went with Hastie to a table. They put the episode out of their minds. But a half hour later a group of women called on Ickes.

Ickes did not look up. "Good afternoon, ladies. What can I do for you?"

"Mr. Secretary, do you know that Negroes are eating in lunchroom?"

"Yes."

"What are you going to do about it?"

"Not a damn thing, ladies."

So much for segregated lunchrooms at Interior.

Weaver also recalls having gone to see Ickes about assignment to the housing division.

"Young man, what do you know about housing?"

"Very little."

"You'll do fine. None of those sons of bitches know anything about it either."

Weaver's problem then was to determine how to make contractors and labor unions give jobs to blacks. Contractors would say, "We'll hire them if the union will take them in," and the unions would say, "We'll take them in if the contractors will let them work." Weaver and Hastie devised a plan that required the employment of blacks as at least 5 percent of skilled, 15 percent semiskilled, and 40 percent of unskilled workers. Failure to meet those requirements was prima facie evidence of racial discrimination. Weaver's staff checked weekly payrolls to see if the goals were met under the first affirmative action plan ever used by the federal government. "It was my idea and Bill's law," he says.[3]

Hastie was the only one of the several black lawyers at Interior who was on the department's legal staff. He called his "a minor professional post," but when he left the department Margold called his departure "a distinct loss because he isn't replaceable."[4] Appointment of blacks to governmental positions, Frederick S. Weaver had written, was one thing, whereas meaningful duties for them was another. Hastie's assignments had enabled him to help protect Native Americans' oil rights, prevent commercial exploitation of Alaskan reindeers, and facilitate the construction of irrigation projects and the provision of federal support for vocational education. He represented the department in litigation in Washington and elsewhere. Most important of all, Hastie played a part in shaping the future of the Virgin Islands, which President Herbert C. Hoover had called "an effective poorhouse."[5]

According to Gordon K. Lewis, a mercantile oligarchy ruled the impoverished masses in the Virgin Islands. "All of [this]," he writes, "could be seen, generally, as the ineradicable contradiction between democracy and empire." Hastie helped to lay the legal basis for the eradication of this contradiction through his work on the Organic Act of 1936. The Department of the Interior had relieved the Navy of administering the Virgin Islands in 1931, fourteen years after the United States bought them from Denmark for twenty-five million dollars. The islands were purchased in order to prevent Germany from establishing a militarily strategic Caribbean station on St. Thomas. As the German threat faded, so did both American interest in these possessions and the prospects for first-class citizenship for Virgin Islanders.[6]

Before the enactment of the Organic Act, the Supreme Court had ruled both that the Constitution did not require the granting of full citizenship rights to the people of unincorporated territories, and that Congress enjoyed complete command over them. Congress decreed the Virgin Islands to be an unincorporated territory; for ten years (from 1917 to 1927) Virgin Islanders were "nationals," not "citizens"; and for an even greater period the islands were plagued by additional problems: vagueness about the local colonial councils' jurisdiction and the shifting of gubernatorial responsibility among President, Congress, and the Department of the Navy. Indeed, between 1917 and 1931 the Navy itself was a gargantuan problem. Chosen by President Woodrow Wilson to administer the new possession, the Navy, a segregated service, was by temperament, training, and example unfit for the assignment. Naval officers were incapable of seeking or accepting the cooperation of the colonial councils or the support of the people. Yet they had a free hand to work their authoritarian, prejudiced will; in Washington the Navy Department provided no supervision, and few people knew anything about the Virgin Islands.[7]

Rothschild Francis, editor of *The Emancipator* on St. Thomas, tried to inform higher authorities about the dictatorial ways of the Navy. As he engaged a series of naval governors in verbal contest, other native islanders engaged their troops in physical combat. Francis, Hamilton Jackson, Ralph de Chabert, and other polemicists also helped to pave the way for a feat that Hastie achieved at Interior. President Harry S.

Truman said Hastie wrote the Organic Act. This struck Hastie as "fulsome." The legislation resulted from the work of a group that included other lawyers in the department, lawyers and legislators from the Virgin Islands, and members of the staffs of the drafting and territorial committees in the Senate. "So I was an active participant in that process. But . . . nobody was *the* draftsman of the act, though I think and hope that my influence and my ideas may have had some importance in getting a reasonably satisfactory Organic Act drafted."[8]

Hastie did not want credit for his work on the Organic Act. It was enough that its supporters succeeded in preparing "a good and workable new charter . . . that would get acceptance in the Congress." But Roy Bornn, then commissioner of public welfare, says that Virgin Islanders involved in the process believed that Hastie was the principal draftsman. Hastie handled the matter almost exclusively at Interior. But he was self-effacing about it all, adds Bornn, for he was "very modest, very unassuming, very calm, quiet—and yet such a dynamic personality."[9]

The Organic Act became effective on 22 June 1936, nineteen years after a temporary government had been authorized by the act of 3 March 1917. The latter measure vested all civil, military, and judicial powers in a governor appointed by the President with the advice and consent of the Senate. It specifically authorized the appointment of an army or navy officer. And it gave full effect to the last Danish legislation concerning governmental organization in the islands, the Colonial Law of 1906. Under the Colonial Law, there were two local legislatures, called colonial councils, which were empowered to comment on any legislation before the Danish Parliament concerning the then Danish West Indies. The colonial council for St. Croix consisted of five appointed and thirteen elected members; that for St. Thomas–St. John four appointed and eleven elected members. The governor set the regular dates for the legislatures to meet, and when he was of a mind to do so he dissolved the legislatures. The right to vote was restricted to men, native or five-year residents, who were at least twenty-five or older. They had to be of "unblemished character" and could never have been convicted of an "ignominious act." In addition, they had to earn an annual salary of $300 or more or else own property yielding a minimal rental of $60 on St. Croix and St. John or $120 on St. Thomas.[10]

The Organic Act renamed the legislatures municipal councils, cut their membership in half, and provided for the election of members at large as well as by districts. It abolished the governor's power to dissolve the legislatures and appoint legislators. His veto could be overridden by a two-thirds vote in the legislature, but that vote was subject to presidential veto, which was final. In joint session the municipal councils constituted the Legislative Assembly, which enacted laws applicable throughout the islands. Legislation enacted by the assembly or by the councils was subject to congressional annulment. At long last Congress granted to Virgin Islanders their basic rights: due process of law and equal protection of the laws; counsel in criminal prosecution; protection against double jeopardy and self-incrimination; write of habeus corpus; prohibition of slavery; proscription of ex post facto laws and bills of attainder; nonimprisonment for debt; and freedom from religious or political tests as a condition for holding office.[11]

The Organic Act abolished property and income requirements for voting, extending the franchise to all residents, women as well as men, who were twenty-one or older, able to read and write English, and citizens of the United States. Between 1917 and 1927 native Virgin Islanders had been Danish subjects under United States protection, but they were citizens of neither country. On 25 February 1927 United States citizenship was conferred on some islanders of native and Danish stock and on their children born between 17 January 1917 and 25 February 1927. This made many Virgin Islanders in Puerto Rico, the Canal Zone, and the United States citizens with no country. Final clarification of citizenship status was made on 28 June 1932 when citizenship status was given to Virgin Islanders living in those locales or in any American insular possession or territory.[12]

By virtue of his work at Interior, Hastie's future was to be further intertwined with that of the Virgin Islands. He got to know Lieutenant Governor Lawrence W. Cramer. In 1937, when President Roosevelt was impressed by the British example of appointing native lawyers to judicial posts in their predominantly black colonies, he asked Cramer and his wife what they thought about his naming a black to the district court in the Virgin Islands; the Cramers, enthusiastic about the idea, suggested Hastie for the appointment.[13]

Hastie did not want the appointment for himself, but he did want it for some black. "Upon whom could we concentrate as a candidate to be pushed?" he asked Charles Hamilton Houston. Ickes, however, wanted Hastie for the position. "He has more than made good [as an assistant solicitor]," the secretary wrote about Hastie. "He is not only an excellent lawyer but a man of fine character and sensibilities who, in my judgment, is qualified to be judge of any United States District Court anywhere." So Ickes recommended that Roosevelt appoint Hastie to the district court. He had already made the recommendation to Homer S. Cummings, the attorney general, who had not responded.[14]

Opposing Hastie, Cummings contended that the appointment of a colored continental would offend black Virgin Islanders. Therefore he held up Hastie's nomination for several months. Finally, on 5 February

1937, Roosevelt sent it to the Senate. After the President appointed Hastie, opposition came from Senator William H. King of Utah despite his promise to Ickes that he would not raise it. King, chairman of a subcommittee of the Senate Judiciary Committee, said that native Virgin Islanders disapproved of the nomination of a colored man. He doubted that Hastie would maintain "a judicial point of view" about interaction between black islanders and the government. He questioned Hastie's competence as a trial attorney. Supporting Hastie at King's hearings on the nomination were Thurgood Marshall (National Bar Association), William L. Houston (president, Washington [D.C.] Bar Association), Leon A. Ransom (Howard University Law School), Edward P. Lovett (Washington [D.C.] Bar Association), Victor H. Daniel (native Virgin Islander), and Rufus G. Poole (Hastie's white colleague at Interior). Providing moral support were Hastie's mother and his wife, the former Alma Syphax, whom he had married in February 1935.[15]

Allied with King against Hastie was Ickes's antagonist, Senator Millard E. Tydings of Maryland, chairman of the Territories and Insular Affairs Committee. "He gouges, bites and scratches," Ickes said of Tydings, "and sometimes even hits in the clinches." But Ickes knew how to deal with Tydings: He sent a black member of his staff to alert blacks in Maryland to Tydings's treachery; their protest—in the form of letters, telegrams, and visits by several delegations—straightened Tydings out. On 19 March 1937 Hastie was confirmed.[16]

By virtue of his work on the Organic Act, Hastie had standing even with Virgin Islanders who had reservations about his appointment. Of special importance to the nonelite was the broadening of the electoral base of politics. "That caused a complete revolution in the political, economic, and social conditions of the entire Virgin Islands," said attorney Alphonso A. Christian, Sr. "It was very unsettling to a lot of people. But it gave a tremendous amount of opportunity to the masses of the people."[17]

It unsettled people who felt that women should not be allowed to vote. This view was chauvinistic (women belonged at home performing proper chores, not at polling places exercising political power) and racist (their votes would contribute to majority advantage in an overwhelmingly black area). But the women in the teachers' association on St. Thomas would have none of that. In 1934, after the electoral board rejected their applications to vote because of their sex, twenty-three women contested the decision in court. Anna Maria Vessup, Edith Williams, and Eulalie Stevens, who met all other qualifications for voting, were selected as plaintiffs in the lawsuit against the electoral board in 1935. Robert Claiborn, their attorney, contended that the

Nineteenth Amendment to the Constitution invalidated the provision of the Danish Amalienborg Code of 1906, which disenfranchised women. Judge Albert Levitt agreed, and on 28 December 1935 he ordered that the twenty-three women be permitted to vote in the next election. Four months later he heard the petition of 166 women on St. Croix against sexual discrimination in voting. Both before and after Evelyn Richardson, Marie Moorhead, Jane Joseph, and others won, there were lawsuits to secure women's suffrage in keeping with the Constitution, specifically the Nineteenth Amendment.[18] Hastie's craftsmanship on the Organic Act finally laid the issue to rest. "Unquestionably," concludes Darwin D. Creque, "the Organic Act of 1936 revolutionized the political, social and legal status of the Virgin Islands."[19]

Before 1936, whether appointed by the governor or elected, legislators protected the interests of the middle and upper classes from which they came and to which they owed their selection. The Organic Act gave the masses no say about the selection of their governor; that remained a presidential prerogative. But it enabled them to elect their legislators. The Virgin Islands Progressive Guide originated in 1937 to help them make the most of that opportunity. Its young native-born founders—Aubrey C. Ottley, Valdemar A. Hill, Roy P. Gordon, Carlos A. Downing, Oswald E. Harris, and Henry V. Richards—promptly began their program of voter education and registration. Only 5.5 percent of Virgin Islanders had ever voted; many were apathetic; and the first elections under universal suffrage were just one year off. The Progressive Guide's candidates, like its founders and other members, had no political experience. But they had political godfathers, namely, James A. Bough, the native-born United States District Attorney, and Hastie.[20]

Begun as a civic organization, the Progressive Guide quickly became the first political party in the Virgin Islands, a transformation that Hastie guided its leaders in achieving. The Progressive Guide was pro-labor and favored economic reform and self-government. Disclaiming an interest in politics as such—"We don't care who sits in the legislature as long as we get the results we are after"—the organization made its goals clear. It wanted a change of legislators, not because all of them were incompetent, but because all were "seeing in one and the same direction with rare exception." That direction was not favorable to the people. For example, the legislators tolerated discrimination that assured white continentals, even those with "a questionable past" and those without relevant job experience, salaries enabling them to live luxuriously while natives lived hand-to-mouth. Let a native replace a continental on a job and the salary was automatically cut. The Progressive Guide advocated the cessation—or reversal—of this preferential treatment. To earn a living many natives migrated to the mainland, a practice that the Progressive Guide deplored. "Right now it would take a Sinclair Lewis to adequately portray the pathos of struggling Virgin Islanders leaving . . . to endeavor to scrape a living in machine-minded New York, while indolent continentals drift to the balmy Isles, and live luxuriously. There is no reason why this should be so."[21]

Salary discrimination was but one of the economic problems confronting the Progressive Guide. Another, on St. Thomas in particular, was the influx of aliens, cheap labor whose exploitation by politicians and employers displaced Virgin Islanders from jobs. Then came the hordes of relatives, whose presence not only deprived Virgin Islanders of work but also undercut efforts to raise wage standards. The alien problem had to be solved, Aubrey Ottley warned. "Either that, or the natives are heading for perdition."[22]

Propelling them toward perdition, the Progressive Guide argued, was an ill wind blowing in from Puerto Rico. For on St. Croix as well as St. Thomas Puerto Ricans were "taking control of the most important function of a community—the business end." Their success was attributed to their business acumen, to their cooperation (with each other), and to the selfishness of Virgin Islanders, which made mutual support impossible. "We do not wish the Puerto Ricans ill (God forbid!), but we positively desire the Virgin Islanders to also make use of the source of profit and improvement that the seemingly clairvoyant incomers have been able to spot with unerring good judgment." How else could the latter control their homeland?[23]

The problem of aliens was one of two spooks troubling the Progressive Guide. The other, which it called "the ghost that haunts St. Thomas," was the want of economic independence. The fact that the island was not self-sufficient was forgotten or disregarded as relief money flowed in from "those blessed Alphabetical Agencies in Washington," but if people on the mainland could not foretell how long that assistance would continue—and they could not—that was all the more reason for Virgin Islanders not to take relief for granted. Both aid from Washington and "adequate and permanent industries" in the Virgin Islands were needed. Success in securing both depended largely on whether local politicians cooperated with rather than fought Congress.[24]

The Progressive Guide saw salvation in improved voter and public education. While conducting the former type of instruction, it encouraged the latter for adults as well as children. Its salute to the class of 1938 at Charlotte Amalie High School was given on 25 June in the same fashion as were its commentaries on the weightiest topics, that is, in a front-page editorial in its mimeographed weekly bulletin the *Progressive*

Guide. Blue and gold (senior high school) and red and white (junior high school)—the Apollo Theatre had been brightened by those colors the night before (Friday), when Judge Hastie gave the commencement address for the fifty-two students graduating from high school and the thirty-two completing junior high school. Hastie minced no words in telling the people who packed the theater that St. Thomas fell far short of the mark in some critical respects. He said that teachers are paid too little, physical education is not taught, overcrowding prevents many students from entering high school, school buildings are wretched, and equipment is inferior. More money is spent on public education in the states than in the Virgin Islands, where the schools have to try to fill the void resulting from the absence of a YMCA, community center, and social services agencies. "The schools have to be more than schools usually are in other places where there are more than schools; indeed, the schools have to be community centers whence those desirable things which civilization and our democratic form of government find necessary may be spread."[25]

It was clear, the *Daily News* commented, that Judge Hastie saw his "obligation to the community as a living thing," which he planned to fulfill "regardless of the feelings of any individual whom circumstances may have made vulnerable." It was not only as a godfather to the Progressive Guide that he fulfilled that obligation. In doing so, for example, he challenged the Civilian Conservation Corps (CCC), which the organization praised as one of the most beneficial projects of the New Deal. The corpsmen learned job skills, furthered their education, earned allotments for their families, made their parents proud of them, and performed useful tasks.[26]

The CCC program hit a snag when the camp director, P. W. Hartmann, advised "supervisory and facilitating personnel" to refrain from writing newspaper articles. Hartmann's memorandum (10 March 1939) was the topic of discussion three days later when Hastie, Dr. John S. Moorhead, and Alvaro de Lugo met with him and other ranking CCC officials. Hastie, Moorhead (the municipal physician for St. Thomas–St. John), and de Lugo (postmaster of Charlotte Amalie) failed to persuade Hubler to their view that the memorandum was an infringement of basic civil liberties. Hubler refused to consider rescinding the memorandum and substituting other means of ensuring that employees' statements were not seen as CCC policy. The personnel to whom the memorandum applied were not CCC enrollees but National Park Service employees who supervised and administered the CCC's national parks projects. Bringing all this to the attention of William Trent, Ickes's Adviser on Negro Affairs, Hastie said the interference with free speech was especially regrettable coming on the heels of an American Civil Liberties Union reference to the Virgin Islands as a place where civil liberties were protected. He was certain (and Trent confirmed) that the secretary would want to be informed in order to take appropriate action.[27]

"While the issue itself is not racial, most of the employees affected by the objectionable memorandum are colored Virgin Islanders," Hastie wrote to Trent the day after the meeting with Hubler. The point was noteworthy. This was not the first time the CCC had tried to whip the natives into line. Roy P. Gordon and Valdemar A. Hill, Sr., helped to establish the Virgin Islands Progressive Guide when they were CCC employees. Accusing them of participating in politics in violation of the Hatch Act, their supervisor suspended them without pay. On appeal Ickes overruled him and put them on leave with pay.[28]

Between those two CCC episodes the Progressive Guide made its bid for legislative seats in the elections of 1938. It won three seats from the People's party, sometimes called the Blue party or the Blue-blooded party because it represented the elite. Two years later it swept the People's party completely out of office. In 1931 Governor Paul Pearson had charged Virgin Islanders to nurture "divine discontent" about their lot. The St. Thomas *Daily News* believed that the Progressive Guide's founders had been inspired by Pearson. Their organization, it added, "may become a powerful youth movement." With Hastie "very instrumental" in its development, the Progressive Guide was to become that and more.[29]

Hastie would rue the day his godchild became dominant in politics. In the thirties, however, he was not thinking of that. He had more immediate concerns, such as coping with racial discrimination that cut several ways. Although some opposition to his nomination had cropped up in the Virgin Islands, he was warmly welcomed upon his arrival there on 27 April 1937. On 7 May more than two hundred blacks attended a reception for him at the governor's mansion. Some seventy-five persons honored him at a dinner on 20 May at the Grand Hotel. Although appreciative of this, Hastie urged his admirers never again to discriminate against whites.[30]

Whites, of course, discriminated against blacks, including Hastie, to whom the Contant Tennis Club refused the membership traditionally given to the district court judge on St. Thomas. Through Roy Bornn the black West Side Tennis Club extended membership to Hastie. "Well, I certainly want to be a member of *your* club," Hastie told Bornn, adding that he would gladly have joined the Contant as well. On the court Hastie met another member of the West Side Tennis Club whom Bornn describes as "a very finished and very lovely young lady." She was the daughter of one of the wealthiest men in the islands, Herbert E. Lockhart, Sr. Her name was Beryl. Bornn introduced Hastie to her. "But I didn't know what I was doing at the moment," he says. "We often laughed about it, Bill and myself."[31]

Hastie and his wife, the former Alma Scurlock of Washington, organized the Go To College Club, which encouraged students to attend colleges in the United States rather than in Europe. Since Virgin Islanders were to be under American jurisdiction, Hastie said, they should know America, including black America. Accordingly, he urged that they enroll at Howard University. He persuaded a number of Virgin Islanders to change their opinion of Howard, which they had considered inferior to European schools. They had "a touch of the white fever," says Geraldo Guirty, a reporter. In other words, it was more important to go to a white school than to go to a black school. Many who considered Howard University inferior to European schools, says Enid Baa, a librarian and historian, suffered from "mistaken identity" and neither knew nor cared whether they were black, white, or something else. But Dr. Roy A. Anduze, a physician, claims that Hastie understood the danger of such unawareness in race-conscious American society.[32]

Hastie set a fine example of conduct in race relations, including interaction among blacks. "Until he came here," says Christian, "persons of his complexion or lighter were almost considered a different race from persons who were darker. He attempted to re-educate the people in that respect, telling them that as long as you have any kind of colored blood in you, you're colored or black or whatever you want to call it."[33] The point was lost on such Virgin Islanders as the one who rhapsodized, "What happy memories the sight of Germans on the streets of Charlotte Amalie brings to those who have enjoyed pleasant contacts with a large-hearted, friendly race!" This effusive article appeared in the St. Thomas *Daily News* twice within several days of the arrival of the Nazi sailing schoolship, the *Horst Wessel*, in St. Thomas harbor on 5 May 1938.

"Newspapers tell us of the fantastic happenings in Germany," the rhapsodist held, "but these young men dispel any illusions one might nourish of a race gone berserk or changed overnight into dragons and nightmare terrors." Bright-eyed and smiling, the men of the *Horst Wessel* reminded the writer of a day, come and gone, before the birth of most of them. "Not many people can understand what the Germans meant to St. Thomas for more than thirty years." For example, the Germans once employed many workers in coal wharves and warehouses on Hassel Island. German sailors not only wooed but also wed native-born women. "It is history now, but in the lives of many yet alive, the period was filled with stirring social incidents, riding parties, picnics, sailing, swimming, tennis and dancing the clock around. How merrily the cadets entered into the fun often shared by officers and civilians of many races and nations!"[34]

Dancing had again been scheduled—for Friday the thirteenth—at the Grand Hotel, though not quite "the clock around" (only from 9 P.M. until 2 A.M.). Invitations had gone out to the *Horst Wessel*. Plans for the dance, which was held to benefit the local high school, were made by a committee on which Hastie's wife served. Having been told by the management that a select group would be welcome at the hotel, the committee excluded from its list of invitees a number of native teachers but included the foreign sailors. Hearing this, Hastie left his chambers that afternoon and went home. "He made his position clear to me," says his former wife, who, better comprehending the implications of the committee's action, agreed that they should invite its members to meet at their home. On the evening before the dance, they met with the committee, criticized its action, and urged withdrawal of the invitation to the Germans. When the committee refused to exclude the sailors, Mrs. Hastie resigned from it.[35]

The story made news on the mainland as well as in the islands.[36] Meanwhile, Maud Proudfoot, chairman of the dance committee, and Elsa A. Lindqvist, the sponsor, had given their version of the episode. There was no color discrimination, they maintained. The committee "simply followed an established list," adding names as they were proposed. And it simply followed tradition in inviting the men of the *Horst Wessel*. "In the good old days when foreign warships frequented our port, dates for bazaars and benefit entertainments were usually set with a view to capturing the patronage of visiting ships, thus garnering some foreign money for a local cause." To have invited the Nazis was therefore "the natural thing to do." Mrs. Hastie, knowing all this, had not objected before the eleventh hour.[37]

St. Thomians had long regarded light or white skin as a badge of superiority, the *Progressive Guide* declared. "They loathe even the mere implication that they are colored, and the term Negro when applied to them is an abomination. And so St. Thomas has been witnessing a sort of Mason and Dixon division among its people who are of the same bred [sic], with a much too much of a social stratta [sic] sandwiched in." Other native newsmen agreed.[38]

Proudfoot and Lindqvist had complained that subsequent advocacy of democratic attitudes within the black race had amounted to a "crusade" for racial hatred. That they viewed such advocacy as fostering racial hatred amazed Hastie. For his part, he favored the elimination of invitation lists for dances intended to benefit high schools. If all "respectable citizens" were invited, St. Thomas would be "a happier and a better place." It was he who had raised the issue of race. "I did so in connection with the *Horst Wessel* invitation, pointing out that the members of the committee were Negroes and, as such, should have too much race pride to attend a social gathering with representatives of a government so intolerant of all non-Aryan races as is the present government of Germany." The invitations to the Nazis were withdrawn, but the Hasties, hearing that some of them planned to be at the dance anyway, did not attend.[39]

Before going to the Virgin Islands and while there, Hastie promoted a drive toward freedom that had begun two hundred years before he went to work at Interior. The slave rebellion on St. John in 1733 had been crushed by Danes with the assistance of French reinforcements from Martinique. More than a century passed before the second rebellion by slaves occurred on St. Croix in 1848. Danish, Spanish, and British forces suppressed the rebellion—but not before Governor General Peter von Scholten emancipated all slaves in the Virgin Islands. Various motives have been ascribed to von Scholten—regard for his native-born mistress, fear of a loss of Danish lives, belief that a successful rebellion was inevitable—but Hastie thought that the governor general had "seized the events" to justify action that he had long wanted to take.[40]

In 1939, halfway through his term on the federal bench, Hastie resigned in order to become dean of the Howard University Law School. President Roosevelt would have tried to persuade him to serve the remaining two years of his appointment "were it not for the fact," he told Hastie, "that there is a very important field at Howard University which will give you a wider scope in working toward the advancement of the Negro, which I know is your ideal of service. I am sure you are capable of rendering the kind of help needed there."[41]

"From the time that Hastie came down to the islands, he really had a very telling impact," says Alphonso A. Christian, Sr., who was to become a judge on the Territorial Court there. In the Department of the Interior he had not worked on just another piece of legislation; he had written a constitution for the Virgin Islands. And in the district court he had not been just another judge; he had advanced the day when Virgin Islanders would take their fate more into their own hands. More than a godfather to the Progressive Guide, he had been a founding father to the Virgin Islands—and this in an era when few whites and no blacks performed comparable service to the nation. But he could not remain in "a place off from the mainstream"; his interests, as he himself realized, "were basically on the mainland."[42] And as President Roosevelt surmised, help was needed there, Hastie was the man to provide it, and the Howard University Law School was a suitable base of operations.

CHAPTER NINE
Stimson's Stables

The search for a dean of the law school at Howard University had been under way for some time before the job was offered to Hastie. Lloyd K. Garrison, a member of the search committee and dean of the law school at the University of Wisconsin, made the case for Hastie. He said that Hastie would provide exactly what was needed. And that, said Garrison, "aside from legal capacity, is a cool head, wise judgment and the greatest possible tact, patience, courage, and poise." Those traits commended Hastie for another assignment not long after he accepted the position at Howard. This assignment was in the War Department, whose tenet was encapsulated in the title of a chapter in Walter White's autobiography: "Fighters Wanted—No Negroes." That blacks had to fight for the right to militarily defend their nation, said Hastie, was "one of the greatest ironies" of World War II.[1]

The fight was unavoidable because decision makers in the department were determined to minimize the impact that the utilization of blacks would have on military operations and social relations. In this they were guided by memoirs such as those of Major General Robert Lee Bullard, who had soured on black troops during World War I. Bullard swore that black troops were undependable, but that was not his only objection to them. "The Negro," he added, "is a more sensual man than the white man and at the same time he is far more offensive to white women than is a white man."[2]

In addition to personal recollections like Bullard's, the War Department relied on studies such as that prepared by field grade officers enrolled at the Army War College, whose commandant recommended its adoption to govern the use of blacks "in the next war." This study declared blacks to be incapable of leadership. Superstitious, jolly, docile, subservient, convinced of their inferiority—all this "by nature"—blacks had full faith in white officers, none in their own officers. They could be used in combat, but only under white commanders. About one of the authors (then a general) Hastie would say, "A man so misinformed and having such convictions about the Negro cannot possibly judge wisely the considerations involved in the issue now in controversy."[3]

But such men made the policies that were to frustrate Hastie. They decided, for example, that although blacks would be assigned to combat as well as service units, the enlisted men in those units would all be black. The assignment of officers would be 50 percent greater than in comparable white units. White officers would serve all units; black officers only black units and overhead installations. Although segregated, blacks would be allowed to participate in national defense. But this was not generally known. The policy was not classified as secret either when it was first established in 1937 or when it was revised in 1940, but only a few top officials knew its provisions. The commandant of the War College in 1937 had convinced the War Department that publicity would facilitate politically inspired attacks.[4]

Acceptable to some blacks, this policy for segregated service was unacceptable to Hastie, who declared that blacks would soldier, dig ditches, do anything to aid national defense. "But we won't be black auxiliaries."[5] In July 1940, before he left Howard University Law School to work in the War Department, the American Defense requested his signature—and his good offices in obtaining his colleagues' signatures— on its petition for compulsory universal selective military training. Replying that he and his colleagues believed the petition should oppose racial or religious discrimination in the selection of men for training or service, Hastie wired back: *Will sign petition thus amplified.*[6]

Warren A. Seavey of Harvard Law School, agreeing that "something drastic should be done," forwarded Hastie's letter to Secretary of War Henry L. Stimson. This prompted Hastie to tell Seavey that although Stimson's office knew about the situation, "I think it is very important that the War Department realize that responsible and patriotic citizens throughout the country are concerned about dangerously reactionary attitudes within our armed service."[7] He believed that most blacks would defend the nation, but with sharply decreasing enthusiasm. "They are wondering what is necessary to make a nation which will continue vicious discrimination to its own detriment in the present emergency change its attitude."[8]

Meanwhile, events were sweeping Hastie into a more direct collision with Secretary Stimson. In September 1940 the Selective Service Act made a man's induction into training and service contingent upon his physical and mental fitness, his acceptance by the armed forces, and the availability of housing, sanitary, medical, and hospital accommodations as determined by the secretary of war or that of the Navy, as appropriate.⁹ Thinking that these provisions governing induction might turn out to be loopholes through which Jim Crow would escape, A. Philip Randolph, Arnold Hill (acting secretary of the National Urban League), and Walter White called on President Roosevelt, Assistant Secretary of War Robert P. Patterson, and Secretary of the Navy Frank Knox on 27 September. The three black leaders took with them a comprehensive memorandum of concerns.¹⁰

Although Roosevelt disliked White's idea of integrating troops in the North, he liked Hill's idea of appointing a black civilian aide to the secretary of war. But when Roosevelt's policy was announced by his press secretary, Stephen Early, on 9 October, it was essentially the mobilization plan that had been adopted but not publicized in 1937. The sole additional provision affirmed the policy of segregation. Early added insult to injury by implying that White, Randolph, and Hill had approved the policy. They, in turn, released the memorandum they had left with the President and demanded that Early be fired. Early was not fired, but he did retract his statement.¹¹

Steve Early did something else a few weeks later. He drove his knee into a New York City policeman's groin. The officer was trying to keep intact the police line set up to protect Roosevelt as he left Pennsylvania Station after having given a campaign speech at Madison Square Garden. The 1940 elections were near, the President needed black votes, and the policeman was a black.¹² Adverse publicity forced the White House to place a call that interrupted the poker game Hastie and others were playing in Robert Weaver's basement. Prepare "a really rousing speech" for the President, they were told, one that will repair the political damage when the President delivers it in Baltimore. Weaver said that talking would not turn the trick. Asked what would, he consulted the others before making suggestions that Roosevelt approved at once. Two days later Benjamin O. Davis, Sr., was appointed the first black general; Campbell C. Johnson was appointed assistant to the director of the Selective Service; and Hastie was appointed civilian aide to the secretary of war.¹³

Hastie hesitated to take the job. "I was reluctant not because of any lack of interest or because it was not an important area, but I was rather skeptical as to what [could be done by] a person with no authority of his own whom I was sure the military did not want serving in the Secretary's office. . . ." He did not want to appear to be an appeaser. Thurgood Marshall anticipated no problem on that count, provided the

War Department made it clear that Hastie's duty would be to fight discrimination.[14] For advice Hastie turned to Felix Frankfurter, who had vouched for him with Roosevelt and Patterson. Frankfurter suggested that the War Department announce that Hastie, despite his firm opposition to segregation, had been persuaded to provide a voice for blacks inside the department. He also suggested that Patterson privately express to Hastie regret about the department's policy. Meanwhile James Rowe, a member of the White House staff, tried to persuade Hastie ("against his better judgment") to take the job.[15]

Stimson's letter of appointment lacked a pledge of integration, so Hastie openly affirmed his own. Selecting Truman K. Gibson, Jr., as his assistant, Hastie set to work on 1 November 1940.[16] He expected the Army to maintain segregated units during wartime. His goals were to minimize discrimination despite the segregated pattern and to prevent segregation from spreading, especially to new kinds of units and to training schools.[17]

At his first meeting with Hastie, Secretary Stimson greeted the "rather decent negro" whose elegance under trying circumstances would impress him[18]—but not for long, because the two men were worlds apart. Stimson, for example, not only disapproved of efforts to discharge Jim Crow from the Army but saw those efforts—and even Hastie's appointment—as black exploitation of Roosevelt's pre-election vulnerability.[19]

At a later meeting with Stimson, and with Under Secretary of War Robert P. Patterson present, Hastie voiced concern about the restrictions placed on the assignment of blacks to any given camp. The policy, as Hastie saw it, was "definitely against" the stationing of too many black soldiers at the same camp. A related practice was that of limiting the size and location of black units. Most such units were small and isolated in white cantonments. Arguing that these practices were detrimental to the pride and morale of black soldiers, Hastie favored the creation of another division or several regiments for blacks. He also favored the transfer of small units to posts where large numbers of blacks were stationed.[20] This led Patterson, in a memorandum to Stimson dated 16 January 1942, to remind the latter, "You mentioned the suitability of Negro soldiers for operations in the Tropics."

In the meeting with Stimson and Patterson, Hastie suggested the racial integration of troops; however minimal, integration would foster an end to prejudice and brutality against black soldiers. Neither Stimson nor Patterson endorsed the suggestion. Also rejecting it, the military high command accused Hastie of trying to foment social revolution contrary to both majority will and the army's mission, which was to defend the nation, not to settle racial issues. General George C. Marshall, the chief of staff, told Stimson that the Army should face squarely the question of utilizing blacks, but in full realization that "experiments within the Army in the solution of social problems are fraught with danger to efficiency, discipline, and morale."[21]

Hastie urged the Army to increase the proportion of blacks to 10 percent by incorporating small black units into larger white units, establishing additional black units, and assigning blacks to all kinds of new units. General Marshall gave the stock response: The nation's safety, not its synecology, was the army's responsibility.[22]

Even good men, said Hastie, tried to "compromise between fairness and intolerance." No sociological experiments, the Army said, even as it regulated the lives of millions. "[While] exercising the most extreme authority, literally of life and death over them, the Army is up to its neck—more often, I think, over its head—in sociology."[23]

Rationalization and perpetuation of this contradiction were, in large measure, grounded in geography.[24] Because warm weather and abundant land were present in the South, a great deal of military training occurred there. But another feature of this region, as Hastie noted, was neither geographically defined nor regionally confined. It was not only in the South that many field grade commanders and even more company grade officers, who were essentially civilians holding commissions, also held racist opinions. Many Southerners were in regular and reserve military service even in time of peace. As time passed, these officers became increasingly powerful.

Some officers were fair-minded, Hastie added, "but a tragically large number sought to enforce in the Army the traditional southern treatment of the Negro. And too often the officer from other sections of the country was [a] ready convert, if conversion was necessary, to the point of view which we characterize sometimes unfortunately and unfairly as 'southern.'" If not controlled by the War Department, this "situation of intrinsic difficulty" would worsen "beyond separate units and evils incident thereto." But the department vested responsibility and discretion in the local commander and was as likely to remove him because he insisted on fair treatment for his black troops as it was because disorders resulted from mistreatment that he approved.

It was not into the South that the black 77th Coast Artillery moved to defend a shipbuilding and industrial area. When the unit took up positions in Marcus Hook, Pennsylvania, between Philadelphia and Wilmington, the townspeople welcomed it. They gave parties, paid visits to the camp, and invited the soldiers into their homes; women asked the GIs how things were going. Observing all this, the unit commander posted a warning: "Any cases of relations between white and colored males and females whether voluntary or not is considered rape and during time of war the penalty is death." The War Department rescinded that warning after black organizations protested. But why did the commander act? "I was afraid some of these Negroes might get ideas," he said.[25]

Those ideas were as taboo in Roosevelt's army as in Hitler's army. That was one reason why black GIs showered Hollywood moguls with thanks for their very own pinup girl, Lena Horne. "They couldn't put Betty Grable's photograph even inside their foot lockers," she would tell her Broadway audience on 25 July 1981. "I was safe. . . . They could put my picture anywhere."[26]

The Army, however, was careful about where photographs of black GIs themselves appeared; photographs amounted to publicity. The War Department, said Hastie, publicized little concerning their war efforts and even less about the results of those rare instances when democracy was given a chance to work. Antiblack congressmen cut appropriations for the Office of War Information to prevent such publicity. Gordon Parks remembers their animus. Parks, a photographer whose interest in the Air Force had been inspired by Hastie's fight against racism in that branch, hoped to accompany the 332nd Fighter Group overseas in December 1943, but its commander, Colonel Benjamin O. Davis, Jr., told him that his papers were not in order and that he would have to learn why in Washington. Parks went to his home office, the Office of War Information in Washington. Assured that all was in order, he returned to the 332nd only to be informed by Davis that the Pentagon had notified him that Parks could not accompany the 332nd.[27]

Anger, regret, and disbelief welled up in Parks. "This is the first Negro fighter group. It's history. It has to be covered. Can't you protest in some way, Colonel?"

"There's nothing, absolutely nothing I can do. The orders are from the Pentagon. They cannot be rescinded. I'm terribly sorry."

Parks's friend Ted Poston had been right. "There's some Southern gentlemen and conservative Republicans on Capitol Hill who don't like the idea of giving this kind of publicity to Negro soldiers."

Those congressmen had a kindred spirit in Brigadier General Alexander D. Surles, director of the Bureau of Public Relations in the War Department. Opposed to any action that suggested "special status" for a group of newsmen, Surles resisted Hastie's efforts to publicize the black soldier by arranging for General Marshall to address a group of black newsmen late in 1941. When Marshall met the newsmen on 8 December, he told them that the War Department was seeking authorization to complete the organization of a black division. He also said that there was thought of establishing Reserve Officers Training Corps units at three black colleges, and that black units were being organized in all branches of the Army. Most significant of all, however, in saying that the department was not satisfied with its progress against discrimination, Marshall added in an aside, "And I am not personally satisfied with it either."[28]

The personal reference by Marshall lifted the group's hopes. Within the hour Colonel E. R. Householder shot them down. Householder had commanded a company and a battalion in the 25th Infantry. ("If I may be pardoned for saying so, I have letters from individual soldiers which prove that I did so with some success.") Then in the Adjutant General's Department he headed the Miscellaneous Division, which handled "the innumerable inquiries and complaints received from individuals and societies purporting to represent the Negro." Many inquiries were initiated by Hastie, through whom the black press consistently channeled prepublication reports about brutality against black soldiers. Disenchanted by the lack of responsibility and self-restraint shown by the black press, Householder's most telling blow against the hopes raised by Marshall was delivered when he took up accusations that the Army and War Department practiced discrimination. Attempts by the Army to solve racial problems "would result in ultimate defeat," Householder asserted. "The Army is not a sociological laboratory. . . ."[29]

The discussion that followed the presentations by Householder and other officials was less than spirited. It was as though no one would pick up Householder's gauntlet. Finally someone did. So the War Department, said Roy Wilkins of *The Crisis*, was telling blacks that current racial practices were best for morale. "They don't know whether that is true because they have never tried any other system to see whether it would work." Integration was working in Reserve Officers training camps and would work in regiments.[30] Claude A. Barnett, founder of the Associated Negro Press, suggested that soldiers be allowed to volunteer to serve in an integrated battalion. What better time than during war was there to put aside racial antagonism? But Brigadier General Benjamin O. Davis, Sr., from the Inspector General's Office, reported high morale and few complaints from black soldiers as things stood. (Davis constantly oozed optimism about race relations, undermining Hastie by arming the War Department in defense against his charges.)[31]

Some of the newspapermen chose to accentuate Marshall's comment and minimize or ignore Householder's remarks. But one black newspaper noted Hastie's predicament: He was an able man without influence or authority in a department clinging to discrimination and segregation. Marshall's news conference was "an obvious attempt to appease belligerent Negro editors who have taken a critical view of the whole panorama of national defense." Wrong on the second count, the newspaper was right on the first. Of course Hastie was competent but impotent. Generally blacks in such positions never did exercise power. And when segregation was challenged, a typical part of the power holders' response was the one Householder gave: "I think it is a problem which probably hasn't been entirely settled among your own people."[32]

Householder's colleague Lt. Col. James W. Boyer castigated Hastie for being determined "to advance the colored race at the expense of the Army." None of the many advances satisfied Hastie, and "he intends to go from one disputed point to another." His loyalty to the NAACP exceeded his loyalty to the War Department. "Incident after incident could be recounted wherein he has demonstrated willful persistence in breaking down the Department's long considered policies." Boyer contended that Hastie, with whom he had often "dealt on a most pleasant basis," created rather than solved problems. However, the Personnel Division credited Hastie with having won some concessions. "If this action is continued," it noted, "the whole program may get out of hand."[33]

Boyer, Householder, and others in the Adjutant General's Department took their cue from Major General E. S. Adams, the adjutant general, who said that blacks preferred segregation in the Army. He declined to discuss that "well-established principle" with black reporters because "too much is going on in the War Department these days to give time to little things like these." The Adjutant General's Department, offering another explanation for advocacy such as Hastie's, dragged out a perennial white explanation of black discontent: "It is well known, of course, that the Negro population has been a focal point of subversive agitation." Subversives were agitating inside as well as outside the armed forces. "While the events which have so far transpired have been scattered, there appears to be underlying all such events a pattern of centralized stimulation."[34]

Hastie detected subversion of another kind. In a memorandum to Stimson dated 22 September 1941 he claimed it deprived the nation of "that passion for national ideals" that was indispensable to national survival. It stemmed from the inconsistency between words and actions. "So long as we condone and appease un-American attitudes and practices within our own military and civilian life, we can never arouse ourselves to the exertion which the present emergency requires."[35]

Hastie himself was an emergency for the War Department where he was subverted in classic bureaucratic fashion. General Marshall told black newsmen on 8 December 1941 that Hastie was "well informed" about developments concerning black troops, but long before then it had become common practice to route information with a cover note stating "Not to be shown to Judge Hastie." For example, Hastie was kept in the dark about the decision to move the 54th Coast Artillery out of Camp Wallace, Texas, action favored by white but opposed by black citizens. He complained to Patterson on 15 July 1941 that the deception sabotaged his office.[36]

The deception continued. On 11 October 1941 a press release informed Hastie that the Bureau of Public Relations had scheduled a conference of delegates from womens' organizations to arrange more effective participation by women in the defense program. Hastie asked whether black organizations had been invited. They had not, he was told; only large federations would be represented; the conference was to be small. Still proceeding informally, Hastie said that the National Council of Negro Women, a large federation, would begin a weeklong convention in Washington on 13 October, the place and date of the bureau's conference. Even on last-minute notice "authoritative spokesmen" such as Mary McLeod Bethune would accept an invitation to the conference. They received no invitation. Their exclusion, Hastie told General Surles, "will seriously hamper our efforts to promote essential unity at this time." Once again the War Department had manifested "a disposition to make [the black] an unwelcomed appendage to the Program of National Defense rather than an integral part of it."[37]

Hastie himself was an "unwelcomed appendage" to a department more committed to undermining than to undergirding him. Ranking officers, for example, deceived him in the spirit of claims such as that made by the Adjutant General's Department, namely, that "the pattern of centralized stimulation" had created no rebellion by most blacks in uniform because they had used "good common sense." The fact of nonrebellion, however, did not obliterate the reality of black military life portrayed by John Oliver Killens.[38]

> "We ain't soldiers, tell him, Corporal Solly. You a Race man. Talk to this antique fool." In Killens's novel, Scotty was putting Buck straight. "Don't you know this regiment is a service outfit, ass? I thought you at least had that much sense. Any old fool knows that colored people join the service and white folks join the Army."
>
> "They training us how to serve the white soldiers," Bookworm added.

". . . [Y]ou might as well let the facts hit you in the face," Solly said. "Hitler and Tojo and the governor of Georgia are on the same damn team. All three of them're against you and me. . . ."

There had to be some way to get some relief from the trouble being heaped on them, Bookworm thought. "Maybe with that Double V-for-Victory jive, we could take our case to the N-double-A-C-P. . . ."

James G. Thompson started the jive. "Will things be better for the next generation of Negroes in the peace to follow? Would it be too much to demand greater liberty and opportunity for people of my race, in exchange for the sacrifice of my life?" In Wichita, Kansas, awaiting the call to war, Thompson answered the questions. "Let colored Americans adopt the Double V—VV—for a double victory: the first over our enemies from without, the second over our enemies within." Having received Thompson's letter, the Pittsburgh *Courier* headlined the Double V idea, recruited its eight thousand dealers, distributed car stickers, buttons, and blotters proclaiming "Democracy: At Home—Abroad." Once ignited, the idea spread like wildfire among blacks. Some would reverse the order of the two V's, or at least place them side by side.[39]

Adam Clayton Powell, Jr., a minister in Harlem, spoke for many blacks when he declared, "As long as Negroes are kept in the glory-hole of ships, in the isolated 99th Pursuit Squadron, in the Jim Crow regiments of the army, and get no closer to the assembly line in defense industries than as porters in the toilets, the war is futile and the peace is already lost."[40]

Hastie made his views known at the Morgan State College commencement in Baltimore on 10 June 1940. "In this present war, we have got to insist as never before that we're anxious to do our part, but only if equal justice is meted out to us during this period with the prospect of the same thing continuing afterwards." At a meeting sponsored by the National Negro Stop-Hitler Committee in Chicago in November 1941, he said, "I believe it is of the utmost importance to the American Negro that Hitler be stopped." Rights that blacks took for granted would be nonexistent under Hitler, who held slavery and death in store for blacks. So foreign threats could not be disregarded. But neither could domestic trouble. "I believe that while training an army to fight Hitler, we should not let up one bit in our effort to prevent discrimination in the army right here."[41]

At a conference in Harlem on 10 January 1942, Hastie asked seventy delegates from eighteen national black organizations whether blacks were fully backing the war effort. By a two-to-one vote the delegates affirmed his belief that such support was lacking.⁴² Jim Crow's myrmidons in the War Department knew what the Double V protest was all about. They received reports from the Bureau of Intelligence. Classified as secret and approved for his receipt by Archibald MacLeish, director of the bureau's Office of Facts and Figures, one such report reached Hastie's desk in June 1942.⁴³

This secret report disclosed that almost half (41 percent) of the blacks in Harlem cherished the second V that they had added to the "V for Victory" campaign: V for Victory over racism in the United States. Negroes and poor whites in New York City differed about the importance of the second V. They were asked whether it was more important to improve democracy in the United States than to beat Germany and Japan. Thirty-eight percent of the black respondents answered in the affirmative (significantly, 12 percent declined to answer), but only 5 percent of the white respondents agreed. About five out of every ten blacks (50.5 percent)—but nine out of every ten whites—placed primary emphasis on beating Germany and Japan. According to this report, blacks were not disloyal. "On the contrary, [blacks] are deeply devoted to American ideals, asking only that these ideals be realized in relation to themselves." One ideal was equal opportunity in the armed forces. Nearly half (45.5 percent) of the black respondents thought the Army was unfair, and over two thirds (67 percent) thought the Navy was unfair. Although it had been recently lifted, the ban on their becoming seamen was resented by blacks. They criticized the Army's failure to protect black soldiers against violence by white MPs and civilians, but even more its segregation of black troops. The report continued: "The line of reasoning indicated in comments made to interviewers was that there shouldn't be colored soldiers and there shouldn't be white soldiers, there should only be American soldiers."

Roosevelt gave blacks reason to believe, in 1942, that he was empathetic. In August he suggested the national slogan "Minorities Are Vital to Victory." The President was playing on the theme of the NAACP conference to which he sent his message on June 14: "Victory Is Vital to Minorities." But as the report noted, the crucial attitudes were those held by whites toward blacks in and out of the military: "Negroes cannot put their shoulders to the wheel in the war effort if they are not permitted to do so."⁴⁴

Nevertheless, the report added, a third (36 percent) of the whites in the Northeast and West and over half (53 percent) of those in the South blamed blacks themselves for inequities in opportunities. Furthermore, whites knew which opportunities blacks least wanted but were totally unwilling to grant the ones—equal jobs and wages and, in the South, nonsegregation—blacks most wanted. Whites were much more inclined than blacks to believe that conditions had improved for the latter since the war began, but they were far less inclined to believe that conditions for blacks would improve after the war ended. Nearly half of the whites thought blacks had as good a chance as whites to get war jobs and to get ahead in the military. There were regional differences in anti-black sentiment. "In all parts of the country, however, large numbers of people were unsympathetic to blacks," the report read. "Not only were they cold to their aspirations; many evidently felt that rights long since granted to Negroes should be revoked."

It was, in Chester Himes's phrase, "as sure as water is wet" that Hastie would cross swords with revokers in the War Department. He did so in the summer of 1942, when he opposed plans to segregate housing and dining facilities for officers and civilians at the army air base near Tuskegee, Alabama. His recommendation that the Army ban such segregation elicited from Colonel John R. Dean, secretary of the general staff, the old excuse that the Army could not jeopardize the war effort by balking at "accepted social customs."[45]

Hastie found that argument specious. "The Army, despite all protestation that it is not concerned with sociological [patterns] seems to be undertaking to impose a segregated pattern on the leisure time activities of colored soldiers in England, despite the willingness of the British to welcome and accept Negro troops." White soldiers resented the friendliness of the English, especially women, toward black soldiers. It was reported that, in keeping with Lieutenant General Dwight D. Eisenhower's request, blacks in the Women's Army Auxiliary Corps would be shipped to England "to provide companionship" for black soldiers. That, Hastie told Patterson, was not the purpose the corps was intended to serve.[46]

Eisenhower's request typified the Army's penchant for what Hastie called "sociological adventure." Hastie thought educational programs in race relations were required to curb the very social experimentation that the Army and the War Department disclaimed. But Stimson disagreed, asserting that officers and soldiers did not need instruction "on matters which I believe can be safely left to their own good sense and fairness as American citizens." He believed that white soldiers respected their black comrades.[47] And hearing that racism in the Army would be the topic of a speech by Archibald MacLeish, a top official, Stimson spent an hour explaining to him how "this crime of our forefathers had produced a problem which was almost impossible of solution in this country. . . ." This, the secretary added, was especially the case during time of war.[48]

MacLeish might have been influenced by Stimson's remarks. In his nationally broadcast speech at the National Urban League's dinner on 11 February 1942, MacLeish said that blacks would contribute to the war effort because they knew the difference between slavery and freedom. Blacks, he continued, also knew that their mistreatment in the United States was attributable to imperfections in democracy. This rationalization of racism was unacceptable to *The Crisis*, which held that lynching was not an imperfection of democratic life. It felt that until government took effective action against lynching and all other forms of racism, no one could blame blacks for not jumping up and cracking their heels together about the "war for democracy." No less than *The Crisis* and MacLeish, Hastie and Stimson saw this war in quite a different light and placed a vastly different interpretation on the assertion, "Liberty is the only thing you cannot have unless you are willing to give it to others."[49]

Hastie and Stimson did not see eye to eye on policies that were important for reasons practical, symbolic, or both. One such policy governed the blood bank. At first blacks were disqualified as donors. After they were permitted, in January 1942, to become donors, their blood was kept separate from that donated by whites, and it was injected into white soldiers only if requested by those soldiers. Major General James C. Magee, the army surgeon general, while admitting that the reasons for this policy were not "biologically convincing," insisted that they were "commonly recognized as psychologically important in America." The American Medical Association, the American Association of Physical Anthropologists, the American Red Cross, and the Army Medical Corps itself acknowledged the lack of scientific justification for the segregation of blood. But Magee said that 95 percent of all soldiers were white and preferred blood from whites. Give it to them. As for black soldiers, give them plasma from black donors if they object to receiving it from white donors.[50]

In August 1942 Hastie asked Surgeon General Norman T. Kirk to reconsider the suggestion that the Red Cross be advised to discontinue its blood-segregation practice while, as far as possible, honoring soldiers' requests for direct transfusion from donors of their own race. Hastie's query was answered by Assistant Surgeon General C. C. Hillman in a form letter. The Army agreed that "there is no biological incompatibility between plasma of different races when administered therapeutically," Hillman told Hastie. But many whites did not want black blood injected into them. Whether that disposition resulted from bias or ignorance, "it nevertheless exists," Hillman added, and could be disregarded only at risk of failing to obtain plasma. Why didn't blacks seize their "wonderful opportunity" to help by donating blood? What evidence that would be of their "superior concern over National welfare as compared with racial interest."[51]

"This language seems to me to be an admission that the whole basis for the policy is an unwillingness to defy the prejudiced sensibilities of a portion of our population," Hastie wrote later in assuring the New Jersey Urban League executive secretary that the Red Cross undoubtedly had lied in saying that a "high ranking colored officer" had represented black soldiers in asserting (as the Red Cross put it) that "they would be happier if they knew that the blood to save them when they were wounded came from their own people."[52]

Hastie's friend Charlie Drew provided expert scientific opposition to segregated blood banks. Dr. Drew had conducted research about preserving blood and using it in transfusions. This work, performed at Columbia University in 1935 on a General Education Board Fellowship, had been followed five years later by his initiation of the "Blood for Britain" project, which concerned the battlefield use of blood plasma. In 1941 Drew had been the director of the American Red Cross Blood Bank. In his opinion, the policy requiring segregation of blood was "indefensible from any point of view." Taking a similar stand, the American Association of Physical Anthropologists underscored part of the irony inherent in the policy. It pointed to many Southern families' reliance on black wet nurses and asserted that the white infants ingested not only "the nutritious elements in the milk of those colored women [but also] many of the same substances which were circulating in the blood stream [sic] of the women who were suckling them."[53]

The irony did not escape the Philadelphia *Tribune*. So there is to be no mixing of blood, it observed, adding, "It is perfectly all right for 'white' blood to be pumped into colored soldiers; and it continues as an American custom, wherever the white man goes to mix his blood with native women, whether in the Pacific or in Africa." Red Cross officials admitted that a mixup in labels would leave them (and everyone else) unable to distinguish between "black" and "white" blood. Hastie, like others, argued that the policy hindered the war effort and reflected a Nazi view of race. He told Stimson that all blacks were insulted by policy made in deference to "those who insist that our country treat the Negro as a loathsome being even as all non-Aryans are regarded under the ideology which we are fighting to the finish with force of arms. Even the saving of the lives of soldiers is weighed against the appeasement of the sentiments most alien to our professions, and found to be wanting."[54]

Hastie's offer of his resignation in protest accounted for the change in policy from one of outright rejection to one of segregated acceptance of blood from black donors. Stimson did not accept the resignation. He thought he understood Hastie. "I really think what he wanted was a talk and some encouragement and these Patterson and I gave him," the secretary noted in his diaries. But before the year was out, his patience had been exhausted. Some of Hastie's complaints, he felt, were "well founded and probably remediable," but some were "trivial [and] of the impossible class to solve which represent the hopeless side of the insoluble problem of the black race in this country."[55]

There in his Augean stables (i.e., the War Department) Stimson had discovered the so-called black problem. "But this perhaps is a misnomer," Langston Hughes has written, "for if Negroes are a problem to white Americans, whites are even more of a problem to Negroes. . . ."[56]

CHAPTER TEN
Domestic Enemies

Hastie was well informed about the problems that whites created for blacks in the Army. Take black officers, for example. Although the War Department had never collected data about their performance in World War I, it had drawn disparaging conclusions that Hastie considered the primary reason for its refusal to allow blacks to become officers. In fact, said Hastie, the Army had not wanted to commission black officers in 1916. Forced to train seven hundred of them at Fort Des Moines, Iowa, it took its revenge by selecting unsuitable candidates, many of whom became incompetent officers and therefore living "proof" that blacks could not carry their commissioned weight. Even those who were competent had the doors to Regular Army commissioned ranks slammed in their faces and the retention of reserve army commissions denied them. When World War II began, blacks constituted only one half of 1 percent of the Officers' Reserve Corps. Only five blacks were Regular Army officers, and three of them were chaplains. By December 1942 a policy of operating integrated (except for the Air Force) Officer Candidate Schools (OCS) produced more than a thousand black officers. Other blacks won battlefield commissions, and still others were commissioned from civilian life. Black women were obtaining commissions in the Women's Army Auxiliary Corps and as nurses.[1]

Blacks who became officers were at a severe disadvantage. Generally they served with or commanded only black troops. Black nurses and doctors took care only of black patients. Black officers were assigned to combat divisions, but units staffed only by them were unlikely to see combat. Closed to those officers were virtually all administrative and other positions whose occupants dealt with both black and white troops. Of course, none were to command white officers, and promotion was possible only when the War Department certified an opening for a black. To the assistant chief of staff Hastie complained that black officers found it very hard to obtain field grade promotions or even to advance in the four regiments that were to have only black officers.[2]

For all the difficulties, however, Hastie believed that black officers were far better off in the second than in the first world war. Greater progress was in store, he wrote, because current officer candidates were future colonels and generals. "Whether these thousands of young men and women remain in the Army or return to civil life, they can no longer accept without question the traditional and stereotyped conceptions of racial inferiorities and differences." For this reason Hastie worked to improve OCS opportunities for blacks. The odds against them were great. For example, commanders often refused to notify blacks about OCS opportunities. Some were biased, others were unconvinced of the War Department's sincerity about admitting blacks to OCS, and still others did not want to lose their best soldiers. To make matters worse, many of these soldiers chose to remain sergeants rather than becoming second lieutenants (commonly considered a black officer's permanent rank) and confronting greater hostility from white officers, enlisted men, and civilians.[3]

Other features of life and duty discouraged black enrollment in OCS. For example, it was hard to place those who were technically or professionally trained. When Hastie tried to pave the way for lawyers in the Judge Advocate General's Department, he was told by JAG that the place for them was in black divisions. The problem was that in divisions commanded by whites, only white officers were allowed to head special and general sections. One JAG incident, however, brightened Hastie's day. At commencement exercises at Fort Washington (near the District of Columbia) Hastie, who was the speaker, spotted three black officers in the graduating class. Awarding Joel Adams his certificate, Hastie first uttered a formal "Congratulations" and then whispered, "Well, how do you do, man? I'm so glad to see you."[4]

Black officers' difficulties were not confined to military posts, as William E. Raynor and his friend Bob learned. While double-dating in the South, the sailors saw a man running to join other blacks who had been waiting in the rain for the bus. When he boarded it, they saw his captain's bars and ribbons from World War I. They saw him pay the seven-cent fare, pause as he passed empty "white" seats to cross the segregation line and stand, declining with a smile the seat offered by a little old black woman. Then they themselves were standing, knowing not when or why, only that they could not sit among empty seats that he was forbidden to take.[5]

Their Southern dates were embarrassed, and maybe the captain was, too. He got off at the first stop a half mile down the road. In doing so, he paused, said nothing, gave no smile, only a quick look into each sailor's eyes. A tremble of the lips and he was gone. Betty Lou loudly insisted that the sailors sit down, and Agnes yanked Raynor's sleeve. But he and Bob had forgotten them; the moment was filled with "the pain and humiliation" in the captain's eyes. As Bob said, "It was the first time we'd ever seen the inside of a heart."

What if the sailors' sensitivity had been characteristic of the top military men in the War Department and in the field? What if Major General Richard Donovan had typified more, and Major J. D. Frederick fewer, of them? Donovan, the commanding general of the VIII Service Command at Fort Sam Houston, Texas, had been outraged by the murder of a black MP, Private Raymond Carr, by a Louisiana state trooper, Dalton McCollum.[6] On 1 November 1942 in Alexandria, Louisiana, four state troopers tried to arrest Carr while he was on duty as an MP. He ran. Two chased him, pistols blazing; they overtook him, and McCollum shot him. Carr was alleged to have used abusive language, a misdemeanor. He was alleged to have attacked McCollum with the only weapon he carried, a stick. Donovan wanted McCollum punished. Under his orders Colonel K. F. Hanst asked State Police Assistant Superintendent L. A. Newsome to arrest McCollum. But Newsome said he would not arrest a state trooper "for shooting a nigger." Carr was a black American soldier, Hanst said. To Newsome "he was still a nigger."[7]

Appealing to Governor Sam Houston Jones's "inherent sense of fairness and right," and informing him that no action had been taken against the murderer, Donovan asked Jones "to join me in the position that any soldier, in uniform, regardless of his color or racial extraction, is entitled to fair and just treatment," and that civilian policemen could not molest, arrest, or attack with pistols on-duty MPs whose conduct could be challenged at military headquarters.[8]

Wanting "every effort" made to punish McCollum, doubting that "local racial sentiment" would permit indictment by a grand jury, and reinforcing his appeal to Governor Jones, Donovan suggested that the attorney general be asked whether McCollum could be tried for violation of federal law. If not, then emergency legislation should be obtained as a federal shield for soldiers. In line with his suggestion, the adjutant general drafted a letter and sent it to Hastie's office. Hastie's assistant drafted a substitute, which Stimson, enrgaged by the impossibility of bringing McCollum to justice, signed and sent to Attorney General Francis Biddle.[9]

Hastie liked Donovan's suggestion. Biddle did not. "It seems to me that this is a case for the state to indict for manslaughter or murder," he told Stimson, "and that it becomes a little silly for the Federal Government to prosecute . . . on the ground that the man who was killed was deprived of his life without due process of law." But he had sent an investigator to Louisiana, and when that man's report was in hand, Biddle would suggest that Stimson ask Governor Jones for vigorous prosecution of the case. "I note that Major General Donovan has already requested such action."[10]

Hastie commended Donovan, whose "forthright action and recommendations may well have provided the necessary impetus for basic remedial action." He had no such praise for Major J. D. Frederick, who figured in his investigation of the murder of Private Albert King, a black GI.[11] King was gunned down by Sergeant Robert A. Lummus on 21 March 1941 at about 4:30 A.M. at Fort Benning, Georgia, after a fight involving Lummus, white soldiers, and Private Lawrence J. Hoover. While the fight was going on, King left the bus and his companion, Hoover.

A court martial exonerated Lummus of what Hastie called a "callous and wanton shooting. . . ." Lummus had shot King five times; first in the lower abdomen and then, as King lay prostrate, "in the back near the left kidney . . . in front of the left ear . . . at the upper margin of the left ear . . . [and in] the left side of the neck. . . ." He testified that King threw up his hands after the first bullet hit him. Rubbish, said Hastie, adding that a shot in the stomach from a heavy-caliber weapon causes a man to crumble or clutch his stomach, not raise his hands. Hastie concluded that King lifted his hands before, not after, Lummus shot him.

After its findings had been made, a board of inquiry in this episode was told by Major Frederick that King had escaped from MP custody and was being recaptured at the time of the shooting. How could King have escaped custody and been subject to recapture, Hastie wanted to know, when he had not been arrested? (Only Hoover had been arrested.) And how could King have been subject to recapture when he had not been in custody? Frederick told the board that King was unauthorized to be at the scene. King had a valid pass to leave the post, Hastie countered, and was on a highway inside the camp, apparently on his way back to his quarters.

Frederick disapproved the board's initial finding that King had been killed in the line of duty. On his orders the board reversed itself. Hastie protested, telling Stimson that the amended finding should be nullified and the order requiring it investigated. He reasoned that this nullification "would render partial justice in this wanton slaying of an unarmed soldier."

Too few commanders and staff members were like Donovan, and too many were like Frederick and Major General William Bryden, deputy chief of staff, whose report about disorders near Gurdon, Arkansas, in 1941 Hastie rebutted.[12] The 94th Engineer Battalion, consisting of blacks from Detroit and Chicago, went into bivouac on 10 August. The next night two or three hundred of them unjustifiably demonstrated in Gurdon. Or so Bryden reported on 15 October. The next day Hastie contested Bryden's accuracy and thoroughness. What Bryden called a "group demonstration" had been, in fact, "a mass exodus" by panic-stricken men from the black part of town back to their bivouac area. Contrary to Bryden's assertion that "white people were insulted, threatened and placed in fear of their safety," Hastie concluded that although profanely noisy, the soldiers had no intentions other than to return to bivouac, where, it was rumored, a soldier had been beaten. His purpose was not to excuse the soldiers' disorderliness but to provide "a more revealing picture of the facts" than the one given by Bryden.

Bryden reported that on 12 August state troopers ordered the battalion's guard away from the bivouac entrance, "using force to exact compliance with their demands and displaying firearms." Hastie argued that the troopers attacked the bivouac. "It is hoped that neither the brevity [thirty-seven words] nor the restrained language of [Bryden's] description minimizes in any way the gravity of such an assault upon a Military encampment." Agitators in the battalion caused the incident, Bryden asserted, adapting (Hastie said) the characterization that military investigators had applied to the black GIs who complained that their sentries were not armed.

On 14 August the third incident occurred.[13] Just before noon state troopers informed the provost marshall, Second Army, that soldiers from the battalion, disorderly and without military supervision, were moving along the highway. The troopers asked if they should take charge. The Provost Marshall replied that they should do so until his own men arrived. Armed troopers cleared the soldiers from the highway, striking an officer in doing so. "This incident was ended peaceably by arrival of the Military Police," Bryden added. But Hastie gave a different version.

According to Hastie, the soldiers were under military control, moving to a bivouac site at a greater distance from Gurdon.[14] Second Lieutenant Donald A. Curry was in command of the second platoon when Company C took a ten-minute break, on Highway 67 near the Little Missouri

River, between 10 and 11 A.M. A state trooper passed in a patrol car, turned around, and drove back. "Get those God-damned niggers off the road or I'll come back and get them off for you," the trooper snarled at Curry. "And as for you, boy, you can go to hell." Forty-five minutes later state troopers stopped the soldiers, cursed them, and ordered them off the highway. One state trooper named Mason tore off Curry's glasses and struck him, drawing blood from his mouth. "Don't hit him!" another trooper yelled. "Kill him!"

Bryden claimed that the provost marshall asked state troopers to take charge of the disorderly troops until MPs arrived. According to the record, said Hastie, the men were not disorderly and the provost marshall's response to the telephone call from the state troopers was that he would assemble MPs and meet them at the scene. He did not request them to take control in the interim. Moreover, contrary to Bryden's statement that the MPs arrived after the incident, two truckloads of MPs under the command of Captain J. E. Morgan were parked at the side of the road during the incident and did not budge. After the incident Morgan persuaded the state troopers to move off to the side of the road, and the soldiers entrucked to move to their bivouac area.[15]

Bryden blamed the battalion for the trouble. Removal of its commander improved discipline. The "proper relationship" between Arkansas police and the military was established. That, said Bryden, was that. Hastie believed that his disposition to fault the battalion was "ill advised." The men who were disorderly should be punished accordingly, "but it is a grievous and serious matter that troops can be assaulted with impunity in bivouac and on the march, and more so that the Army should accept such evidence with complacency. . . ."[16]

As Hastie recommended, Under Secretary of War Patterson referred the matter to the attorney general. The Department of Justice concluded that there was insufficient evidence of criminal responsibility attached to the state troopers, especially since power had been delegated to them. A report from the inspector general undergirded a decision contrary to Hastie's purpose. The Justice Department closed its files on Gurdon on 12 February 1942, the Wednesday on which Private Felix Hall was last seen alive at Fort Benning, Georgia. Hands and feet bound by baling wire, his body was found on 28 March, with one end of a rope about the neck, the other around the trunk of a small tree at the brink of the ravine in which his body hung.

The assumption was that Hall was murdered by a person or persons unknown.[17] But the identities of other persons who had murdered black soldiers in 1942 were known. One was A. J. Hay, a policeman in Little Rock, Arkansas, who pumped five shots into Sergeant Thomas B. Foster, who lay on the ground, helpless. "The policeman, having completed this routine chore, proceeded to fill and smoke his pipe," Hastie informed Stimson. "His lack of concern is understandable since a [federal] grand jury did not see fit to indict him." Hastie also informed Stimson about a killing in Mobile, Alabama, one of the places where police powers were exercised by bus drivers, one of whom fatally shot Private Henry Williams on 15 October 1942. Another known murderer was H. W. Steinert, the policeman in Columbia, South Carolina, who killed Private Larry Stroud with a shot through the back of the head. A coroner's jury took Steinert at his word when he said he fired because Stroud stooped down as if to pick up a stone. Also known were the policeman who killed Private Thomas Broadus in Baltimore and the guard who killed Private James W. Martin, a prisoner in the Camp Lee, Virginia, guardhouse, claiming that his warning shot had gone astray.[18]

Hastie also apprised Stimson of assaults that fell short of murder in 1942.[19] In Norfolk, Virginia, Sergeant Ravell Smith and Private Lewis Branson occupied seats in the colored section of a bus. Whites wanted the seats. The soldiers refused to move. City policemen arrested, handcuffed, and beat the soldiers. The month before, in Montgomery, Alabama, a driver resented a lieutenant's taking a rear seat on a bus he intended to fill with white passengers. The two policemen who responded to his call beat and jailed the lieutenant. A broken nose, blackened eyes, and bruises were suffered by Lieutenant Norma Greene, an army nurse.

"This matter has reached the stage where the Army can no longer afford to dismiss the situation by saying that these cases are beyond its jurisdiction," Hastie stated when suggesting that Stimson ask Biddle to join in an appeal for legislation to give federal courts jurisdiction over violent crimes against or by on-duty military personnel. The War Department should shed its unwillingness to risk offending negrophobes in the South. "This continuing wave of violence may lead to rioting at any time," Hastie continued, "and certainly it is raising havoc with the spirit of Negro soldiers, many of whom have reached the stage that they would much rather fight their domestic enemies than their foreign foes. . . ."[20]

Bullied and beaten, cursed and killed, the black soldier found travel between camp and town, in Hastie's words, "a succession of hideous and maddening experiences." This affected more than the soldier's morale. "His fighting spirit is aroused and his will to violent action becomes strong and implacable. But he isn't thinking about the Germans or the Japanese." He thinks, said Hastie, about "an enemy more immediate and more maddening. . . ."[21]

Hastie and that enemy fought in the summer of 1941. On 8 July four black second lieutenants arrived in Washington, D.C. Traveling from Fort Ontario, New York, to Fort Eustis, Virginia, they disobeyed the conductor's and a railroad agent's orders to move to the coach reserved for blacks. On to Richmond they rode. But from Richmond they were supposed to ride in accordance with state segregation laws, argued Arthur H. Gass, manager of the Association of American Railroads. The judge advocate general considered the argument and said that if accommodations were (as he assumed) equal for blacks, then at issue was segregation rather than discrimination. He said that the railroad had met its legal obligation, and that no conflict between state and federal law existed. The officers should be ordered to obey state law. The judge advocate general suggested that Hastie's opinion be obtained, but he himself agreed with Gass.[22]

Hastie disagreed. The contention was that there was no discrimination if the separate quarters for blacks and whites were equal, but the reality was that the accommodations for blacks usually were inferior and inadequate. Their day coach, dining car, and smoker often were the rear part of the baggage car. Placed close to the engine, the car was filled with smoke and cinders. The conductor and news vendor occupied two of the few available seats. Soldiers had to stand when all seats were taken; other coaches may have had many empty seats, but they were off limits to blacks. These views were in an NAACP memorandum that Hastie undoubtedly wrote.[23]

In another memorandum, which Hastie definitely wrote, he contended that because state segregation law was an unconstitutional interference with interstate commerce, the officers were not obligated to obey it. *Plessy v. Ferguson*, cited by the judge advocate general in defense of the Virginia statute, applied to *intrastate* commerce. Had the officers been provided with Pullman accommodations, the dispute would not have arisen, for racially separate Pullman cars were not provided in the South or even required by the Virginia segregation statute. Hastie thought it uncommon and undesirable for adverse conclusions about the officers' conduct to have been drawn without their having been asked to comment on the facts. Their comments about traveling by day coach rather than by Pullman and about "the relative accommodations" for black and white GIs would have been relevant.[24]

Hastie and the judge advocate general both wanted the attorney general's opinion. Awaiting that opinion, the quartermaster general concurred with Hastie's suggestion and notified field personnel that the Association of American Railroads did not intend to discriminate against military personnel.[25] The attorney general thought it "not altogether unlikely" that the Supreme Court would invalidate such an issue if confronted with it. Still, he backed the judge advocate general "pending final determination of the question by the Supreme Court." That was on 19 December 1941. At the time Hastie thought federal legislation would be the means of overcoming such state laws.[26] Meanwhile, every black soldier faced trouble in the area of public transportation, for there was, in Hastie's opinion, a legion operating "on the theory that whenever someone must suffer extra privation, it should be the Negro."[27]

This legion included local military commanders who joined other whites to force segregation upon black soldiers, often in violation of army policy. Not that military authorities were blameless: They had yet to get tough with the Interstate Commerce Commission and the Office of Defense Transportation, both of which had the power to combat discrimination. For example, the Office of Defense Transportation could control the allocation of rolling stock and equipment sought by bus and railroad lines. Its director, Joseph B. Eastman, needed no other power to squash discrimination. A year earlier, at his request, the NAACP had sent him a memorandum containing its views on discrimination in public transportation. Eastman had not done anything asked of him.[28]

If he were to try to set aside Jim Crow laws himself or to ask Congress to do so, Eastman told Walter White, it would appear, especially to the South, "that I was taking advantage of the war to accomplish a social reform having only slight relation to efficient transportation." Besides, the attempt would fail, while disrupting his operations. Rather than benefiting, blacks would suffer. "Time and events are playing into your hands. If wise policies are pursued, at the end of the war the position of the Negro will be much improved." Why jeopardize these prospects? Of course, Eastman knew that whites opposed gradualism. "But we are dealing with emotions, not reason. I am not persuaded, therefore, that what you ask me to do would be in the best interests of the colored people, whose advancement you so rightly and zealously seek."[29]

Under Secretary of War Patterson received the same memorandum that White sent to Eastman. It contended that discrimination eroded morale and fostered wasteful duplication of facilities. Patterson did not refute that argument, as Eastman did. Instead he passed the buck: "This matter lies entirely within the province of the Office of Defense Transportation."[30]

Not entirely, for Patterson did not take the black soldier into account, while Hastie did: "He sees Jim Crow and those who impose it as his enemies." Those like the Southern official who, chided for being disrespectful of the military uniform, shot back, "Listen, to us down here a 'nigger' is just a 'nigger,' I don't care how you dress him up."[31]

Hastie knew that—and more. "The Negro soldier would rather die fighting that man . . . and the ideas that man stands for in America than to go to the trouble of traveling half way around the world to die fighting a stranger because he has the same ideas." Although unpleasant, this truth had to be understood if friction between the black GI and racists were to be effectively handled. "As a soldier he fights his enemies whenever half a chance presents itself."[32] At the time that Hastie issued the warning, violence flourished. "By early summer," wrote Ulysses Lee, "the harvest of racial antagonism was beginning to assume bumper proportions." But in 1943 disorders occurred more often, involved more troops, and most likely resulted from initiatives by black soldiers than in years past.[33]

Hastie coupled warnings with suggestions for the prevention of trouble. At his urging and on Provost Marshall General Allen W. Gulliou's recommendation, black MPs were assigned to camps where large numbers of blacks were stationed. This included Fort Bragg, the scene of rioting in August 1941.[34] The next summer Hastie recommended that blacks with military, legal, or police experience be commissioned in Zone of the Interior MP battalions. All who had applied for commissions had been rejected; one had been given written notice that only whites would be commissioned. Hastie's recommendation would be deferred, Assistant Chief of Staff Donald Wilson responded. As younger officers became available or qualified to replace them, overage officers on troop duty would be reassigned. The black MP battalions would provide ideal reassignment positions that Wilson did not want to preempt by commissioning civilians.[35]

In his reply Hastie focused on overage officers. He said that Wilson's information about the availability of replacements was outdated. In the three (out of four) regiments that were stateside and staffed by a number of overage officers, there were 24 such officers. Ten were in a regiment that had 106 officers and was 60 percent overstrength, Hastie said. There were 5 overage officers in one of the remaining regiments and 9 in the second, but each of these units had over 120 officers, fewer than authorized but enough to permit reassignments, especially since OCS would graduate 100 or more infantry officers each month.[36]

Immediate reassignment of overage officers seemed practicable to Hastie. A major, three captains, and three or four lieutenants would be an adequate nucleus for each of three battalions. Other vacancies could be filled by officers drawn from elsewhere, including pools of former officers having appropriate military or civilian experience. He hoped his suggestions would be considered in the pending study concerning the utilization of black MP officers.[37]

In August 1941 Hastie suggested informal group discussions involving GIs in the same camp. Interracial problems might be solved that way. "It is apparent, I believe, that we do not solve such problems by trying to keep colored and white soldiers away from each other." The idea, killed by General James A. Ulio, head of the Morale Branch, was essentially that which was put into effect in the European theater eleven months later in keeping with General Eisenhower's directive that racial discrimination be "sedulously avoided." Another of Hastie's recommendations—that the Bureau of Public Relations and the Special Service Division prepare an educational program to inculcate wholesome racial attitudes in soldiers, both black and white—was also adopted. In addition, he recommended that blacks be employed as special service officers, that President Roosevelt broadcast an appeal for racial tolerance, that Secretary Stimson publicly encourage civilians and military personnel to amicably coexist, that local commanders and public relations officers be ordered to work more diligently toward that end, and that Military Intelligence share information to help thwart organized attempts to direct violence against blacks in the Army and defense work.[38]

Adoption of Hastie's recommendations would have been tantamount to the War Department admitting that racism created serious problems in the military. Rather than making that admission, Under Secretary Patterson accused "certain sections of the press" of capitalizing on isolated incidents "to promote, in the Army, social gains which have not been attained in the country as a whole and . . . using the Army as a means of promoting such gains among the civilian population."[39] This agitation, said Patterson, impeded the department's efforts to establish an Army that could defend the nation while fitting into its "accepted social order."[40] His castigation of the black press was reinforced by Assistant Secretary of War John J. McCloy, whom Hastie understood to have said that cessation of this protest would better serve the purposes of the protesters. In response to Hastie's criticism,[41] McCloy said that the black press should not divert the black's mind from the preeminent task, namely, winning the war. Too few blacks, soldiers and civilians alike, seemed "vitally concerned" about the first part of the Double-V campaign to suit McCloy.[42]

"May I urge that . . . you point out the basic issues of this war," Hastie wrote, "and the impossibility of foreclosing those issues at home while we stir people up to fight for them all over the world."[43]

"Frankly, I do not think that the basic issues of this war are involved in the question of whether Colored troops serve in segregated or mixed units," McCloy answered, "and I doubt whether you can convince the people of the United States that the basic issues of freedom are involved in such a question."[44]

McCloy's appraisal of the mood of white Americans was correct. That, however, was more a commentary on him and them than on Hastie, who might well have been reminded of Frederick Douglass's statement: "Those who profess to favor freedom and yet deprecate agitation are men who want crops without plowing up the ground; they want rain without thunder and lightning. They want the ocean without the awful roar of its many waters."[45]

The press that McCloy and Patterson attacked and Hastie defended was described by the Bureau of Intelligence as "ardently democratic, being primarily concerned with the actual practice and extension of democracy in the war effort." Perhaps for that reason and admittedly for another—fear of creating a void that less credible sources, including the enemy, would fill—the bureau opposed suppression of the black press.[46] Meanwhile, Hastie tried to correct a serious problem, namely, the frequent unavailability of factual information to the black press. He offered to relay information from the department to the black press and suggested the issuance of special and supplemental press releases to keep blacks informed about the war effort.[47]

But not all of the would-be suppressors were in the War Department. Many field commanders banned black newspapers in post exchanges, libraries, unit reading rooms, and service clubs. They ordered their mail officers to withhold black newspapers, an action that intelligence officers sometimes took on their own. Many urged censorship or other control of the black press.[48] Years later Hastie disclosed that some black publishers had been in danger of criminal prosecution for having publicized injustice in the war effort.[49]

But censorship and prosecution were not the only reprisals available to sabotage the work of the black press and, given his reliance on it, Hastie himself. Wanting to increase the recruitment of blacks and receiving black editors' complaints, in October 1942 Hastie asked General H. B. Lewis, the acting adjutant general, to encourage the Bureau of Public Relations to place advertisements in black newspapers.[50] When asked why the War Department did not use black newspapers in its recruiting campaigns, Lewis said that their circulation was below that specified as acceptable by "competent advertising counsel" retained by the department in its efforts to be "perfectly fair and just." It is unlikely that the disingenuousness of this response to his query escaped Roy Garvin, national advertising editor of the *Afro-American* newspapers, who undoubtedly knew that even publications with acceptable circulation were disqualified if they championed particular racial or religious causes.[51] Carrying the rationalization one step further, Patterson said that only by adhering to the formula could the department protect itself against accusations of bias. Neither he nor Lewis mentioned the impossibility of black publications satisfying criteria adopted, applied, and justified by whites.[52]

What could be done about this discriminatory policy? What Hastie did was to ask Walter White to protest it. The policy was not changed, but a compromise was reached. At Stimson's request, Hastie, White, and General Surles discussed ways to reduce racial friction and to increase the recruitment of blacks into the Army and into special training. Hence the increases in the number of blacks inducted, the number given special training in integrated schools, and the number of black units. The Bureau of Public Relations began the publication of *Army Talk*, a pamphlet, and produced *The Negro Soldier*, a motion picture shown in theaters throughout the nation, including the South, after the NAACP filed an *amicus curiae* brief in federal court against the War Activities Committee of the Motion Picture Industry. Hastie and Thurgood Marshall wrote the brief.[53]

Hastie gave the black press much credit for making change possible, even inevitable. Many articles, he said, were inflammatory but not overdrawn. Things far worse than those reported were happening. "You could not read accounts of what was happening to colored soldiers and civilians without being angry and indignant." The black press lit fires of "immediate indignation" without which constructive change would have been impossible. Years afterward, when Hastie was honored by the National Newspaper Publishers Association as a "pioneer champion" of equal opportunity in the armed forces, something had remained unchanged with him. "The one thing I would fear most would be loss of the indignation against injustice."[54]

Hastie himself lit fires of indignation, at times with help from the black press, the NAACP, and other sympathetic souls, including some in the War Department. He never sought help from President Roosevelt, thinking it inappropriate for an aide to the secretary of war to directly or (through Mrs. Roosevelt) indirectly appeal to the President. But he found Stimson to be more a liability than an asset. The secretary—an honest man, a dedicated public servant, and "a patriot in the best and the highest sense of the word," Hastie explained—simply did not understand racism.[55]

Thirty years later Hastie recalled a conversation with Stimson concerning military racism. They had reached the decisive point in their differences, the Rubicon symbolized by the airstrip at a base in Alabama. "I will always remember one sentence, both what he said and the words he used, which to me reflected the problem of getting him to move in this area. He said, 'Mr. Hastie, is it not true that your people are basically agriculturalists?'"[56]

CHAPTER ELEVEN
Invincible Man

Stimson wanted blacks to farm; Hastie wanted them to fly. But what was Hastie to do? Stimson's belief in the inability of blacks to qualify as combat pilots was common among military policymakers.

During World War I blacks had been excluded from the Air Service of the Signal Corps; in fact, they had not even been permitted to apply for enlistment because no black squadron was being organized. Their requests to become air observers had also been denied. Plans of the Office of the Director of Military Aeronautics that would have allowed them to pull airfield fatigue and police duty had been disapproved. Although a few blacks had been construction workers in the Air Service, no blacks had flown or maintained airplanes. After the war ended, military authorities had turned down requests that the organized reserves include black units. There had been no such units in existence and no black officers had been commissioned in the Air Service.[1]

In 1939 an amendment to Public Law 18 required the Civil Aeronautics Authority (CAA) to designate for the training of black pilots one or more of the civilian schools that were eligible borrowers of training equipment from the War Department. Subsequent negotiations between the CAA and the Army Air Corps resulted in the CAA's consenting to train black pilots under its program. This agreement created the erroneous impression that henceforth the Air Corps would conduct such training. This impression was compounded when the CAA established Civilian Pilot Training (CPT) units at Tuskegee Institute and five other black colleges: Delaware State College, Hampton Institute, Howard University, North Carolina A & T College, and West Virginia State College. At some Northern colleges and universities a few blacks enrolled in additional CPT units. Cadet boards, however, continued to turn away black applicants upon their successful completion of CPT programs. The general staff was partially of the opinion that the Air Corps should be required to take its share of black enlisted men during wartime. But could this be done while preserving segregation?[2]

Tension surrounding this question was heightened by the passage of the Selective Service Act of 1940, which seemed to ensure the inclusion of blacks in the Air Corps. Though the law forbade their exclusion from the military services, the Air Corps circumvented it. In October 1940 the War Department announced the training of black pilots, mechanics, and technical specialists. Hastie secured correction of the announcement when he disclosed that the only flight training open to blacks was that given by the CAA. And yet the Air Corps could no longer go its own way—that is, not completely. In December its plans called for the recruitment of blacks into the all-black 99th Fighter Squadron. While auxiliary personnel, including mechanics, were to be trained at Chanute Field, Illinois, pilots were to be trained at Tuskegee Institute, which had acquired one airfield and obtained the use of another for its CPT program. For the basic and advanced training of the 99th, a third airfield was to be constructed (along with all other living and training facilities) at Tuskegee Army Air Field, a four-million-dollar tribute to Jim Crow.[3]

According to one report, Hastie at first opposed the experiment at Tuskegee but later withdrew his objections. Hastie, however, said that he was overruled in this, the first instance of Air Corps defiance of army policy that required officers' training to be integrated. But the Air Command, he added, was only one of the parties in an "unholy alliance." Since Tuskegee Institute would not be involved in the program at the army air base, he could not immediately discern the school administrators' motives for allying themselves with the Air Command. And then Hastie, who advocated that blacks be trained along with whites at schools under military contract to provide primary training for cadets, realized that Tuskegee Institute wanted to provide such training for all black cadets. Therefore, the school assisted the Air Command in finding and negotiating the acquisition of land for a training center for blacks. There were already in existence three military flight-training centers: one in Texas, a second on the West Coast, and a third at Maxwell Field, forty miles from the Tuskegee Institute. Whites were trained at these centers. Blacks would not have to be trained at any of them if the Air Corps managed to establish a fourth exclusively for them. The Air Command wanted a segregated center, and Tuskegee wanted a racial monopoly. Each got what it wanted.[4]

Hastie's analysis antagonized Alexander B. Siegel, a Tuskegee trustee to whom Walter White sent a copy. Siegel defended the institute on grounds that since "no power on earth" could prevent whites from abusing blacks in integrated situations, blacks should be segregated for their own protection. It did not matter if they were segregated, so long as they were able to contribute to "our chief business at this time [, which] is to win the war." Furthermore, segregation would foster in black cadets the peace of mind and calmness that were essential to success. Hastie's rebuttal was that had the cadets' best interests been the overriding consideration, training would have been offered in the Far West rather than in the Deep South. Moreover, integration was far more conducive than segregation to their developing the qualities mentioned by Siegel; army officers' training had demonstrated that. Arguments such as Siegel's were predictable, said Hastie. "The defender of segregation invariably becomes a sophist when he tries to justify his position."[5]

Hastie's high hopes had been that the Air Corps would be spared the segregation that contaminated the other military services. That, of course, was not to be. Given no choice but to accept twenty-five hundred blacks as part of the mobilization plan adopted by the War Department in October 1940, the Air Corps organized them into ten (separate) aviation squadrons. As Hastie pointed out in a pamphlet the NAACP published in 1943, the ten squadrons had no white counterparts and no assignments other than common labor. "These units," he wrote, "were set up in order that it could be said that Negroes were in the Air Corps."[6]

Hastie's protest against segregated dining and sleeping facilities in the Tuskegee experiment had prompted Robert A. Lovett, assistant secretary of war, to say that everything was being done according to "regularly established practice." Unmollified, Hastie replied, "As I see it, the question is not what is or has been done in other instances, but whether this is the correct thing to do under the peculiar circumstances of the present case." The segregated program was costly in money, time (six months' delay in starting) and adaptability to military needs (pursuit training only). Hastie could think of several reasons to integrate training: morale, acquaintance, mutual respect, and knowledge of nationwide air defense. "But why in the name of common sense should all of this elaborate special machinery be set up to train Negro flyers?"[7]

A policy memorandum of 30 January 1941 had decreed that within a year all officers at the ground and flying schools at Tuskegee air base, except for the post commander, executive officer, and squadron commander, were to be black. In reply to the query that Hastie made in July 1942 about compliance with that policy, General Henry H. Arnold said that a year had not actually expired because the school had not opened until October 1941. Arnold, chief of the Army Air Corps, assured Hastie that original plans to replace white with black administrative officers at Tuskegee had not been changed.

Within several months Hastie again inquired about the elevation of blacks to such positions. Holding subordinate positions or none at all were black officers in finance, chemical warfare, medicine, and athletics, some of whom had been prominent in their communities. Given the responsibility of responding, the Southeast Army Air Force Training Command (SAAFTC) at Maxwell Field, which controlled Tuskegee, asserted that Hastie was referring to a prewar plan for which no subsequent directive requiring implementation had been issued. Besides, the number of officers and enlisted men at Tuskegee far exceeded the number envisaged when the plan was prepared. SAAFTC claimed to have searched in vain for black officers who had the requisite technical skills. Major General George E. Stratemeyer, chief of the air staff, made that argument in December 1942, telling Hastie that the Air Corps felt unjustified in experimenting with inexperienced replacements. Undoubtedly he shared SAAFTC's view that Hastie's motives were racial rather than military.[8]

Hastie's concern was that "reluctance to accord authority" to blacks accounted for the failure to carry out the original plan. He often witnessed at Tuskegee "the spectacle of white officers, one or two of outstanding ability but most of them unexceptional," gaining top administrative posts and promotions while blacks, even those of high reputation in their professions, had opportunity for neither. Hastie argued, "Certainly racial discrimination is shown when not one of the half dozen Negro graduates of the Adjutant General's School, now stationed at Tuskegee, has been assigned to an Adjutant's duties in Post Headquarters."[9]

Although his critics would not concede the point, Hastie was concerned about military efficiency as well as racial justice. Did the racial assignment of ground crews to pilots enhance the air forces when it resulted in black pilots having no crews because none were trained and available? How was national defense advanced when the Air Corps, seeking applicants for a program intended to train ten thousand meteorologists, rejected many solely because they were black? And did the similar disqualification of blacks in other specialities—armament, engineering, and aviation medicine, to mention only a few—guarantee victory over the Nazis? These questions went to the heart of military efficiency. Unlike his detractors, Hastie saw no affirmative answer to them.[10]

If there were men in the War Department who would rather risk losing the war than utilize all of the nation's human resources—and Hastie was convinced there were—he could not help but oppose them. And if there were men of "integrity and good will" in the department— he likewise felt there were many—that was encouraging. It was not, however, sufficient to meet the need for justice envisaged by blacks. But even the good men in the War Department seemed not to understand this. One of them, a general who worked with Hastie against racism, once stormed into his office, angry and perplexed. The general said that as he had crossed the street near the War Department, he had almost been hit by a truck whose black driver had yelled, "I should have run over you, you son of a bitch!"[11]

Recounting the incident, Hastie said, "I tried to explain to my friend that to the truck driver he represented all of the wrongs the Army had perpetrated and was still perpetrating upon the Negro. More than that, it is almost impossible for any white person, at least until he becomes personally known and respected—and sometimes not even then—to escape the odium which the white community has earned through generations of mistreatment of the Negro."

With that mistreatment uppermost in his mind, Hastie had written his memorandum to Stimson on 8 January 1942. In the conflict with foreign enemies, which was in some respects the most poignant phase of the bittersweet war against bigotry, individual campaigns were insufficient. "On every essential front we must be moved to aggressive, far reaching and uncompromising action." Progress had been made, but if the nation was to profit by the potential inherent in one tenth of its people, "we must go much further and faster," he added in his memorandum to the secretary, concluding, "I have not found in the War Department the group will and understanding which are prerequisite to such acceleration."

Six months later the Air Corps still showed no evidence of acceleration. In fact, the Air Command planned to train specialists and technicians at the Tuskegee air base in order to avoid enrolling them in its other technical schools. In opposition, Hastie told Lovett that this would be both inefficient and unnecessary. The Army was training black technicians in integrated schools. Indeed, the Air Corps itself had trained its first group of black mechanics and technicians under integration at Chanute Field, Illinois. Despite Hastie's urging the adoption of that approach, the first instance was also the last. The Air Corps endeavored to establish segregated training at Tuskegee and elsewhere, but as a result of its dalliance it produced an increasing number of black pilots for whom there were no crews.[12]

Late in 1942 Hastie learned that the Air Corps planned to pattern Jefferson Barracks in Missouri after Tuskegee air base. Ground crews, including officers, were to be trained there, according to a press release and the newspapers in St. Louis. He asked Assistant Secretary for Air Lovett whether such plans existed and was told that there were no plans to make Jefferson Barracks the second segregated air base. Telling Stimson later that Lovett's lack of candor "cannot be considered an excusable inadvertence," Hastie was moving to the point of informing the nation that the Air Command viewed the training of blacks as an experiment. "The tragedy is that by not wanting the Negro in the first place and by doubting his capacity, [it] has committed itself psychologically to courses of action which themselves become major obstacles to the success of Negroes in the Air Forces."[13]

Stimson, the recipient of Hastie's memorandum, compartmentalized freedom for blacks: It could be economic and political but never social. Stimson viewed as "twin devils" the advocate who would keep blacks in "their place" and the one who would propel them "from complex reality to unattainable reality" in one swift move. Having told Hastie the pleasure he derived from a visit to Tuskegee, Stimson noted in his diaries, "I did my best to win him and to reassure him as to our attitude and I felt I had not made a success of it." By October 1942 he had decided that Hastie was disappointing, virtually useless, and unrealistic.[14]

A feather in Stimson's cap, Tuskegee was a thorn in Hastie's side. It was an affliction made worse by plans for Jefferson Barracks, which demonstrated further that Hastie was an "outsider inside." And not really inside, for the top brass never accepted him. They resented his criticism, damned his NAACP ties, and animadverted his office. "Dear Bill," Felix Frankfurter had written in 1940. "The Secretary of War was, if I may say so, very wise in calling you to his aid, and the country is fortunate that you were able to accept." But the secretary and other officials made Hastie's job, as it was said, "a little like sweeping back the sea with a broom. . . ."[15]

Hastie had seen ingrained racism in the Army, "outright fascism" in the Navy, but in the Air Corps he saw some hope for democracy. New, numerically significant (comprising nearly a third of the Army), and increasingly independent, the Air Corps had discriminated against its black squadrons, but at least the one formed for combat (the 99th Fighter Squadron) had not been assigned to perform odd jobs along with the others. *Time* magazine trumpeted the "practical impossibility" of integration, but Hastie pointed to the Army's OCS program as a model of integration for the Air Corps. The magazine, like the Air

Command, underestimated "the essential decency" of the young GIs being called to fight and die "for a free world." Patterson and Marshall were inclined toward Hastie's purposes; Colonel Walter Beddell Smith, secretary to the general staff, advocated positive change, though not the eradication of segregation; and for his insistence that the War Department take the initiative in improving the black soldier's experience in the South, Major Hodding Carter, the former newspaper editor from Mississippi, was relieved of his duties as assistant chief of staff for public relations and "quietly reassigned overseas to edit an army newspaper." But these men were in a minority in the War Department, where ranking military officials, Hastie added, "treated me as an outsider, though courteously." Arnold was "entirely out of sympathy with my efforts," and Lovett "always seemed politely disinterested in them." Hastie attributed mainly to them the attitude that the army air forces took toward him. In a word, "Hostile!"[16]

Some progress was made while Hastie was in the War Department. Conditions on several bases and even in surrounding communities were improved, but Hastie felt that, on the whole, they remained "really very bad" throughout the war years, and "the basic resistance of field commanders and of many persons in the general staff in the War Department never let down." The air forces, he felt, were reactionary and his differences with that branch were irreconcilable. Racism drained the nation of the "spiritual strength" without which military victory was impossible. He had hoped that formal entry into the war would rectify that national mistake, but still lacking was firm commitment to the very ideals that the nation had gone to war to protect. Hastie therefore decided that he would be more useful "as a private citizen who can express himself upon such issues than as a member of the War Department under obligation to refrain from such public expression." His resignation took effect on 31 January 1943.[17]

The War Department counterattacked with Major General George E. Stratemeyer's rebuttal of the detailed memorandum that Hastie sent to Stimson. Hastie would have liked nothing better than to have reality conform more to Stratemeyer's assessment than to his own, but he found nothing in the general's memorandum that convinced him that he had erred. He was right. The rebuttal was impressively unresponsive to the allegations he had made. Stimson accepted Hastie's resignation with regret, for Hastie had given "very faithful and effective service." He hoped that Hastie would feel free to provide advice as time went by—or so he said.[18]

At a press conference Under Secretary of War Patterson commented on Hastie's "great ability and fidelity," said that he "respected him

Hastie as an infant. *Courtesy Beck Cultural Exchange Center.*

Charles Hamilton Houston. *Courtesy Moorland-Spingarn Research Center, Howard University.*

Roberta Childs Hastie. *Courtesy Beck Cultural Exchange Center.*

From top to bottom: Hastie, W. Mercer Cook, W. Montague Cobb. *Courtesy Hastie family photo album.*

Hastie is second from the right in the first row. Two other men important in his life are also shown: in the back row, fourth from the left (with the man in the suit to his left) is Charles R. Drew; to the right of the man in the suit (and with the small man in the black shirt to his left) is W. Montague Cobb. *Courtesy Moorland-Spingarn Research Center, Howard University.*

VARSITY TRACK TEAM. 1923
Drew   Bates   Lyons   McKay
McLaughry Roundy Jones Dunbar McCormick Shepherd Strong Dodd
Hastie Lamberton Brown Clark Nelligan Cobb Darling Giles
Courtesy Amherst College Archives.

Hastie (middle of fourth row) and the faculty and administrative staff at Bordentown Manual Training School. Principal William R. Valentine (center of second row) in vested suit, with pen in lapel pocket. *Courtesy Hastie family photo album.*

Hastie at left; Ralph J. Bunche at right. *Courtesy Hastie family photo album.*

Roberta Childs Hastie and Mary McLeon Bethune (with sign) picketing with the New Negro Alliance. *Courtesy Moorland-Spingarn Research Center, Howard University.*

General George C. Marshall, Army Chief of Staff, addressing black newspapermen. *Photo by U.S. Army Signal Corps.*

The legendary figure in this photograph is Captain Benjamin O. Davis, Jr., third from the left, with hands clasped. *Photo by Southeast Air Corps Training Center.*

William Hastie. *Courtesy Hastie family photo album.*

Walter White, of the NAACP, visiting the Hasties in the Virgin Islands. *Courtesy Hastie family photo album.*

William, Karen, Billy, and Beryl Hastie. *Courtesy Fritz Henle.*

During a ceremony held on February 23, 1948, President Harry S. Truman presents a plaque commemorating the hundredth anniversary of the emancipation of the slaves on July 3, 1848. Governor William H. Hastie looks on at Emancipation Garden. *Courtesy The Daily News, Charlotte Amalie, St. Thomas, Virgin Islands.*

President Truman and Governor William H. Hastie, with Magens Bay, St. Thomas, in the distance. *U.S. Navy. Courtesy Harry S. Truman Library.*

thoroughly," and claimed that "excellent progress" had been made in utilizing blacks in all of the armed forces. The moment that the press conference ended, reporters hurried to Hastie's office, but he declined to discuss his resignation before its effective date. Even then he would not make sweeping accusations; he would not tarnish the many "fine officials" at the Pentagon.[19]

"Bill Hastie has done an heroic thing," wrote Louis Lautier. At Hastie's request he and Truman K. Gibson, Jr., had remained in the Office of Civilian Aide. "In twenty years on the Washington scene, I have known of no other colored man who has quit a government job as a protest against biased policies."[20]

W. E. B. Du Bois saluted Hastie. One of the two types of advisers on race relations, he wrote, "transmits to the public with such apologetic airs as he can assume" his department's refusal to act on any suggestions to ease racial unfairness; the second kind analyzes the situation, offers advice, and "works on hopefully" as long as his department reasonably cooperates. When it refuses to cooperate, he refuses to continue. "It is, of course, this second type of official alone who is useful and valuable," Du Bois continued. "The other is nothing. Hastie belongs to the valuable sort and will not be easily replaced."[21]

Thinking of Hastie as "my ace in the hole," Lieutenant Monroe Dowling joined the Army. He would not have done that had he even suspected that Hastie was dissatisfied. "Of course you were never a person to take anyone into your confidence, but I had no idea conditions were as bad as your resignation indicates." Things were not too bad with Dowling, but since all officers in his outfit outranked him, they could take the choice assignments—"leaving the hell for me—for example I work nights from 10:45 P.M. to 7:30 A.M."—giving him no voice in decision making, which occurred in the daytime. All this was tolerable when Hastie was in office, but now Dowling's promotion might not come through on July 1. Dowling asked whether they could talk the next time Hastie was in New York.[22]

"Dear Mony," Hastie replied, saying that things were too hectic to allow more than a brief note but that they should have a long talk when he next visited New York. Unable to resist the temptation, he added a postscript: "Helen can really be sure where you are late at night now."[23]

Three years after Hastie's resignation, Patterson, who had become secretary of war, said publicly, "Democracy in the Army during the war was due largely to the effective, brilliant and sincere work done by Mr. Hastie." A senior Air Corps officer, said Hastie, once "very positively informed me" that his branch would not be integrated within the next two hundred years. Closer to the mark was the field investigator who described the aftermath of Hastie's departure as "a kind of quiet social revolution . . . that in time . . . will have an incalculable effect upon the civilian population." In the opinion of James C. Evans, who followed in his footsteps in the War Department, Hastie was "the prime mover" in that revolution.[24]

In the wake of Hastie's resignation, the Army and even the Navy began to experiment with integration. And, says Alan M. Osur, because of the resignation the army air forces "began to move with a speed and determination never previously observed in the area of race relations." Training schools were desegregated, with the exception of Tuskegee air base, and even there facilities, administrative positions, and flight instructors' ranks were desegregated. Blacks were admitted to schools of aviation medicine, and those who were physicians became eligible for army medical reserve commissions. Plans to make Jefferson Barracks a segregated post were scrapped.[25]

Blacks, whom Stimson saw as men of the land, became men of the skies. Lena Horne, their pinup girl, remembers those who were Tuskegee Airmen.[26] "They had this impossible dream of becoming pilots, fighting for their country. The country created a special air force for them." At first the Tuskegee Airmen were not given a chance to prove that they could fight. The 99th had its full cadre of pilots by early August 1942, but it was not designated combat ready until late September. Several months passed before the squadron, commanded by Lieutenant Colonel Benjamin O. Davis, Jr. [who succeeded Captain George Roberts], was sent to the Mediterranean theater of operations. Assigned to escort bombers and attack ground targets, pilots of the 99th were not involved in the glamour of air battle. Their lack of aerial victories overshadowed their other invaluable contributions, and they would have been relieved of combat duty and given routine convoy assignments had it not been for General Dwight D. Eisenhower's vote of confidence and an army report that favorably compared the 99th with white squadrons.

Beginning in August 1943, when Davis was given command of the 332nd Fighter Group (the other all-black unit, which was stateside), and lasting until July 1944, the 99th was shifted from one white fighter group to another. On 3 July, six months after the 332nd Fighter Group was dispatched to Italy, the men of the 99th rejoined their former commander, Colonel Davis, when their fighter squadron became one of four that comprised the 332nd. The 100th, 301st, and 302nd were the other squadrons.

When finally given an opportunity to fight, black pilots made the most of it. In January and February 1944 they shot down seventeen German planes (a record for an American squadron), probably destroyed four more, and damaged six. At Normandy, in June 1944, they affirmed their reputation as the best escort service for bombers. In November the 99th set a record for missions flown (twenty-six). Nine months later, flying propeller-driven P-51s on the longest (sixteen-hundred-mile) mission by the 15th Strategic Air Force, they were challenged by eight new jet planes over Berlin and shot down three. On that mission they kept intact a record unequaled by any other escort pilots, namely, completion of the mission without the loss of a bomber to the Luftwaffe.

Usually called Tuskegee Airmen, these black pilots were called something else by the white crews of the bombers that they escorted on 200 combat missions. Because they flew planes whose tails were painted bright red, and because they never lost a bomber to German aircraft, the bomber crews called them the Red-Tailed Angels. Hastie had opposed the segregated program that prepared them to fly the 15,533 sorties and 1,578 missions on which they won the following awards: Legion of Merit (1), Silver Star (1), Soldier Medal (2), Purple Heart (8), Distinguished Flying Cross (95), Bronze Star (14), and Air Medal and Cluster (744). But having wanted for the Tuskegee Airmen what they themselves most wanted—a chance to fly and fight—Hastie was proud of their combat record. Appreciating what he had done to help make possible their impossible dream, they called him "the father of the black air force." He deserved a broader appellation, for his resignation spurred the armed forces so that they not only integrated military life but also spearheaded the desegregation of civilian life. To author Lee Nichols Hastie described the integration of the armed forces as "the most remarkable and heartening sociological demonstration of this century."

Hastie's resignation kindled this demonstration. "It was courageous," says Secretary of Transportation William T. Coleman, Jr. Characteristically, Hastie exemplified the grace under pressure that Ernest Hemingway called courage. But there was more to the resignation. "It was also symbolic," Coleman adds. Indeed. It reminded the nation that not all black men were invisible; many were invincible.[27]

For his invincibility Hastie was honored by the NAACP, which presented its Spingarn Award to this "champion of equal justice" in 1943. "Though young in years," the citation read, "his record of achievement is notable measured by any standard, however absolute." The record included a decade of outstanding service with the NAACP and at Hastie's other base of operations, the law school at Howard University. Returning to the law school, Hastie was destined to further justify the Spingarn Award Committee's accolade: "His every act . . . has established a standard of character and conduct which [we are] honored to recognize."[28]

"These are days for strong men to courageously expose wrong," Adam Clayton Powell, Jr., a Harlem minister, said when Hastie resigned.[29] Hastie saw himself in more modest light. Then, as later, he believed that individuals should "speak out and take a stand for the things in which they believe" lest we become "a country of frightened people." But he did not think that his resignation was a courageous act, "because I wasn't faced with hunger."[30] He could return to the law school at Howard University.

Never one to surrender, Hastie returned to Howard and continued his battle against military racism. "While he was in the War Department," says Archibald T. LeCesne, his law students at Howard University, "black GIs who thought they were being wronged would say, 'Goddamn it, I'm gonna write to Hastie.' And many of them meant it."[31]

They still meant it after Hastie left the War Department, but relatively few wrote in the vein adopted in November 1943 by Alexander P. Haley. Perhaps the difference was attributable to the fact that Haley, a coastguardsman, did not think that blacks were being wronged. For example, two and a half years earlier—when not one of the Navy's 5,026 blacks was an officer, not more than a handful held a rank higher than messman, all were called "the chambermaids of the Navy" (virtually all were messboys, stewards, or chefs), and Hastie endorsed the view that the Army was a national disgrace and the Marine Corps were the "height of Lily Whiteism," but neither was as racist as the Navy[32]—Haley had offered an opinion to a black newspaper. As given by the newspaper, Haley's defense of naval policy was as follows:

1. It is the navy's [sic] job to defend our country.
2. Navy efficiency and co-operation are at a maximum when there is no friction.
3. Best way to eliminate friction is to keep all colored people out of the navy [sic] except as cooks and waiters.[33]

Secretary of the Navy Frank Knox, who rejected President Roosevelt's suggestion that colored bands be used on at least some ships, contended that attempts at integration would adversely affect morale, harmony, and efficiency. Writing to Hastie in 1943, Haley was less his kindred spirit than Knox's. Telling Hastie that he had wondered why blacks seemed content to be "chambermaids," Haley added, "Finally, it dawned upon me—it was the boys themselves . . ." They did not want to take on anything requiring "any mental ability whatsoever," and thought that he was "crazy" to forgo sleep in order to learn, grow, advance—as he had, with the approval of his white shipmates. He expected to attend the Coast Guard Academy—his captain had written a recommendation in behalf of "one of the most respected men on board this ship"—and to gain a position in which he could help other blacks. He wondered whether Hastie would let the people recruiting blacks into the naval services know that.³⁴

Although Hastie had not dealt with the Coast Guard, he did know something about conditions topside in the Navy. One story making the rounds had President Roosevelt, in conference with ranking army and navy staff members, asking what was being done about the growing number of complaints from blacks about national defense. The army man told him what was being done. The navy man told him, "We file them in the wastebasket." Nevertheless, Hastie sent a copy of Haley's letter to the Coast Guard. "I hope and believe this will serve a useful purpose," he told Haley.³⁵

Haley had written the letter because he thought correspondence with Hastie might be inspirational. "More important, however, is the fact that I want to tell somebody who matters and is in a position to understand my reaction, how another colored American feels about this so-called 'race problem.'" Though it may not have been real to Haley, the problem was very real to Tony Weaver, who was with the black 332nd Fighter Group at Selfridge Field, Michigan. It was there that Judy Edwards had been killed in June 1943 when his plane crashed. Gordon Parks, the photographer, learned this when he read the note from his best pal, Weaver. At Oscado there were no mortuary facilities for blacks so Weaver was detailed to accompany Edwards's body to Steubenville, more than three hundred miles away. "How could anybody do anything like this?" he asked Parks, wondering whether physicians at the hospital had examined Judy Edwards, "since this would have required them to touch him, too." The "so-called 'race problem'" was just as real to Hastie as it was to Weaver, and this accounts for his involvement in the army Scottsboro case, the nightmarish experience of Privates Edward R. Loury, Jr., and Frank Fisher, Jr.³⁶

"While I was having intercourse with her, she pushed me on the shoulder," Loury said. "I stopped and stood up. I saw a soldier and a girl approaching us. They came to a few feet from where we were lying, saw us, and then turned around and went back the same way they came. . . ."

Houses of prostitution in New Caledonia were off limits to black soldiers. So when the car reached Prostitution Hill, Edward R. Loury, Frank Fisher, Jr., and their buddy got out. The three had hitched a ride after leaving a carnival near Nouméa, the capital of New Caledonia, a French overseas territory some 750 miles to the east of Australia. The night of 2 May 1943 was clear. Seeing a man and a woman emerge from the bushes near a parked jeep, the soldiers walked over to them. The man with the young native woman was a white officer. As the soldiers drew near, the woman got into the driver's seat of the jeep. She did not speak English and they did not speak French, but Lieutenant Robert L. Engels spoke both languages. He served as interpreter in the age-old transaction: sex for money.

Subsequently disputed were the questions of whether the discussion was amiable, whether Louise Mounien got out of the jeep willingly or unwillingly, and whether she bargained about price. But never disputed were the assertions that she willingly accompanied the four men to a more secluded spot on Prostitution Hill, had sexual intercourse once with each of the soldiers, and accepted three dollars in payment. She neither resisted nor cried out, not even when another soldier and a woman came near.

The intruding couple also saw Engels. Did they recognize him? Did he think so? As soon as the blacks took amiable leave of him and Mounien, the lieutenant overrode her objections and drove to the MP station. Four days later he and two other officers visited the unit to which Fisher and Loury were assigned. The men stood in company formation and the officers looked them over. Then the company was restricted to camp. After Fisher's arrest, Captain Norman Jones put Loury under company arrest. He said that Loury was sure to have been involved in anything that Fisher had done.

Fisher had been nineteen years old, Loury twenty, when they volunteered for army service in September 1942. Nine months later they were convicted on charges of having raped Louise Mounien. The theater commander at Melbourne, Australia, reviewed and approved their convictions and sentences: dishonorable discharge, forfeiture of all pay and allowances, and life imprisonment. They were serving time in the United States Penitentiary at McNeil Island, Washington, when a petition for clemency was submitted to Secretary of War Stimson by Congressman Vito Marcantonio, president of the International Labor Defense, and Hastie, chairman of the NAACP Legal Committee.[37]

Marcantonio and Hastie offered four justifications for clemency. The first was that rape, "a crime which can be committed only by compulsion against a woman's will," had not occurred. Mounien consented to sexual intercourse. Asked whether anyone hit or threatened her, she said no. "She remained passive lying on her back upon the ground after each act of intercourse until the departing soldier had rejoined his fellows and the newcomer had arrived where she was lying. . . ." The travel distance was about fifteen yards; Marcantonio and Hastie thought it "inconceivable" that when that far away from the next man in line, "any woman would lie quietly awaiting successive acts of ravishment."

The only complaint Mounien made was that the soldiers were rude in departing "without saying Bon Soir." Perhaps other men in her life said good night. Someone had delivered a more telling message—gonorrhea, which a surgeon at the local hospital detected when he examined her within twenty-four hours of the action on Prostitute Hill. On both 5 and 8 May a Dr. Ginieys reported that she had gonorrhea, with which neither Fisher, Loury, nor Engels was infected, according to Captain Salvatore L. Pernice, local military chief of the Urological Service, who examined them on 7 May. The lesions at the neck of Mounien's uterus discovered by Ginieys could not have resulted from intercourse within the twenty-four hours preceding his examination of her; so none of the men with her on 2 May infected her. Ginieys also reported that he found no medical evidence of violence or rape. None of these reports was presented to the court-martial.

Mounien told Captain John F. Saxon, the investigating officer, that she had been in Nouméa only two months but "had known many American soldiers." Her landlord remembered their steady nocturnal procession to her door. This, said Marcantonio and Hastie, is "a young woman of loose morals, accustomed to daily cohabitation with casual acquaintances among military personnel . . . The picture of her as a frightened innocent, terrified by the very approach of the petitioners, does not make sense. . . ."

The second argument for clemency was that military authorities were prejudiced against Fisher and Loury. Captain Saxon had read the report on Mounien's character by French civil authorities, had known the medical findings, and had read the MP testimony that within an hour of the incident the woman seemed not to have been in tears or hysterical. He recommended that no charge of rape be filed. Military authorities ignored his recommendation and withheld from the court-martial the evidence supporting it.

Lack of opportunity to examine Saxon's evidence or to gather its own was one of several glaring shortcomings of the defense that had been provided for Fisher and Loury. Counsel first interviewed them on Sunday afternoon for one trial (Fisher's) on Monday and the other (Loury's) on Tuesday. The two officers appointed as counsel advised them against testifying. Determined to testify, Loury took the witness stand after a recess. The judge advocate, the prosecutor, and several jurors questioned him. His counsel neither questioned nor helped him in any way.

Counsel certainly did not help him when it consented to the admission of self-incriminating statements obtained from the accused. The day before, while representing Fisher, counsel had heard the men who obtained the statements admit that they had intimidated Fisher and Loury.

"If you won't play ball with us, we will have to play ball on your head," Lieutenant David S. Teeples had said in the MP station, demanding that they admit their guilt. He brandished a blackjack. "Do you know what this is?"

"Yes," Loury answered, "but you're not going to use that on my head. Other people aren't supposed to hit the head of my mother's son."

Teeples was unimpressed. He said that his assignment was to break the case. "If I have to shoot you and hang you to a tree to be sure we have the man who did this, I'll do that." He added that all he wanted from Loury was a statement that Fisher had threatened Lieutenant Engels with a .45 pistol.

At 2:30 A.M. Loury was taken from a downstairs office to a cell. Later that morning a doctor ordered him taken to the hospital; the effects of an overnight chill made Loury miserable. At about 11:30 P.M. he was taken upstairs and questioned. Teeples still would not send him to the hospital. Instead, he threatened to return Loury to his cell, thereby saving ammunition.

"I asked him what he meant," Loury recalled. "He said that at sunrise they would have to shoot me. I told them that they might as well start shooting. . . . Then they tried to get me to sign the statement, but I refused. . . ."

The questioning stopped. Loury slept. Teeples and Sergeant Donahue woke him. Each grabbed one of his arms. They said he would be shot that night. "Your people won't know anything about you," said Teeples. "They will think a Jap got hold of you."

Back to his cell at 3:00 A.M. "It was cold there and I had no blankets." At 3:30 A.M. back upstairs. Sign the statement! No! Back to the cell—still not to the hospital. Two blankets. No questioning the next day, but again that night. Sign the statement! No! The fifth day. "You're in this very deep now," said Teeples, "and there is no way out for you."

"I was feeling very bad and didn't want to be bothered anymore," Loury said about this confrontation in his cell. Hearing that, Teeples wrote out a statement. Loury refused to sign it or to write his own. Eventually he agreed to give a statement, and Teeples helped write it. Upstairs, before Major Moffat, Loury signed it. He did not swear to it, but at the trial Major Moffat swore that he had done so voluntarily.

Moffat was not questioned by defense counsel. Teeples and Donahue, curiously, were never called as witnesses.

As a third argument for clemency Marcantonio and Hastie cited substantial conflict between the testimony of Engels and that of Mounien. Engels said Fisher was armed; Mounien said neither Fisher nor Loury was armed. Mounien said she cried out; Engels said she did not. Lastly, the fourth argument was that even if the convictions were lawful, the life sentences were excessive. Military justice tended to parallel civilian justice, said Hastie and Marcantonio. Under the penal law in New Caledonia the maximum punishment for rape was twenty years at hard labor. Why, then, were Fisher and Loury sentenced to life behind bars? "All persons concerned in the proceedings against them were white Americans, who as such had lived in an environment where social attitudes concerning whites and Negroes having sexual relations with the same woman are strong and prejudicial . . . No reasonable person can believe that white soldiers would have been convicted or even tried for rape in the circumstances of the present case."

The plea to have Fisher and Loury exonerated and returned to military duty went to Under Secretary of War Robert P. Patterson. Marcantonio had entered the battle at the prisoners' request, Hastie at Marcantonio's invitation.[38] The petition, said Patterson, raised questions relevant to clemency but not to guilt. Fisher and Loury were young, had little education, were deficient in "family background and training" (both were orphans), had no criminal record while civilians, had enlisted voluntarily, and had soldiered satisfactorily before that night on Prostitute Hill. On 31 March 1944 the sentences were reduced to ten years for Fisher and eight for Loury.[39]

At the first legally available opportunity, a year later, Hastie and Marcantonio petitioned Secretary of War Stimson for clemency. ("Before filing the petition," Hastie wrote to Marcantonio, "I will get in touch with you with reference to the matter of publicity.") They rebutted Patterson's opinion of 31 March 1944,[40] saying, "It accepts the incredible as true, wholly ignoring the facts which destroy credibility."

Patterson had relied on the version of the encounter given by Engels, the "acquiescent bystander" who gave it to exculpate himself. During the course of the investigation and the trials, Engels gave four conflicting versions. Engels, said Patterson, did not try to protect Mounien, but he did not rape her either. He was discharged under other than honorable conditions for having disgraced the uniform. "His craven action neither excuses nor mitigates the acts of the prisoners. On the contrary, it emphasizes the boldness of their action and their determination to have the girl." Fisher's paying her was seen by Patterson as an attempt to absolve himself. Fisher was alleged to have said, "Officer, would you call this rape? We paid her." Hastie and Marcantonio said that his discharge from the Army showed how much confidence Engels, who made the allegation, enjoyed among his superiors. On the joint statement that Marcantonio sent to him on 17 September 1945 for comment before presenting it to President Roosevelt, Hastie wrote about Fisher's paying Mounien: "Not his attitude in giving but hers as evidenced by taking that is important." On the statement, which was identical to the draft petition that he had sent to Marcantonio on 13 April 1945, Hastie also noted, "Difficult for one who has not seen America thru Negro's eyes to understand."

Hastie placed the second comment in the margin next to the paragraph in which he and Marcantonio contested Patterson's interpretation of the remark about payment. The rebuttal: "Every Negro man knows that any sex relation he may have with the paramour of a white man, if discovered, is likely to be followed by a charge of rape. In the case of a transaction between a Negro enlisted man and the paramour of a white officer, the anticipation of such a charge is even stronger."

If asked, Fisher's question was a mocking, taunting, and, as it turned out, overly optimistic assumption that no rape charges could arise. The question was not an admission that he had had Mounien without her consent. To call it an admission that justified Fisher's conviction, said Hastie and Marcantonio, underscored "the flimsy character of the evidence of the prosecution. If actions ever speak louder than words, they do so in this case." They said that Patterson had never answered the argument they had made in their petition the year before.

Trying to offset the argument that Mounien's promiscuity made "incredible" her tale about having been frightened, Patterson said that counsel disregarded "the well-established rule of law" that in a rape case the victim's unchastity is no defense. Hastie and Marcantonio "respectfully submitted that in this particular your opinion ignores our argument and without answering it suggests an entirely different matter which we have never urged."

Their being blacks did not influence the trials of Fisher and Loury, said Patterson. Their sentences exceeded the maximum for rape under civil law in New Caledonia but fell short of the maximum (death) under the Articles of War. Moreover, the importance of controlling crime by soldiers on the friendly island was great. Hastie and Marcantonio had dealt with those points in the brief they submitted to Patterson in 1944. The adjutant general had invited the military command to argue against clemency on grounds that granting it would damage military-civilian relations on the island. Concerning the acceptance of the invitation Hastie and Marcantonio commented, "It merely emphasizes the anti-Negro prejudice of the military command in New Caledonia."

New Caledonia was not the worst of it all. Deploring the mutilation and murder of black soldiers, Hastie said in the summer of 1943 that there were countless suggestions about ways to increase the likelihood of victory in the war. Buy war bonds, shun the black market, and, miners, return to work—all this and more Americans were being urged to do. "But the President, the top military authorities and all officials in Washington to whom the people would listen have been silent on the entire subject of violence against the Negro soldier."[41]

CHAPTER TWELVE
Call Dupont 6100

When Hastie left the War Department, he did not intend to return to Howard University as its law school dean if peace seemed to be a year or more away. If, however, peace seemed nearer, "I may return now because we must get ready for a tremendous expansion and some real discussion of legal education."[1]

At the law school Hastie was awaited by such students as Pauli Murray and Billy Jones. Ever since Hastie had represented Thomas R. Hocutt in the case against segregation at the University of North Carolina in 1933, Pauli Murray had been impressed by his steadfast leadership in the field of law she planned to enter. And Billy Jones remembers having felt that his anticipation of Hastie's return was rewarded. "Andy Ransom had been an effective acting dean," he says, "but when Hastie returned, there was a breath of fresh air. You could feel it in the things he did. It made you think of Charlie Houston."[2]

To understand Hastie's law school, it is necessary to understand Vice Dean Houston's law school. In fact, to understand Hastie it is helpful to understand Houston, "my friend, my distant kinsman, my mentor and senior colleague in both the practice and the teaching of law." Their fathers, who had been clerks in government service in the 1880s, were such close friends that the two families considered themselves cousins. After young Bill moved to Washington, D.C., the Houstons became even more like family to him. Twelve years old then, Bill was ten years Houston's junior. He admired Houston. "Indeed . . . I followed his footsteps through college and law school, into the practice of law with him and his father, and into law teaching and, under his inspiration and leadership, into the struggle to correct the appalling racism of American law."[3]

When Bill Hastie first knew him, Houston was spoiled, self-centered, reclusive, comfortable, and undisturbed by racism. "If he was not pleased with the status of the Negro," Hastie wrote, "he was not greatly moved by it and had no passionate concern to change it."[4]

Then the twenty-one-year-old Houston went off to World War I, earned an army commission, and encountered an enemy for which he was unprepared: racial bigotry. A naive, self-satisfied youth became a wise, tough man "with a hatred for American racism that he never wore on his sleeve but would always retain as a motivating force." Houston went from the Army to Harvard Law School, where he became the first black editor on the *Harvard Law Review*. He earned his LL.B. (1922), S.J.D. (1923), and a Sheldon Traveling Fellowship that enabled him to study civil law at the University of Madrid.[5]

Houston's military experience probably made the difference between his becoming a successful lawyer somewhat concerned about racial betterment and his becoming, in Hastie's opinion, "the effective leader of the essential first stage of the 20th century struggle to make the concept of America as a nation 'with liberty and justice for all,' a reality rather than a hypocritical platitude."[6]

Hastie's explanation of Houston's most impressive traits is self-revealing: ". . . In serious pursuits he was tough, combative and unsentimental, demanding excellence of himself and of his students and professional colleagues, though in social relations he was warm, always approachable and held in great affection." Houston worked hard and expected others to do the same. Students seeking his financial help received it; those seeking his toleration of complaints about assignments received his standard response, which was that there was to be "no tea for the feeble, no crepe for the dead."[7]

No criticism or condemnation made Houston retreat from a position, however unpopular, once he believed it to be correct. Blacks and liberals were enraged by his endorsement of President Roosevelt's nomination of Senator Hugo Black to the Supreme Court. Black's brief membership in the Ku Klux Klan in Alabama was unforgivable, they said. But as an NAACP lobbyist Houston had come to respect Black, who was chairman of the Senate Committee on Education. He backed Black's nomination and persuaded Walter White to do the same. Black's judicial record vindicated that faith.[8]

Houston believed black lawyers were indispensable in the fight against racism. He did not trust the average white lawyer, the beneficiary of the racism that had to be destroyed. The black lawyer's raison d'être, Houston said, was to be a social engineer committed to the advancement of his people, and the only justification for the Howard

University Law School was its performance of "a distinct, necessary work for the social good."[9]

Between 1930, when he became vice dean of the law school, and 1935, when he left to take charge of the NAACP's legal program, Houston worked to assure that performance. "His students soon learned that he was interested only in first rate achievement, not in excuses for shortcomings," Hastie wrote. "If students were sometimes driven to swearing at their Dean, they also swore by him."[10]

Hastie's students did not swear at him, but they did swear by him. For he, like Houston, made certain that students received more than instruction at Howard. And Hastie, like Houston, exemplified all the sterling qualities that law students, regardless of race or color, would have been well advised to develop, and that their law students, because of race and color, were required to develop. The requirement was necessitated by a mean national spirit that made applicable to William Hastie the observation Ossie Davis was to make about Godfrey Cambridge years later. Davis said that Cambridge won acclaim but not acceptance in this country. "America was not ready to love Godfrey. It was prepared to kiss him on both cheeks, but not on the lips."[11]

More so for Hastie than for any other black lawyer, the irony was even sharper. None had risen higher in public accomplishment, but while white America refused even to kiss him on the cheeks, it slobbered over John Cook.[12] Caricatured by Cook on the "Amos 'n' Andy" radio show, black lawyers were ignored by scholars. Houston urged the collection of information about them "before it is too late."[13] They had been so disregarded in works about American life and law that one wonders whether the oversight was intentional. "Someone has written that whom the gods would destroy they first deprive of a sense of history," Hastie once said. "I think this is true. For history informs us of past mistakes from which we can learn without repeating them. It also inspires us and gives us confidence and hope bred of victories already won."[14]

Hastie's students were fortunate, for the hope and confidence they required were personified by him and his faculty. Professors at Howard's law school were made in the mold of Robert Morris, the black lawyer who teamed with Charles Sumner in 1849 to make the first judicial attack on segregated education.[15] They were the vocational descendants of black lawyers who had sat in Congress and of those who had founded the National Bar Association (NBA).[16] And they were the professional untouchables, for the American Bar Association (ABA) had excluded them from its founding in 1878 until 1912, when three were admitted—and a resolution was adopted requiring future disclosure of applicants' race. While on the federal bench in 1939, Hastie himself was rejected by the ABA![17]

ABA membership would have pleased Hastie as a breach in discrimination against black lawyers. But the ABA could never have had a meaning comparable to that of the NBA.[18] After all, NBA founders were trailblazers who would have agreed with Edward Brathwaite that persons who create nothing "must exist on nothing."[19]

When Hastie was admitted to the bar in 1931, the local NBA counterpart, which Houston and six others had founded on 25 May 1925, was the Washington Bar Association. "It was in its infancy, small but active and vociferous as becomes a healthy infant," Hastie would recall. It was also very much needed. Of the hundred or so black graduates of law schools (mainly Howard University Law School), few made a living as lawyers, not even in local or national government. "I know, because in [1933] when I became a junior member of the legal staff of the Interior Department, my appointment was front page news in the weekly black press," said Hastie. Hardships assured by racism were aggravated by the Great Depression.[20] Nevertheless, some blacks earned a living in the law. Hastie called them "professional pioneers". One was Louis R. Mehlinger, a founder of the Washington Bar Association. Hastie was a junior at Amherst when Mehlinger's brother, James Edward, committed an unpardonable sin in Mississippi. James Mehlinger contracted to buy the farm that his father had rented for five years. For that display of independence he was killed by Klansmen. His father, whom they shot, subsequently moved the family to Washington, D.C.[21]

Louis Mehlinger attended a black college in Mississippi where he learned "manual training." He taught carpentry in Florida before moving to Washington in January 1907. After illness and the cessation of night courses at Howard University cut short his dental schooling in 1908, Mehlinger became a secretary in the Treasury Department. Later he became an army intelligence officer. Discharged in March 1919, he returned to Treasury to reclaim his job but was promoted instead. A higher salary ($1,800) enticed him to leave the position (accounts adjuster) to go to work for Emmett Scott, the secretary-treasurer at Howard.

Mehlinger studied law at night. "Howard didn't have a proper law library then. I got my training from the Library of Congress. At 9 o'clock every Sunday morning, I was down there waiting for the door to open. And I stayed there all day until they closed the place at night."

Mehlinger completed law school in 1921. But a black law graduate, even magna cum laude (in a class of twenty-six), had no chance of joining a major law firm and virtually no chance of making a living in solo practice. Many blacks in need of lawyers could not afford them, and many who could afford them preferred white lawyers. The only black lawyers at the Justice Department, Perry W. Howard, the attorney general's special assistant, secured a law clerk's job for Mehlinger. Twenty-six years would pass before blacks were admitted to bar review courses in Washington, so Mehlinger made the most of the department's library. He passed the examination in December 1921. Two months later he was admitted to the bar.

As an assistant attorney who argued cases in the Court of Claims, Mehlinger traveled throughout the country. Usually he was treated well. But he recalls a trip to Ithaca, New York, in 1932 to join a Court of Claims commissioner at a hearing. He waited a long time in the United States Attorney's office, then asked a secretary if he could go into the courtroom to continue waiting for the commissioner.

"Are you his messenger?" the secretary asked.

For years Mehlinger was the only black who worked as a lawyer at Justice, except for Howard and his successors appointed by President Roosevelt. "Perry was engaged in politics, ran around a lot, and I did all the law work in the office." Mehlinger rose steadily to become a senior attorney. Judge Joseph C. Waddy, who was an elevator operator even while he was a practicing attorney, said that Mehlinger's was "a rather minor job but, at the time, it was a plum for us."

Running elevators was an occupational probability for black law graduates in those years—and an occupational blessing, since it was often the only job they could get. So it was with William B. Bryant. His grandfather had fled from Alabama in 1910, just ahead of whites who were dedicated to the proposition that he should be lynched. The rest of the family followed. Born on 18 September 1911 in Wetumpka, Alabama, Bryant grew up in northeast Washington, attended Dunbar High School and Howard University, and decided to go to law school.[22]

"What do you want to go to law school for?" his grandfather asked. "Negro lawyers don't amount to anything." Bryant knew no lawyers. "At that time, there wasn't anybody I could point to and say, 'No, you're wrong.'" But at Howard he found a black lawyer who amounted to something.

Charlie Houston disapproved of Bryant's holding a night job while in school: "You can't do this and get the law." But Bryant was his own man. "I told him his job was to put the law in the classroom and my job was to get it and that I wasn't asking any special favors from him or the school."

That was that. Bryant earned some of his best marks in Houston's classes and undoubtedly pleased his model by graduating at the top of his class in 1936. A dozen years passed before he practiced law. In the interim he ran an elevator in an apartment building at a salary of thirty-two dollars a month, copied records of the District Recorder of Deeds (on a WPA project), and worked at the Frederick Douglass Home on a bookbinding project. He worked for the National Youth Administration (as a mail messenger) and the Office of War Information (as a researcher), and became an army lieutenant colonel. In 1948 he became a lawyer. He and Wesley Williams opened offices, and for two years Bryant "made very little money, but I enjoyed myself."

Hastie once wrote, "Saint Francis of Assisi is said to have prayed: 'God grant me the serenity to accept the things I cannot change, the courage to change the things I can and the wisdom to know the difference.' But at times it may be better for the Omnipotent One to give men the wit and the will to continue to plan purposefully and to struggle as best they know how to change things that seem immutable."[23]

So many things seemed immutable. In Washington most black lawyers were prevented from rising above domestic relations, personal injury, and criminal litigation. Even those who had a measure of success were hard-pressed. "There was hardly a night you could spend at home if you wanted to keep up a practice," says District of Columbia Superior Court Judge William S. Thompson. "You had to do it. You had to hustle. You could be the smartest in the world legally, and, if people didn't know you, you were on your way to the poorhouse."[24]

Some of their guides to the poorhouse were blacks who, like Dean Kelly Miller at Howard University, preferred white lawyers to handle their important business. "I am not outraged because I happen to be a lawyer," Hastie had said about that. Just let me ask if it is advisable to place colored students in the care of a man who thinks that way. Is the dean preparing his students for unimportant work? Hastie did not suggest that color should outweigh competence in Miller's selection of counsel; he denounced Miller's premise that competence and incompetence were separated by a color line. "As long as there lurks deep down in the consciousness of Kelly Miller or any other Negro a feeling that white is a little better than black, just so long is that man a hindrance to human advancement."[25]

When Hastie had become dean, among the hindrances was the exclusion of blacks from the white bar association in Washington and from the library that it operated in the District Municipal Building. After attorneys H. L. Brown and Pete Tyson brought suit in 1941 to compel either open admission to the law library or its removal from federal property, Attorney General Robert H. Jackson ordered the admission of members of the bar regardless of "race, color, religion, or sex." The District of Columbia Bar Association removed its library from the court building rather than obeying Jackson's order.[26]

Hastie's determination to remove hindrances several years earlier had produced headlines: "Hastie's Attempt to Have Bar Associate with Negro Congress Creates Row"; "National Negro Congress Endorsement Causes Breach"; "Negro Congress Is Red Inspired, Red Directed, Red Controlled, Says Howard"; "Bar Association Splits." The controversy involved factions led by Hastie and Perry W. Howard. At issue was the question of whether the Washington Bar Association should support efforts by the National Negro Congress to win federal legislation to stamp out segregation, to aid sharecroppers and tenant farmers, to provide social security protection for domestic and farm workers as well as for other workers, to guarantee American neutrality should war occur, and to assure blacks full voting powers. Hastie's speech in favor of the bar association's assisting the National Negro Congress touched off debate at a special meeting in March 1936. He read the majority opinion of the committee appointed to advise the association. The committee's chairman, Howard, reported the minority's opposition to that recommendation.[27]

"Slander," "crooks," "reds"—fulmination had preceded the association's adoption of the majority report by a fourteen-to-twelve vote. Thirteen members denounced the vote as unpatriotic and unwise. Meeting at the Mu-So-Lit Club the next Tuesday night, they withdrew from the association and formed the Harlen-Terrell Lawyers' Association. Among them were some of the most prominent black lawyers in the city.[28]

One who had backed Bill Hastie was William L. Houston, founder of one of the few black law firms in Washington. It became Houston and Houston when his son, Charlie, was admitted to the bar in 1924, and Houston, Houston, and Hastie in 1931. The firm was noteworthy for its training of a number of judicial marvels, including William Houston's law partners. While "Amos 'n' Andy" vilified black lawyers, Bill and Charlie revolutionized American life and law. The radio show and the revolution underscore W. E. B. Du Bois's affirmation: "It was the black man that raised a vision of democracy in America such as neither Americans nor Europeans conceived in the eighteenth century and such as they have not even accepted in the twentieth century; and yet a conception which every clear sighted man knows is true and inevitable."[29]

True, but not inevitable. So Hastie had returned to Howard.

Of the young, eager day students who had enrolled at Howard's law school since October 1941, a number had withdrawn while others had been called away by Uncle Sam. No students in the evening college, which resumed operation in 1941 primarily to accommodate war workers, had been forced out of school by military service. There were one part-time and five full-time faculty members, several of whom carried teaching loads in excess of the maximum normally sanctioned by the Association of American Law Schools. In addition to his administrative duties, Hastie taught ten hours.[30]

Hastie was a stickler for preparation. One day students went unprepared to his class in "Evidence"; they had been preoccupied with an examination in another course.

"Don't ever walk into this class unless you are prepared," Hastie told them. He slammed his book down and dismissed them.[31]

That was both the only time Pauli Murray ever saw Hastie's temper flare and the last time she went to his class unprepared. In January 1944 he asked whether she would like to apply for a Rosenwald Fellowship. She applied for it and for admission to Harvard Law School. That set off the "Harvard Howl."

"Your picture and the salutation on your college transcript . . . indicate that you are not of the sex entitled to be admitted to Harvard Law School," T. R. Powell, chairman of the Graduate Committee, told her. She appealed to President Roosevelt, to Justice Felix Frankfurter (as a former faculty member), and to Dean Lloyd K. Garrison, a member of the Board of Overseers at Harvard. The Dean of Harvard University informed Roosevelt that the "comparatively simple" problem could be solved because "Radcliffe degrees are really Harvard degrees." But women could not enter the law school through Radcliffe.[32]

"One cannot sit under men like Hastie and Ransom and not be inspired to go to Harvard," Pauli Murray told Frankfurter. The top student in her class, she and Hastie had decided that Harvard was the place for her. Harvard could not escape the increasing pressure for fair play toward (women) applicants who were giving a good account of themselves in the legal profession and in the war effort.[33]

By June she was spent. The dispute and her studies had taken their toll. She put the "Harvard Howl" in her dean's hands and asked Powell to direct all correspondence to him.[34]

Hastie learned from Professor Erwin Griswold in August 1944 both the certainty of Pauli Murray's exclusion and at least part of the reason for it. More than ten years earlier the law school faculty, in a close vote, approved the admission of women but was overruled by the Harvard Corporation. Griswold thought it was in 1942 that the issue again surfaced but was tabled; the faculty felt it should not make a decision while many of its members were away in military or governmental service.[35]

Griswold favored the admission of women, and "in normal times" would have voted accordingly. Pauli Murray was turned away for no reason other than sex. "As I see it, that is no real basis for a discrimination, but there is a lot of history behind it, and such things do not change easily."[36]

Griswold recalled the rejection, several years earlier, of another woman who had gone on to do at Columbia some work that he would have liked to have had her do at Harvard. "She was turned down here by the Dean, without any reference to the Faculty," Griswold wrote. "Miss Murray had at least carried the matter further, but I can hold out little hope for her success at the present time."[37]

Hastie appreciated Griswold's "illuminating letter." This, however, was one respect in which Howard did not emulate Harvard. Howard had never discriminated against applicants because of sex or race. It was the first law school in Washington to admit women, and the only one to do so before 1900. Six of those who graduated between 1882 and 1900 were white; five obtained LL.M. degrees. As of 1940 the law school counted a state governor among its white graduates. In that year Hastie reiterated his predecessors' objection to the American Bar Association's designation of the law school as "colored." He objected both because Howard had been singled out for racial designation and because the designation was inaccurate.[38]

Powell was miffed. Pauli Murray had not corrected "my misapprehension" and still refused to acknowledge her femininity. She had not been candid, having written to him, "Without commenting on your assumption as to my sex, since you have not subjected me to medical examination, I am interested in the democratic technique of free discussion and reexamination of traditions and policies in all democratic institutions."[39]

"I think that what you have viewed as a lack of candor in revealing the applicant's sex is essentially the unwillingness of the feminist to concede that sex is a material consideration," Hastie replied to Powell, who had written in light of Harvard-Howard relations.[40]

But she sent copies of letters to the President and to others, trying to drum up publicity, Powell complained. Hastie explained that Murray sent copies to the White House because she was among the "able and ambitious young people" who held Mrs. Roosevelt's personal interest.[41]

"Miss Murray is a very sincere, able and determined person," Hastie continued, "and we of the faculty have done nothing to discourage her from pursuing orderly processes in an attack upon what she believes to be an unsound and unjust provision of University law. . . ."

"Pauli, you don't have a chance," Garrison told her. Why did she want a chance? Perhaps because Harvard was prestigious, she says, or because Howard was "a miniature edition in sepia of Harvard." Perhaps because she had more than held her own against the Howard man,

Francisco Caniero, who was then at Harvard, and was "top man" in her class, or because she had been influenced by Hastie and other Harvard men at Howard. "Perhaps it was because I believe in the equality of human beings without regard to race, color, sex, or creed."[42]

Like her dean, Pauli Murray fought for noble causes. The battle to enter Harvard Law School was not hers alone, she informed A. Calvert Smith in an undated letter; it was that of women who wanted a chance to obtain "the very best legal education." All sorts of traditions were giving way to progress. Women were bearing their fair share of the war effort. They were holding their own at the bar, on the bench, and in administrative positions that required legal talent. She added that "although Harvard might lose in the sense of loss of a tradition, it might gain in the quality of the law school student personnel."

Nor was that all, she continued, for "even if you reject my application and I give up the fight, you will not be rid of applications and similar kinds of arguments on the part of young women." Those at Howard would try to gain admission to Harvard Law School. And even *that* was not all. The moment that young white women learned about her efforts to enter Harvard Law School, they would renew their own, which fitted neatly into the current demand for an "equal rights" amendment to the Constitution.

As Hastie believed, Harvard would have gained by Pauli Murray's presence. But unlike the dean whom she admired at Howard, the law school at Harvard neither knew nor cared to know the precious person described by John P. Lewis in his column in PM. "Pauli looks like a high school kid, quick and alert and gay, and carries herself with an air of irresponsibility that belies everything inside that head of hers," Lewis wrote on 14 June 1944. "She laughs easily, but her eyes are serious . . . [S]he leaves you with an impression of wisdom and courage. This slight, wiry, tousled girl reflects, with a sensitivity I cannot put in words, the humanity of a people who have learned of human nature from the harder, not the softer side. There were tragedy, human dignity, strength, maybe even a touch of destiny about her that, it seemed to me later, were more than hers alone—as if she personified in one small bundle the whole of her people."

But Garrison was right. Pauli Murray did not have a chance. While she was ready for Harvard, the law school was not ready for a female, much less a feminist. Hastie believed in Pauli Murray: in the ability she had shown as a law student and in the promise she held as a lawyer. He also believed in Robert L. Carter.[43]

Carter, a federal district court judge in New York City, remembers the day he first met Dean Hastie. At the time Carter was a student at

Lincoln University in Pennsylvania. "Bill came and delivered a speech. I thought he was one of the most distinguished looking people I had ever seen. He was so rational and impressive. The experience was part of the reason that I decided to go to law school."

Carter next met Hastie at Howard. "I was finishing law school. We were all getting into preparation for the war, and he was very busy." Hastie was very interested in students, though other instructors spent more time with them than he did. "He had a lot of other things he was doing," recalls Carter, who went from Howard to Columbia University School of Law on a Rosenwald Fellowship.

To Professor Edwin W. Patterson, chairman of the Committee on Graduate Instruction at Columbia, Hastie wrote a letter of recommendation in which he noted that "Carter has a penetrating mind which reaches the essence of the legal problem, and independence in thinking which helps him draw his own conclusions without undue deference to the thinking of others." He would have liked for Carter to have gained more experience in writing while at Howard but was confident that he would do well in research. Whether writing or speaking, "he gets his ideas across with simplicity and economy of language."

Carter obtained his master's degree and went into the Army, where additional traits that Hastie liked—broad interest in public concerns and "social intelligence"—were not as likely to be appreciated.

"Boy! The Army. . . ." Carter collects his thoughts, chuckles, goes on. "The first thing I met was all this vehement prejudice and particularly some officers' prejudice against educated blacks. My first assignment was down in Georgia. The captain was an old army man from the South. We were all from New Jersey, New York, and so forth; everybody had a pretty good education—about 10 percent of us had college degrees and a few had higher degrees. And he simply told us that he didn't think that niggers ought to be educated and that we shouldn't think we were going to get any breaks from him."

Carter became a lieutenant. "I had a great deal of trouble in the Army. It seems to me it wasn't of my doing. It was so many things—the Army was so full of prejudice." These included things like his going to the Officers' Club, his pressing charges against two white enlisted men who made racist remarks to him, his refusing to obey orders to live off the base as all other blacks did. "They didn't want you on the base. So I had to fight for everything. It was not a pleasant experience. I really don't regard myself as being somebody who likes to fight all the time, but I don't want to be pushed around."

So the military shipped him around, first to Columbus, Ohio, then to Michigan. He continued to rebel. In Michigan, while defending a black soldier accused of raping a white woman, he showed her to be a prostitute. That did it! Major Arthur P. Hayes brought charges and Carter was given an administrative discharge, one that was neither honorable nor dishonorable and left him subject to be drafted.

Discharged and distressed, Carter went to see Hastie at Howard. "He was very calm," says Carter. "He didn't feel that I had done anything, and that made me feel good." Hastie was counsel on Carter's petition to the review board. During the hearing Carter's temper flared. "Bill gave me a look and I calmed down."

Carter received an honorable discharge. To clear your record, Hastie told him, you should apply for readmission. Carter applied. "The adjutant general said 'No, thanks.' But Bill saved my life. You know that thing would have followed me for the rest of my days."

Carter went on to handle numerous court-martial convictions that the NAACP took to the War Department Board of Appeals. Hastie made certain that Carter carried most of the weight in a case in which they were co-counsel. He wanted to make a firsthand assessment of Carter's work. Afterward he recommended to Walter White that Carter's salary be raised to three thousand dollars. He liked what he saw in Carter: ability, maturity, persuasiveness, and guts. "It is quite possible that time may prove that he is as valuable an addition to our staff as Thurgood was some ten years ago." Carter proved exactly that.

Carter became the kind of lawyer that Hastie was determined to train at Howard. Arranging graduate school opportunities for his students was part of his overall plan, but Hastie did not limit their advanced training to formal education. Like his classmate Pauli Murray, Billy Jones lost an opportunity that Hastie wanted for him. When sent by Hastie to Houston's office, Jones was working as a guard for the government. Charles Houston needed a law clerk but wanted one who could type his own work so that a secretary need not be tied up doing that. Because Jones could not type, Houston did not hire him. When Jones reported this, Hastie, who wore a hearing aid, cocked his head to one side. "Say that again." Jones did.

"It is unfortunate that you were there at the right time but didn't have the right tools," Hastie said. "This is what you've got to do: You've got to get there at the right time and you've got to be equipped." This point was not lost on Billy Jones and would not have been lost on Eugene H. Clarke, Jr., who enrolled in 1940 with forty-four classmates and graduated in 1943 with four or five. "I believe it was Hastie who told our freshman class that we were a land-grant school; therefore they had to accept us, but there sure was no requirement that they graduate us."[44]

Clarke laughed, and was asked if Hastie had been serious about that.

"Hastie was a very *serious* person about the law," says Clarke, a lawyer in Philadelphia. "He could crack some jokes about it, but he didn't play around when he said something about the law."

Even when it might have seemed Hastie was easing up, says Clarke, that was not so. For example, when Spottswood W. Robinson III flunked eight (out of fifteen) of the students in a class, the eight appealed to Hastie for re-examination and the grading of it by "that notoriously easy marker, Ransom." Arranging a reexamination was no easy matter. All classes were taught in the morning, and a student would not very likely repeat a course that was being taught at the same time as were other required courses. The students' request was unprecedented, but Hastie granted it. Only one passed. Clarke believes that is what Hastie expected: His was a demanding law school. If there were, say, forty-five students in a class, the professor flunked thirty. There was never a question of whether students would recite, just how often. "If you want to memorize something," they were told, "go down the hill to the medical school. That's where they memorize things. Up here you learn how to think."

What should they get out of law school? As Hastie saw it, a certain amount of specialized knowledge that might enable them to pass the bar examination and perhaps become "successful" lawyers if possible. But that was merely the foundation on which to build during and after law school. The law changes. Lawyers are the architects of such change. Therefore, more important than the acquisition of information was the development of talent to shape and apply the law "for broad social interests no less than for the client who pays a good fee."[45]

Houston had stressed perfection. He was determined to build a black cadre of superlative lawyers. "He used to say that he wouldn't be satisfied until he went up on the hill to some dance and saw all his students sitting around the side reading law books," Thurgood Marshall recalls.[46]

Hastie did not go that far. But he might have been tempted by Oliver W. Hill. Hastie was teaching part-time when he first met Hill in the thirties. Hill says that they were at a fraternity party. The party started at about half past midnight and ended around four hours later. Hill went home to study for his eight o'clock class.[47]

It had been a night to remember. Foxy women, handsome men, music, drink, laughter. And dime dancing: plant your feet, place your hands (one on each of her buttocks), pull *hard*, and restrict all movement (synchronized, of course) to the area below the waist and above the knees (the more rhythmic, the more enviable). The space required was no bigger than a dime.[48] Students were not the only dime dancers, either. What a night. But now Hill was back in the classroom.

"Well, I know our good friend is well prepared this morning," said Hastie. Hill was. He answered Hastie's question, folded his arms across his chest, and leaned back in his chair, the kind with the wide arms to permit note taking.

"I must have gotten too damned comfortable, because the next damned thing I knew the daggone chair had tumbled all the way back and I landed on the damned floor. And Bill said, 'Well, you go home and get your rest tonight.' I had beaten him on the assignment; then I got caught sleeping."

The faculty's accessibility facilitated students' learning to think. "This was one of the beauties of the law school," Clarke recalls. "At that time I might go into some after-hours club and there would be one of my law school profs standing at the bar. I'd say, 'May I see you a minute? Goddamn, you talked about something this morning and I don't know what the hell you were talking about. Now we're just gonna have to stop right here . . . just set your drink down and explain this *to me*."[49]

The law school was small; the students and faculty therefore knew the same people and went to the same parties. "You could sit down and talk with any of these guys," says Clarke. "This is what made them *big* people. They were never too important to allow you to intrude on their privacy to talk to them."[50]

Professors managed to keep things balanced, however. "I'd play poker with [William R.] Ming in Herbert Hardin's apartment Sunday night, all night long, and win his money and walk into class Monday morning and he'd say, 'Go right ahead, Mr. Clarke, first case.' And I'd curse him under my breath." Laughter follows this recollection, a mark of admiration and fondness. "Ming would go to the board and say, 'Well, I don't know too much about this,' and trace the English crown from its inception, you know."

Brilliant like Ming, Professor Leon A. Ransom was always smoking, always rumpled, the underprivileged type, the opposite of the cool, suave Hastie. James M. Nabrit, Jr., was on the order of Ransom, brilliant, with a Southern drawl that made him even more down to earth. Clarke used to baby-sit for the Nabrits. "Never ask Nabrit a question. 'Mr. Nabrit, I don't understand so and so.' His question would be, 'What have you read on the subject?' I'd say, 'Well, I haven't read anything yet.' 'Well, the outside readings have a whole section on that subject. You go read those and when you get finished, if you still do not understand it, then you come back and talk to me."

Hastie was well cast in his role at Howard, as Ruth Harvey recalls. If students fell short of his expectations and were called into his office, they found him gentle but firm, and always willing to help. From time to time, in groups of six or eight or so, students were invited to the Hasties' home to join him and, perhaps, other members of the faculty for an evening of tea and conversation.[51] "I had an experience which particularly endeared Judge Hastie to me," she says. "In my first year I became ill and had to be hospitalized. Upon my return to school, Judge Hastie would inquire of my health from time to time. Without my knowing it, he was kind enough to write to my parents to let them know how I was getting along. I didn't know that he was observing me that closely. Needless to say, my parents were very grateful because they were very concerned as to my progress. They knew that I would not tell them if I were having unusual difficulty."

Hastie was much the same as he had been in the thirties, when Vincent M. Townsend, Jr., was his student. Townsend's description of him is as follows:

> Very friendly and polite. A perfect gentleman. Soft voice, never loud, never bombastic. Scholarly but not in a way of trying to display it. He never looked down his nose to you, never bowled you over with showing you how much knowledge he had, though he had unlimited knowledge. He always put his teaching at the level of the students, so that no student in the class had to feel humiliated, embarrassed, or under pressure if he wasn't as bright as the next student in the class. With Hastie it was a gentle kind of thing where you really couldn't miss the point, and it was a joy to be in his classroom.[52]

Less joyous but unavoidable were the long hours spent in the Buzzard's Nest, the law library on the floor just below the attic in Founders Library, says Ruth Harvey Charity. As though their studies were insufficient hardship, they could see lovers strolling around the reservoir. Strolling, however, was not for them, says one, Archibald T. LeCesne. "We were law students."[53]

"You had to believe in what you were doing just to stay there," says Townsend.[54] As they believed, others also believed. Townsend had worked his way through school as a dining car waiter on the Atlantic and Pacific Railroad and, once he enrolled in law school, on the Baltimore and Ohio Railroad. On the B. & O. he worked in the private car with the traffic manager all summer, living there and reading books. Other blacks with seniority let him work in their place so that he could earn money; they feigned illness so that he could work on weekends, holidays, and during term breaks.

"I'd tell them what Houston and Hastie and others were teaching us and what we were trying to do for the Negro race," Townsend recalls. "I sold that to them. You see how contagious it was. So they felt that they were making their contribution. They were not educated, they had no big hopes, their whole life was a dining car waiter. 'Here's a man that's trying to get somewhere; we'll be here still waiting tables.' And time after time they would take days off to give me trips."

Hastie's students learned by doing. They researched cases; they attended the Supreme Court, which Houston, Hastie, Marshall, and others transformed into a classroom; and they participated in the rehearsals that were held to prepare for trial and to instruct and indoctrinate the students.

The rehearsals were unforgettable, says Clarke. "It was an amazing thing. Thurgood had everything except his morning coat and his walking pants on. He would stand up and make his remarks as though he were talking to the justices of the Supreme Court, and then he would go into his argument in support of his position. He would be interrupted intermittently by faculty, very seldom by a student, because we were too awed to do much, but we would laugh and take sides with whoever was arguing. Everybody was a little afraid to put his hand up for fear he would sound stupid, you know. But it left its impression upon us."[55]

Before she enrolled at the law school Pauli Murray had been impressed by "the big guns"—Houston, Hastie, Ransom, and Marshall—when they discussed with her attorneys strategy to defend her after she had been arrested for bucking Jim Crow on an interstate Greyhound bus in 1940. "They all sat around like a football team, arguing back and forth." The others were ripping into each other. Hastie sat back, listened, and quietly offered an opinion now and then. Cool, judicious. "I just sat there and . . . It was like heaven to me."[56]

Many students went to Howard for the "impression" and left with it. They wanted to emulate the professors in trying, in the words of Edith Hamilton, "to make gentle the life of the world." Eugene Clarke was one of them. The impression left its mark. "It made me want to be a good lawyer, an understanding and compassionate lawyer . . . to be interested in the rights of minorities. And it made me proud to be a member of a profession that had men like these. Of course, it made me proud to be black like they were. It gave me confidence in my ability."[57]

Each trailblazer at Howard knew, as Roger N. Baldwin has said, that judicial victory or defeat depended in part on judges' attitudes, which "in turn depended on the times, on politics, on public feelings and on their own preconceptions embedded in their training and associations. . . ."[58] Whether elected or appointed, said Houston, public officials serve those who put and keep them in office. Blacks could not expect judges to commit suicide, political or professional. "We cannot depend upon [them] to fight . . . our battles. . . ."[59]

There would always be battles, Hastie believed. "As long as selfishness and cruelty and enmity endure in the hearts and minds of men, social injustice will exist," Hastie said. Blacks suffer "inexcusable oppression" in this nation, but they (like all others) enjoy "freedom in the highest sense"—the right to fight for freedom in every sense.[60] At his law school they learned to fight according to the code of battle that governed Hastie and others in the small strike force that spearheaded the struggle. The strike force consisted mainly of blacks and lawyers and functioned under the NAACP's aegis. Its significance, however, exceeded the bounds of legal protection for blacks; being tested was the potential and actual effectiveness of law as a means of advancing order in American society.[61]

As Hastie explained, members of the strike force had to understand the social order as well as the legal system and then persuade the community as well as the courts to alter entrenched racial reasoning. They had to confront the most competent of lawyers over volatile issues in courts, from the lowest to the highest. "In the struggle, no allowance would be made for the deficient background of any advocate or for inferior early training or for inadequate grasp of subject matter," said Hastie. "[T]here could be no substitute for outstanding ability and resourcefulness and years of rigorous preparation."

But even "big league knowledge, ability and drive" was not enough, Hastie added. Of equal importance was the belief that the nation possessed both the will and the means to ensure racial fair play. This belief was often hard for blacks to come by. The strike force had to believe; otherwise its efforts would have seemed hopeless and would have been futile. But, this belief notwithstanding, despair and defeat would have been assured had the strike force failed to fight each battle within the context of its overall campaign. No person or group could fight effectively on all fronts at all times, and even a particular victory counted only insofar as it advanced the campaign. Scarce resources were not to be, and were not, dissipated "in the pursuit of relatively inconsequential victories."

Perhaps above all else the strike force wasted no time or effort in looking for an easy way out. "[T]here is no simplistic approach to worthwhile achievement in human affairs," Hastie said. "And this is doubly so in its application to the disadvantaged individual or group identified as American Negro. Yet, difficulty need not foreshadow despair or defeat. Rather, achievement can be all the more satisfying because of obstacles surmounted."

Hastie wanted his students to know the facts of life and law. To know, for example, that some whites lowered obstacles. He considered himself fortunate to have been in the Supreme Court when Earl Brewer, former governor of Mississippi, did that.

In 1936, in the dead of night, policemen had taken three black youths into the woods, strung them up by their thumbs, and beat them until they confessed to the murder of a white person. The blacks denied their guilt; but the next day, freshened up and coached by the police, and in the presence of several reputable persons assembled by the police, they made the confessions that clinched their conviction in court.

Illiterate, poor, and despised, they were no match for the state of Mississippi. Brewer was. He asked the Supreme Court to spare his state the shame of that conviction. As the white-haired lawyer made his argument, the justices sat on the edges of their chairs; they made no effort to conceal their reaction, and subsequently negated the death sentences with an opinion that initiated judicial insistence on fair play by state as well as federal officials.[62]

Another fact of life and law was that fighting racism was often a thankless task, one made more so by blacks.[63] For example, Conrad Pearson once had a black client in Durham, North Carolina, named Tucker, a trustee in the county home. One day inmates collected enough money to send someone to town to buy whiskey. They chose this man Tucker. (Had he been outside, they could have bought it from him, for he was a bootlegger, which explains why he was inside.) The fellows got drunk. Someone snitched. Tucker, they thought. He will pay, they said. Tucker reported the threat to the captain, who told him to defend himself.

"They were killing hogs and had knives two feet long," Pearson said. "Tucker was standing at the gate, counting the fellows as they came out. And this white boy jumped him. Didn't know he had this knife. Tucker disemboweled him."

Pearson defended Tucker against a charge of first-degree murder. One day on the street a crippled white boy stopped Pearson. "Loan me a dollar," the boy said, "because I'm going to be a witness in your case and I'm going to tell it just like it happened." Pearson gave him the dollar, and the witness came to court, true to his word. The judge called

Pearson to the bench and said that manslaughter, not murder, was the proper charge. "I won't tell you what I'm going to do," he said, "but I suggest that he plead guilty to manslaughter."

Pearson had challenged the exclusion of blacks from the jury. A guilty plea would negate that challenge, which Pearson wanted to press. But the guilty plea was entered. Better that than the death penalty. Not long afterward Tucker again got in trouble—and got a white lawyer.

Pearson became more, not less, determined. He had been trained at Howard, where Hastie and others were not mere professors but protagonists. In Pearson's time they had used their skills to promote equal justice. As Ransom was to report while substituting for Hastie as dean, faculty members participated in cases designed to refine the law to protect the rights of blacks. In 1940–41, for example, they were either the leading or associate counsel in all but one such major case. Students participated in this "laboratory work" so that they would be inculcated with "a desire to enter into the public service."[64]

Commitment to public service was evident not only in Hastie and in his faculty; it was also evident in his students. In January 1943 Howard students began to hold demonstrations against restaurants in Washington that discriminated against blacks. They were coached by Billy Jones and Pauli Murray. Most of them were not law students, Patricia Roberts Harris, a cabinet officer, recalled years afterward. "But it was Bill Hastie who gave us the courage to understand the relationship between our anger and the intellect that can turn anger to effect by using the society's own goals against that society to make certain that those of us who were then excluded would be included."[65]

Harris had been inspired by Hastie before she enrolled as an undergraduate at Howard. His display of conscience and courage in the War Department had made him admirable, just as his representation of Thomas R. Hocutt had first made him inspirational to Pauli Murray. To the latter he had become even more a beacon as she fought to save Odell Waller's life.[66]

"I won't give you a damn thing."

"You ought to give me something for working, don't you think?" Odell Waller asked Oscar Davis.

"No. I ain't goin' to pay you . . . I will see you when the crop is finished."

Waller had returned to Pittsylvania County, Virginia, in the summer of 1940 to claim his share of tobacco that Davis had withheld the previous summer. He had gone to work in Baltimore, hoping to settle with Davis later. In the meantime his wife and mother had continued to work as domestics and sharecroppers for Davis. They had helped to harvest the wheat that Waller had planted before his departure. He returned to find that Davis had evicted the women and denied them the family's one-fourth share of the crop.

Hearing that, Waller grabbed his gun. He found the Davises, father and son, in the yard of their plantation. Davis had threatened him in the past, but that did not deter Waller. "Mr. Davis, I come here to get my wheat."

Davis became more agitated: "'I told you you warn't goin' to carry it away from . . .' and he used some dirty words [Waller testified] . . . and he usually carried a gun and he run his hand in his pocket like he was trying to pull out something. I had my gun and out with it. I open my pistol and commence to shoot at him. I don't know how many times. Mr. Davis hollered and fell."

Davis died two days later, on 17 July 1940. With a lynch mob at his heels, Waller fled to Ohio. He was extradited and tried by a jury of whites: a businessman, a carpenter, and ten plantation landlords. (Jury service bore a price, called a poll tax, of $1.50.) Waller pleaded self-defense. He was sentenced to be electrocuted.

The Workers Defense League, "a Norman Thomas socialist-oriented agency for the poor," had tried to prevent Waller's extradition, says Pauli Murray, then a member of its executive committee. It continued to defend him as the result of field investigations by her and a young writer, Murray Kempton.

At first the authorities would not permit her to see Waller, who was on death row in Richmond. Finally they did.

"Miss Murray, as God is my judge, I didn't intend to kill the man. I thought he was going to shoot me. I'm as sorry as I can be that it happened."

She believed him. "After all, a young fellow (only twenty-three at the time), scared, going up to the white man to get his rights. You can just see what was probably going on in his head."

Governor James H. Price granted five reprieves. His successor, Colgate W. Darden, granted none. Petitions for a stay of execution were filed with the Supreme Court of the United States—in vain.

"The problem was that Waller's lawyer had tried to show prejudice in the selection of the jury," Murray explains. "This was part of his statement. But he did not put supporting evidence into the record." On appeal the case was handled by John Finnerty, a wealthy volunteer and a determined advocate who pursued Chief Justice Harlan F. Stone through Virginia and West Virginia, enlisting police escorts as he traced the jurist's speaking-tour trail, finally catching up with him.

In Stone's hotel room they argued the case and the law. "There's nothing I can do," said the chief justice. "Nothing in the record supports an appeal."

Waller's champions went to Washington to see if the President could do something. A. Philip Randolph and Mary McLeod Bethune led the delegation. "We were unable to see him, so we tried to see Vice President Wallace," says Murray. "He saw us coming, and he literally ran away. Everybody in Washington knew why we were there."

They got to see Wallace's political secretary. "He picked up Mrs. Bethune's cane and played with it. When you came out of there, you felt lower than dirt, having seen the best we had in the race given that kind of cavalier treatment."

They saw Elmer Davis in the Office of War Information. He could not be moved.

"If this were your son, would you feel the same way?" Murray asked him.

"Your point is well taken."

But he could do nothing. Eventually Randolph got through again to the blacks' best friend in official Washington. In the NAACP's offices, members of the delegation listened, two on each of the five extensions. It was almost eleven o'clock at night.

"Mr. Randolph, I've done all I can do. I can't do any more." Eleanor Roosevelt's voice broke. Twice she had gone to see her husband; twice he had said that it was a matter of law, not men. He and Harry Hopkins were busy with the war. "And if I go to him again, he will be displeased with me. And so there is just nothing more I can do."

Meanwhile, the search for Darden was proving futile. Pauli Murray and Waller's mother had crisscrossed the country seeking support. At the March-on-Washington rally in New York City, a huge poster had exhorted: SAVE ODELL WALLER. But where was Darden? They had to succeed before midnight, 1 July 1942.

"Darden had disappeared," says Pauli Murray. "And the clock was ticking away, ticking away. Well, this was a terrible thing. That was a night I will never forget. We couldn't sleep. And the clock was ticking away. And finally it got to be midnight, and of course that was it."

Pauli Murray had vowed that she would become a lawyer if Waller were executed. Governor Darden and his ilk would have a lot more to fear from a black lawyer than a black sharecropper. She would become a civil rights lawyer. And that meant that she would attend law school at Howard University. It was a school whose dean understood what she had gone through.

"Unless we can have the substance of law," Hastie had said about another miscarriage of justice (viz. the Scottsboro case), "it doesn't matter whether a person is hanged by an unauthorized mob or by a mob known as 'the law.'"[67]

For incentive to fight racism, Hastie and others at Howard had no need to look to Alabama or Virginia. They were located in Washington, "where [Pauli Murray wrote] jim crow rides the American eagle, if indeed he does not put the poor symbol to flight. . . ."[68] Washington was a racist city, and Hastie was as much discriminated against as were his students. Soon after Hastie left the War Department, I. F. Stone invited him to lunch at the National Press Club.[69] The instant they reached a table, a waiter told Stone that he had a telephone call. Actually, the manager wanted to draw Stone away from the table to tell him that Hastie would not be served. Stone insisted on service for his guest. He and Hastie remained at their table for more than an hour, but they were not even served water by the black waiters, who apparently were under orders.

Thinking that club members wrote a lot about democracy, Stone wondered if they would practice it by striking down the policy against having blacks as diners or members. Ten of them signed his petition for a special meeting at which he intended to protest that policy, but twenty-five signatures were required. Stone resigned and joined the black organization, the Capital Press Club. Recalling the incident, Stone says, "Here's a guy who . . . ranked a hell of a lot higher than all the goddamn press agents, politicians, and assorted canaille that congregate there, and they were indignant that I dared to bring him in. . . ."

The indignation was appropriate to a club dedicated to the proposition that all black men were created unequal to all white men, regardless of achievement. This dedication was common among institutions in Washington. For example, in one downtown restaurant a waitress told a white patron that he and his black guests (a federal government lawyer and his wife) would not be served because "you've had too much already." A moment passed before he realized what was happening. If challenged later, the waitress would say that she had refused service to drunkards, not blacks, and they would be unable to prove otherwise. The only downtown restaurants that received blacks as patrons were those in the Y.M.C.A., Union Station, and government agencies. At National Airport, at the National Zoological Park, and on other federal properties privately operated restaurants refused service to blacks.[70]

Restaurants and other institutions in the forties offered the same rationalization of racism that Hastie had rebutted a decade earlier when a community center defended its exclusion of blacks from a city wide track meet that it sponsored for federal and local government employees. The center rationalized its action by saying that it was simply following the example set by the board of education. Although black and white high school athletes from Washington competed in the interracial Penn Relays in Philadelphia (Hastie remembered having been on the Dunbar High relay team that finished ahead of the McKinley High team in April 1921), the school board in Washington forbade their doing so in their hometown. Discrimination began in the local schools, and other institutions usually offered that as justification of their own racist policies. To Hastie that was an observation, not an excuse. "It doesn't make the Community Center look any better. But it does show up the essential rottenness of our school system once more."[71]

The system was still rotten in the forties. Black schools were overcrowded, while white schools had room to spare. Buildings, textbooks, equipment, athletic facilities—there was no end to the advantages that white schools had over black ones. The board of education frowned on school visits among blacks and whites and forbade interracial debates. It denied the Office of Price Administration the use of a school to train an interracial group of clerks and compelled the withdrawal of a white girl from a vocational course that was available only at a black school. The board went to such pains to prevent interaction between whites and blacks that in 1947 a presidential committee concluded: "Official freezing of the segregated school system is complete."[72]

Hastie challenged the board of education indirectly by praising the widely publicized efforts of authorities in Springfield, Massachusetts, to promote democracy through projects that fostered understanding and tolerance among students at all levels. As president of the District of Columbia branch of the NAACP, he challenged the local board directly. In November 1944 he sent to it a resolution adopted on behalf of the branch's eighty-five hundred members and their allies. The resolution urged the board to increase "interracial respect and understanding" among students and teachers. The recent observance of American Education Week had been marred by the segregation of teachers: whites at McKinley High School heard a white speaker, and blacks at Armstrong High School heard a black speaker, Dr. Rayford W. Logan of Howard University. Logan read the telegram he had sent to the superintendent of schools, Dr. Robert Haycock, whom he hoped would read it to the white teachers: *The sheer futility of talking to separate meetings of teachers on "Improving Relationships" should be apparent to a child in the first grade,* Logan wired. *I trust that this is the last year that this travesty and mockery will be repeated.*[73]

Hastie's NAACP agreed with Logan. The controversy over those teachers' meetings, its resolution stated, emphasized "the fundamental shortcomings" of the school system. While school authorities elsewhere in the nation were promoting racial tolerance, those in the nation's capital were promoting racial intolerance. Hastie and others urged the board "to get in step with current progressive developments in education toward better living and mutual respect and understanding in communities composed of many racial and religious groups." As part of its plan for the initiation of intercultural public education, the local NAACP's Committee on Education named its eleven members as some of the persons with whom it advised the board and school officials to confer regarding that program. Hastie was one of those members.[74]

Hastie seemed to be everywhere. In 1945, when a Citizens' Committee Against Segregation in Recreation (CCASR) was formed under the chairmanship of his former high school teacher, E. B. Henderson, Hastie served on its legal committee along with Leon A. Ransom and George E. C. Hayes, its chairman. At the time Hastie was chairman of the legal committee of the organization that spearheaded CCASR's formation, namely, the local NAACP branch. In Hastie, Ransom, and Hayes the CCASR had skilled advocates who were to persuade the federal court of appeals in Washington to open amateur boxing matches to blacks. Moreover, Hastie possessed other skills of value to CCASR. He chaired its steering committee and, on behalf of its thirty-two organizations, presented its general statement at a meeting of the District of Columbia Board of Recreation, protesting the board's segregation of recreational facilities. He argued that there was no legal requirement of segregation and made the following telling points:

> *We are entitled to look to governmental agencies in Washington, the national capital, to set an example of decent and democratic behavior for the nation and for the United Nations which we assume to lead.* The one community above all in which the American people and the people of other nations are entitled to anticipate and to find American ideals put into practice is the capital of the nation. It is noteworthy that even as this issue is being debated, civic and business groups are urging that the United Nations, representing millions of non-white people, shall make Washington their international headquarters.[75]

Although the Interior Department's acting secretary, Abe Fortas, had reiterated his department's opposition to segregation on federal land (on which fourteen of the board's units functioned), Assistant Secretary Oscar L. Chapman said that "intelligent colored leaders"

agreed that some things could not be rushed. Fortas did not mention—
and perhaps Chapman had in mind—swimming pools (two operated by
Interior) or golf courses that the department permitted blacks to use,
but with restrictions on showers and other facilities. Hastie believed
that the board clung to antiquated rules. "At best, it is not a persuasive
argument that an evil should continue because it has existed in the
past. . . ."[76]

All this could be taken as part of the Roosevelt "do nothing" policies
toward blacks that led Hastie to threaten to resign, in 1944, from the
Political Action Committee of the Congress of Industrial Organizations,
which worked for the President's reelection. Of concern to Hastie, for
example, was Roosevelt's refusal to establish a Fair Employment Practices Committee (FEPC) until a "March on Washington" threatened by
A. Philip Randolph left him no choice.[77] One controversy that fell to the
FEPC involved the Capital Transit Company, about which the Temporary Committee for Improved Public Transportation (TCIPT) complained to the Public Utilities Commission in Washington.[78] Chaired by
Hastie, the TCIPT coordinated twenty organizations. Its complaint,
which the Public Utilities Commission said had to be filed with the
FEPC, was that Capital Transit refused to hire black drivers. FEPC
scheduled hearings two and a half years after it received the complaint.[79]

Things had not been going well for the FEPC. Its secretary had
advised blacks to be less "introspective" and more concerned about
"much bigger issues"; three Southerners had been appointed commissioners, while a militant black lawyer, Earl B. Dickerson of Chicago,
had not been reappointed; Roosevelt refused to override the comptroller
general, who had held that federal contracts need not contain an antidiscrimination provision; and hearings on discrimination in railroad
employment had been postponed.[80]

Hastie criticized the postponement of the railroad hearings. The
FEPC, he argued, was doing "excellent work," though assuring equal
opportunity to women was still an unmet challenge. When the NAACP
and thirty organizations held a rally for a permanent FEPC, Hastie
presided. In testimony before a Senate subcommittee the next year
(1945), he called for a permanent FEPC. A sense of fairness may suffice
someday, he said, "but I fear that will not be in our time." Although
important, persuasion was no substitute for sanctions. When the FEPC
was about to impose sanctions on the federal manager of public transportation, who had been appointed after President Harry S. Truman
seized Capital Transit in his professed apprehensiveness about a wild-

cat strike, the President ordered the FEPC to withhold its directive to that official. An editorial in *The Crisis* stated: "Truman Kills FEPC."[81]

Hastie counted Truman among the national leaders who favored civil rights—but not to the point of enraging reactionaries. "In this process 'practical politicians' tend not to be greatly concerned that the price they pay is the sacrifice of fundamental principles of a decent and democratic society and the basic rights of human beings."[82] That sacrifice of principles was evident in Washington in the form that Hastie dubbed "the cornerstone of the racist structure in America," namely, segregated housing. Exorbitant rent, quarters unfit for animals, widespread displacement by public works, rampant tuberculosis, infant mortality twice that among whites—these were the black's lot. Agnes E. Meyer wondered how such uncivilized things happened in her city.[83] Many people thought the answer was restrictive covenants, which were compacts signed by realtors and property owners requiring their unanimous approval of the rental, sale, leasing, or other conveyance of property to blacks. Some signers were motivated by racism, others by the mistaken belief that black occupancy or ownership reduced property values.[84]

Federal policies aggravated matters, and after appeals by Hastie and the NAACP to Presidents Roosevelt and Truman brought no relief, Hastie recommended to congressmen that they no longer leave these matters to whimsical and capricious federal administrators, who used federal authority and money to mold policy to suit their racial preferences. Largely because of those administrators, public housing was entirely segregated in Washington, where the National Capital Housing Authority (NCHA) followed the "community pattern" in building homes, and where the National Housing Agency, with which NCHA contracted to build homes for war workers, placed orders according to occupants' race.[85]

Among defense workers, blacks were kept on long waiting lists even though there were vacancies in units reserved for whites. NCHA required the impossible. It demanded six months' advance notice about the housing needs of defense workers by agency and by race. It also required persons in substandard private housing occupied through necessity rather than by choice to obtain, as a condition of vacating the premises, a court order substantiating those conditions.[86]

Hastie understood the resulting hardships. He had as clients the families of Arnold B. Green, Nelson M. Love, and Julius Turner, all of whom had applied in vain to NCHA for housing that was vacant because whites had not responded to the agency's efforts to rent it. The

Green family of five lived with seven other persons in an apartment consisting of two rooms, a kitchen, and a bath. The seven Loves shared one room, and the seven members of the Turner family lived in two basement rooms. The three family heads worked in the Procurement and Accounting Division of the Office of the Secretary of War. "That government should through willful racial discrimination impose such suffering upon its own citizens and employees is as clearly unlawful as it is morally shocking and reprehensible," Hastie said. He requested immediate relief for the families and for all applicants in similar difficulty.[87]

After the NCHA refused to offer Hastie's clients any of the hundred units that had been vacant for months, he appealed to the National Housing Agency and to the Federal Public Housing Authority, but without high expectations. Those agencies, he told the Senate Banking and Commerce Committee, were racist. Their policies were reprehensible. So were those of the Washington Real Estate Board, which prohibited members from selling to blacks land in mainly white neighborhoods; the two newspapers that would not carry advertisements of such sales; and the banks and loan companies with which they did business. But the biggest of all barriers to open housing was restrictive covenants.[88]

Hastie thought of nullifying covenants through court action. Legal precedent, however, held that individuals could discriminate as they pleased, that the Constitution protected contracts effectuating that predilection, and that since private, not state, action was involved, the Fourteenth Amendment did not prohibit judicial enforcement of those contracts. Nevertheless, NAACP lawyers thought they could persuade the Supreme Court to hold that enforcement of covenants by courts constituted state action prohibited by the Fourteenth Amendment. Eager to try, Hastie welcomed the case of Clara I. Mays in Washington. "I believe this will be the leading case in this field," he told Pauli Murray. "Be sure to read it."[89]

Clara Mays needed suitable housing for herself, her ill sister, her sister's four children (aged two to ten), and her nephew, who had been discharged after four years in the Army. A government worker, she bought a home in February 1944. She made the purchase through Jane Cook, a "straw party" for Consolidated Properties, and Geneva K. Valentine, a black realtor. The United States District Court for the District of Columbia gave Mays sixty days to vacate the home.[90]

On appeal Mays's attorneys argued that changes in the neighborhood made the original goals of the covenant unattainable, that the covenant unlawfully restrained commercial transactions, and that it did not bind the parties involved. (Neither Mays nor Cook had signed it.) The covenant offended public policy, the lawyers added, and violated the Fifth, Thirteenth, and Fourteenth amendments, as well as the implementing federal statutes. Chief Justice Groner of the Court of Appeals acknowledged the seriousness of racial problems but said that their resolution—certainly in private affairs—"cannot be through legal coercion . . . but must be the result of . . . a voluntary consent of individuals. And it is to this end that the wisest and best of each race should set their course."[91]

Although Justice Miller's vote made the result two-to-one against Mays, he said that the Supreme Court should advise his court whether the law should be reinterpreted. He thought that Justice Edgerton's dissent had been "persuasively presented." Edgerton had said that the covenant was unenforceable by a court of equity because its purposes could not be realized. Instead of maintaining property values it depressed them. And the neighborhood was no longer all white.[92]

In Edgerton's opinion the covenant was contrary to public policy in that it unreasonably interfered with the right to sell housing, "a necessity of life." Was enforcement of a covenant the same thing as enforcement of a deed? Edgerton thought not. More so than the incorporation of restraints (on sales) in a deed, their incorporation in a contract (covenant) made possible their quick and easy application to "unlimited quantities of land . . . By holding that such a restraint may be imposed in such a way this court is not simply following precedent. It is adding an unfortunate extension to an unfortunate doctrine." A housing emergency existed for 187,000 blacks in Washington, Edgerton added. "We cannot close our eyes to what is commonly known . . . Since restrictive contracts and covenants are among the factors which limit the supply of housing for Negroes and thereby increase its price, it cannot be sound policy to enforce them today. . . ."[93]

JEWS AND CATHOLICS MAY BE NEXT the Washington *Afro-American* announced in its coverage of attempts to take the appellate court's decision to the Supreme Court. At that time Hastie officially joined Mays's legal team, which consisted of James A. Cobb, George E. C. Hayes, and Leon A. Ransom. The *Afro-American* had cut to the core of things. So had Edgerton's dissent: "In accordance with the familiar principle of 'balancing equities,' the fact that an injunction will cause extreme hardship to the defendant [Mays] without *commensurate* benefit to the plaintiff is in itself a sufficient reason for denying an injunction."[94]

Hastie, Thurgood Marshall, and Spottswood Robinson prepared the petition for a writ of *certiorari* in hope of persuading the Supreme

Court to hear an appeal. The petition asked whether a federal court could disallow Mays's purchase of the home solely because of a contract that a former owner—but neither Mays nor her grantor—had signed long before, with the intention of denying blacks the right to win or occupy the property.[95]

Hastie was satisfied with the petition. The brief, however, was "very uneven and not a model for young lawyers," though it probably would be persuasive with the Supreme Court. "Actually," he told Pauli Murray, "I think the best arguments for certiorari are Justice Edgerton's dissent and Justice Miller's concurrence." If the Court agreed to hear the case, the brief would be revised. Spottswood Robinson says that *Mays* was the case in which the arguments against covenants were best presented. He thought *certiorari* would be granted. So did Hastie.[96] "The mandate has been stayed to permit us to file a petition for a rehearing," Hastie told Murray. "It is only a straw, but we have nothing else to catch hold of." On 28 May 1945 the Court broke the straw. It rejected the petition.[97]

The Supreme Court's rejection of Clara Mays's petition was another of the countless humiliations suffered by blacks in Washington. The Court's decision was announced several months after Hastie, as chairman of the NAACP National Legal Committee, had expressed suspicion that "undercover anti-Negro activities" were being "deliberately fomented and inspired" in that city. The milieu was perfect for such activities. In Washington, for example, a "black" dog was denied burial in a "white" canine cemetery. On a winter morning in Washington in 1945, two women set out for a city hospital, walking because they could not get a taxi, reaching a church-supported hospital just as one of them experienced severe labor pains, and having the hospital turn her away—it did not admit blacks, but it did provide a sheet to cover mother and child at the place of birth: its front sidewalk.[98]

The "Impression" at Hastie's law school was given by lawyers who were friends as well as comrades in battle. They enjoyed their work and, in the main, each other. Most were exceptional attorneys, but even among those Hastie was extraordinary. "Thurgood always believed that when we got together at the law school to work on a case or anything else, the first day should be set aside for telling all the stories, hearing all the lies, playing poker, drinking whiskey, and just having a good time," Elwood H. Chisholm recalls. He adds that Hastie took part in the revelry. "Bill was a regular guy, and yet he had dignity that compelled respect. Even as a young man he just radiated it."[99]

Hastie reminded Chisholm of Joe Louis. "If the Brown Bomber knew you, then you could cuss him, cuff him, and in those days when it was a fighting word you could even call him 'black.' But you knew that you were around the champ." Commenting on the wealth of legal talent and admirable personal traits in the group, Chisholm said that there are many fine lawyers and people "but there are just some in whom you perceive greatness. Nothing that they ever say or do in your presence or elsewhere ever belies this appraisal. In fact, everything confirms it."

Thurgood Marshall held an equally high opinion of Hastie but never hesitated to give it the twist that characterized the special blend of comradeship that linked them. Hastie had moved on from Howard by the time that a newspaper photo caption identified him as Marshall. This prompted Marshall to write:[100]

> My dear Governor:
>
> Not knowing whether or not this obvious libel was inspired by you, I am giving you an opportunity to first disclaim it in order that I may decide whether to sue you jointly with the *Ohio State News* or to sue them individually. Of course, if you desire to make a private settlement, I am always willing to exclude one joint tortfeasor, providing he makes it worth my while to do so. If not, I will be obliged to sue you both.
>
> Unless I hear from you immediately, that is, within the next twelve hours, I will be forced to take the necessary legal steps to protect my interests. My reputation was severely damaged by Carter Wesley [their ally in the Texas white primary struggle], but what little reputation he left for me has been destroyed by this latest libel. In other words, I have got to collect from somebody, and since you have a regular government-backed salary, I see no reason why you should not be the one.
>
> With love and hisses, I am
>
>   Respectfully (?) yours,
>
>   s/ Thurgood Marshall, A.B., LLB, LLD, etc.
>
>   Thurgood Marshall
>   Counsel for Thurgood
>   Marshall and his
>   children.

The faculty at Hastie's law school, like their allies with the NAACP, had more than humor and ability in their favor. They also had courage. His days as a law student behind him, Oliver W. Hill opened his practice in Richmond. As an NAACP lawyer he often felt "skittish" but only once "seriously threatened." In August 1939 he and J. Thomas

Hewin, Jr., were helping blacks in Greenwood County, Virginia, to register to vote in the Democratic primary elections. Throughout one Saturday afternoon they argued the case, only to have their petition for a write of mandamus denied. But the judge told them that he would be in court on election day, the next Tuesday, and would grant the relief sought if they detected any abridgement of voting rights. That word spread throughout the county. And on election day Hill and Hewin returned.[101]

"We were riding around the back roads to the various outlying precincts to see to it that Negroes could vote. And we came flying down this daggone old road and made a turn, a blind turn. We looked down that narrow old road and started down this hill . . . and, boy! You never saw so many redneck crackers in all your daggone *days*. They had poles and ropes . . . and we had a 1936 Ford with a rumble seat. And I looked at Tom and I said, 'Look here, this is it.' And he said, 'It sure is.'

"There was just nowhere for us to turn around so we just kept on down and went in through this damned crowd. There was a little old bridge—one-way bridge, you know—and we rode across this blamed bridge. By the time we got started across the bridge we saw what the situation was. Somebody had run over the cliff there in an automobile and they were trying to fish the car out of the water. That was the reason for all the poles and ropes. But by the time we got across the bridge Tom was so weak that he just stopped and sat . . . I tell you, I've never been so frightened in all my days. We really thought we were facing it at that time. I've been scared a lot of times, but nothing ever like that."

Endangerment of life and limb was widespread for civil rights lawyers. For example, Ransom "didn't even know what had hit me" after he walked into the hall in the Davidson County Courthouse in Nashville in 1942. He and Z. Alexander Looby, counsel for the local NAACP, were representing a black whom an all-white jury had convicted. As soon as Ransom stepped out of court, R. D. Fessey, a white former deputy sheriff, attacked him. Several persons who would have gone to Ransom's aid were held at bay by a white former constable who, gun drawn, yelled, "We are going to teach these Northern Negroes not to come down here raising fancy court questions."[102]

Fessey was arrested but released the next day on one thousand dollars bond. He was scheduled for trial on charges of aggravated assault. Mayor Thomas L. Cummings of Nashville vowed that "complete justice will be done." But Fessey was not indicted by the Davidson County grand jury, whose foreman was said to have been one of three armed white men who had attacked three persons in the same courthouse the day after Fessey's assault on Ransom.[103]

In some Southern towns a black lawyer could always try a case if the court granted permission and a white lawyer sponsored him. "As long as your sponsor was standing, you could talk," says Judge Billy Jones of East St. Louis, Illinois. "But if he sat down, that was it: You had to sit down." And he would sit down the second you got out of line. The South was as bad as it was portrayed. "Segregation was *deep!* It was *rooted!* You took your life in your hands."[104]

Or put it in the trunk of a car. Thurgood Marshall did that more than once as he sneaked into or out of town. "That man risked his life more times than the average soldier on the front lines," says Charles T. Duncan.[105]

One time was November 1946.[106] Marshall was in Columbia, Tennessee, to help Looby of Nashville and Maurice Weaver of Chattanooga defend Lloyd Kennedy and William Pillow in the second trial resulting from a race riot in February 1946. Four policemen were wounded during the riot; Kennedy and Pillow were accused of having shot one.

The mood in Columbia was ominous. People stayed off the streets. Police cars swarmed over Mink Slide, the black section. The prosecutor told the jurors that acquittal of Kennedy and Pillow would assure lawlessness and the murder of their wives by blacks. (This prosecutor had threatened to assault Leon Ransom with a chair during the first of the riot trials several weeks earlier.) After the acquittal—of Pillow only—a white man dashed away from court, saying that the jury's failings had to be corrected. Before hurrying away, two reporters from Nashville gave a third, Harry Raymond of the New York *Daily Worker*, some advice: Get out of here—now! A lynching seemed certain. And people remembered the prediction that if Marshall and Weaver remained in Tennessee, they would remain at the bottom of Duck River.

Darkness had fallen when Marshall, Weaver, Looby, and Raymond drove away from Columbia in Looby's car. Just outside of town they saw a state police car and another car parked. "I told Looby to look out the back and see if the state car was following us," Marshall said later. "Just about that time I heard a siren. I pulled over to the side. A car, the highway car, pulled up. Another car came up, and then another."

The police searched Looby's car for liquor in that dry county. Marshall and his companions watched to make sure they did not plant some in the car. The police did not have a warrant to search the car's occupants, and Marshall denied their request to do so. The officers released Marshall and his companions, then immediately stopped them, arrested Marshall for drunken driving, and put him in one of the cars.

Two deputies sat in front and two flanked Marshall on the backseat. They drove away fast in the car owned by Deputy Sheriff Lynch.

The other police cars drove away on the main highway, but Lynch's car cut off onto a side road. Looby, Weaver, and Raymond had been released, but Looby trailed Lynch's car until it reached the magistrate's office in Columbia. Looby headed for Mink Slide to arrange bail for Marshall; Weaver and Raymond went into the magistrate's office. Upon being politely requested by Weaver, representing Marshall, the magistrate smelled Marshall's breath. "What's up here?" he demanded. "Why this man hasn't even had a drink!"

The magistrate freed Marshall from police custody, though not from the other menace—Duck River. Not that it mattered. Black lawyers such as Marshall, Ransom, and Hastie gave no thought to themselves from the battlefield. As Brenda S. Spears has said, the battle itself gave them "reason to rally, strength to fight."[107] Hastie believed it was important for liberal lawyers to analyze developments in order to show how futile and harmful were "all efforts to buy reaction off cheaply." He thought that they belonged "in the vanguard of that struggle."[108] Those who wished to be there had only to call Dupont 6100 and ask for the dean. He would welcome their help in combating circumstances such as those that cost Mary Turner her life.

CHAPTER THIRTEEN
A Fighter for Us

Mary Turner encountered the mob near Valdosta, Georgia. Having failed to find its intended victim, the mob lynched her companions. One black man had escaped his hanging jury; Mary Turner's husband and two other black men died in his place. The lynching caused her to scream, so the mob threw her into the fire, too. And as flames engulfed Mary Turner, her unborn baby fell from her body and was trampled— as white parents lifted their own children into the air for a clear view.[1]

Unlike the mob near Valdosta, the one in Sikestown, Missouri, found its intended victim. It broke into the jail in the racist town and snatched the black man from his cell. It tied him to a car and dragged him feet first through the black section of town. Then it soaked him with gasoline and set him afire. On 25 January 1942 Cleo Wright was lynched in broad daylight. On 13 May 1942, as arranged by United States Attorney General Francis Biddle, evidence about the lynching was submitted to a grand jury. The jurors decried the "shameful outrage" but returned no indictment. In a report to the National Lawyers Guild, Hastie and Thurgood Marshall commended the Department of Justice for finally starting to investigate lynchings but criticized its methods in presenting evidence to grand juries. Given "a rehearsal of evidence" against Wright, this grand jury had concluded that he was guilty of the charges on which he had been awaiting trial. "In this case," the grand jury reported, "a brutal criminal was denied due process."[2]

Cleo Wright had faced accusations that cost many black men their lives. He had been accused of attempted rape, a charge that made the blood of white men boil figuratively to the point that they made the blood of black men boil literally. Hastie had written:

175

I rather suspect that in the minds of most white men the 'race problem' means the problem of 'intermingling.' . . . I know of [a] gentleman who was interrogated [about intermarriage] by a group of whites assembled to consider the race problem, and who startled his hearers no end by suggesting that the richness of a clear brown complexion was the ultimate in feminine charm. Some of the other ideas which the Negro entertains upon this subject would further tend to allay some of the alarm which the subject excites, though they would not be altogether flattering to 'Nordic' vanity.[3]

Mere accusation of a bruising of this Nordic vanity sufficed to send lynch mobs on their bloody missions, often spurred on by fans who joined them in tearing away victims' fingers, toes, teeth, and bones, to be kept as souvenirs, and who swarmed around mutilated bodies put on public display. Sexual outrage did not account for all lynchings; other reasons for this beastliness were accusations or charges, proved or unproved, that the victims had committed murder, had disagreed with a white person, or had taken legal action against someone white. But the sexual accusation was most effective in sparking oppression of every kind, including lynching.[4]

At congressional hearings, at mass meetings, and in his newspaper column Hastie supported federal antilynching legislation, which he thought worthwhile if it did no more than raise public consciousness about lynching and show blacks that effective organization could help them gain political objectives. He argued, moreover, that "if the citizen has any rights which the state must protect, certainly the state must protect his life, to which all other personal rights are incidental." Countering states-rights arguments against federal antilynching proposals, he reminded the Senate Committee on the Judiciary in 1940 about the Mississippi deputy sheriff who led the posse that beat three suspects into a bloody pulp before hanging them from a tree until they were almost dead. He also told the committee about the Alabama law enforcement officers who removed prisoners from jail cells to a location at which a mob awaited them. About each instance Hastie asked, "Is this a State right?" The pertinent state right, he insisted, was to protect citizens against lynch mobs. In fact, that was not just a right but a duty imposed upon the state by the federal Constitution.[5]

Southern states refused to discharge that duty. Lynchers were not merely bloodthirsty hoodlums; they were also instruments of political repression. Terror discouraged blacks from even trying to share the principal means of control over government, which Hastie considered voting. But in Southern states government did not leave disfranchisement of blacks to the mob; rather, it took that responsibility upon itself, fulfilling it through poll taxes, literacy tests, and other devices, particularly the white primary. Nothing, said Hastie, reveals the essence of democracy as does the history of black suffrage: "Its essence is eternal struggle." In that struggle the greatest obstacle to blacks was the white primary. "Not a party primary but a race primary," said Ralph J. Bunche.

This race primary was so effective that it justifies the mentioning of Elbert Williams and Lonnie E. Smith's fate in the same breath. In July 1940 these black men tried to register to vote: Williams in Brownsville, Tennessee, and Smith in Houston, Texas. Both were lynched: Williams by a mob, Smith by a registrar. Politically the figurative lynching of Smith was as effective as the literal lynching of Williams: Neither man voted.[6] And other blacks got the message intended by the mob in Tennessee and the state in Texas.

The white primary originated in Texas, where, until 1903, political parties controlled their activities, including primary elections. Party executive committees and conventions exercised this control until 1903, when the state legislature vested it entirely in the party executive committees in each county. In 1921 the Supreme Court held, in *Newberry v. United States*, that primaries were not elections safeguarded by the Constitution, and two years later the Texas legislature amended the state election law, disqualifying blacks as voters in Democratic primaries and voiding ballots cast by any who somehow managed to vote. In 1927 the Supreme Court invalidated this legislation in a suit brought by Dr. L. A. Nixon, a black physician living in El Paso.[7]

Unable itself to bar blacks from primaries, the state immediately authorized the Democratic state executive committee do so. Empowered to specify eligibility for party membership and for voting in party primaries, the executive committee restricted eligibility to whites. Nixon contended that this amounted to state action that disfranchised him. In 1932 his suit reached the Supreme Court, which supported Nixon's immediate contention but undermined his ultimate purpose. The power that Nixon contested could not be exercised by the party's state executive committee, said the Court. If it existed, that power inhered in the party's state convention.[8]

Within three weeks the state convention adopted a resolution excluding blacks from party membership and from primaries. The challenge by R. R. Grovey, a barber, reached the Supreme Court in 1935. The Court held that exclusion from party membership was not state action prohibited by the Fourteenth Amendment; political parties, it said, were private or voluntary associations. This decision, Thurgood

Marshall wrote, "was a rude jolt to the political aspirations of Negroes. But hope did not die out. . . ." Both sides in Grovey's case agreed that primaries were equivalent to general elections, and it seemed to some that sooner or later the Court would abandon the fiction that placed primaries out of the reach of federal law. That abandonment occurred in *United States* v. *Classic* (1941) when the Court held that in primaries involving candidates for federal office, improper denial of participation was punishable under the criminal sections of the civil rights laws, whether the interferer was acting in a public or private capacity. In *Classic*, said Marshall, the Supreme Court finally "pierced the façade of legality which had shielded primaries from the reach of Federal laws regulating the conduct of elections. . . . [Its] opinion followed and adopted the very arguments which had been rejected in the *Third Texas Primary Case*; and the Court, without a single reference to *Grovey*, practically overruled it."[9]

Classic did not concern a white primary, but it was a godsend to those seeking to bypass *Grovey*. Walter White had thought that federal legislation might be the solution to the problem posed by *Grovey*. Hastie, assisted by Nathan R. Margold, drafted two bills, but neither was introduced in Congress. In 1936 the Republican National Convention refused to adopt a plank calling for loss of congressional seats for states that disfranchised blacks, and President Roosevelt would not speak out against the white primary. Thwarted at every turn in the political arena, the NAACP welcomed *Classic*, a fateful inadvertence whose "great value" Hastie said was that it provided a route around *Grovey* by safeguarding voting rights against private as well as state action. To use *Classic*, the NAACP had to show that primaries were essential to the selection of nominees and that they determined the outcome of general elections (as they had in all but two congressional and gubernatorial elections in more than eighty-five years). With *Grovey*, said Hastie, the Supreme Court meant to destroy "any reasonable hope" of overturning the white primary. That made it necessary to first reverse that decision. "If, at the same time, we can win a victory as decisive as the last defeat, all the better." And that meant applying *Classic* to the facts of the Texas primary.[10]

Classic was a mixed blessing. It promised a way around *Grovey*, but it added fuel to dissension within the NAACP's ranks. Black Texans had been criticizing the NAACP for its reliance on white lawyers (to the exclusion of black lawyers) in the initial cases. As chairman of the NAACP's National Legal Committee Hastie knew this. He also knew the wisdom of taking steps made advisable by *Classic* but opposed by black Texans. On 14 January 1941 they had filed a challenge to *Grovey*.

The lawsuit was *Hasgett* v. *Werner*. Because it did not concern congressional elections, it would not have permitted a broader base than the Fourteenth and Fifteenth amendments. Black Texans had invested a lot of time, money, hope, and energy in *Hasgett*. But *Hasgett* had to be dropped. A case based on *Classic* would have much brighter prospects. It fell to Marshall and White to sell that strategy to the Texans.[11]

In Texas Marshall needed all his forensic skills. "[You're] messing up the case," he was told. One Texan reportedly told him that "[you] had better win the next case or [don't] return to Texas." On behalf of Dr. Lonnie E. Smith, a dentist, the new suit was filed by Marshall, W. J. Durham, Carter W. Wesley, and H. S. Davis, Jr. It alleged that because they were blacks Smith and other qualified voters had been refused ballots by election officials for the Democratic primary in Harris County. Smith sought damages amounting to five thousand dollars, nullification of the white primary as unconstitutional, and a permanent injunction against it.[12]

The lower federal courts ruled against Smith. In the district court Marshall argued that victory in the Democratic primary virtually guaranteed election in Texas. *Classic*, he said, mentioned both *Nixon* cases but not *Grovey*, which it ignored while cutting "all the ground work [sic] out from under it. . . ."[13] He and Hastie would have their chance to tell that to the Supreme Court itself. On 10 November 1943 the Court gave Hastie and Marshall half an hour to make the fourth attack in twenty years on the white primary. The opposing side presented neither counsel nor briefs. Hastie and Marshall argued that the Constitution and federal law, as interpreted in *Classic*, prohibited election officials from interfering with Smith's voting in primaries that were "an integral part" of electoral choice and effectively controlled it in congressional elections. Marshall opened and Hastie closed the argument. Hastie asserted that the record in *Grovey* had not contained facts necessary for "an adequate legal appraisal of the so-called 'white primary.'" Several justices turned toward the author of the *Grovey* opinion, Mr. Justice Owen J. Roberts, laughed, and made comments that reddened his face. They asked Hastie and Marshall no questions. Roberta Hastie and other spectators expected Roberts to reply to Hastie's hammering at *Grovey*, "but at every blow he grimly closed his lips closer."[14]

The Court ordered reargument on 10 January 1944 to allow the attorney general of Texas to argue *amicus curiae*. Hastie did not know what to make of this unusual order, but Charles Wesley in Texas saw it as a sign that the Supreme Court would side with Smith. Wesley expected the opposition to focus on developments in Texas, emphasizing *Grovey* and trying to avoid *Classic*. Accordingly, he told Marshall, "I predict that you will miss Durham, who knows the Texas situation, when you get ready to argue in January, but I am sure you will not dare ask him to return since he didn't get to argue the first case."[15]

Unlike Durham, whom he had replaced, Hastie had not argued *Smith* before it reached the Supreme Court. Although he thought that Durham should be present for the reargument, Hastie believed it best that he himself and Marshall again present the case. "If we could be sure that the Court's questioning of a particular attorney would be limited to the Texas cases and practice, I would agree heartily that Durham should argue. However, in my judgment, the two critical points are: (1) our theory of state action in the light of the Classic Case, and (2) the theory under which we claim that the plaintiff can recover even if there is no state action."[16]

He explained another consideration to Wesley. "Thurgood and I have had the advantage of repeated discussion and argument between each other and with other lawyers on all aspects of these matters. We have criticized our own position and have argued the matters among ourselves again and again. Unfortunately, the problem of distance has made it impossible for Durham to participate in most of these discussions. For these reasons, I believe that if the Court should cut loose on us we will be in a better position to answer questions." He appreciated Wesley's forthrightness and wanted to be straightforward himself. "To me one of the most pleasant aspects of the NAACP work has been the fact that we have a group of lawyers who have both the ability and independence to think for themselves and the mutual respect which makes possible free and frank discussion of any and all differences of opinion."

Durham was present for the reargument, for Marshall wanted to calm the Texans. There was, Marshall said, "a lot of running off at the mouth. We can always stick together when we are losing, but tend to find means of breaking up when we are winning. . . ."[17] But Marshall and Hastie would not break up. They were a good team, though a study in contrast. "I know of instances where he's gone for days trying to find a word," Marshall says. "Me, I'd go minutes. If I didn't find it, I'd put another word down."[18]

The contentions to be made in their brief were distributed to NAACP lawyers for refinement. About one assigned to him Marshall says: "I worked on it and worked on it, and got it in good shape, I thought. Then I cut it down to twenty pages, and eventually I thought I did a beautiful job: I cut it down to six or eight pages." Hastie said it had to be cut even more. "And we argued, and I in very polite fashion said, 'Well, goddamn it, if it's going to be shortened, you shorten it.' And he said 'Will do.'

And over a weekend he cut those six or eight pages to a paragraph. And then he said, 'Now find something I haven't covered.'"[19] Hastie was a stickler for accuracy, says Marshall. "He used to every now and then say, 'Goddamn it, Thurgood, you didn't check that out.' And it would be just a little thing about that big [holding thumb and forefinger a fraction of an inch apart]."[20]

Having checked everything out in *Smith*, Hastie and Marshall awaited word from the Supreme Court. On 3 April 1944 word came. Roberts had found his tongue by then. Thinking of *Grovey*, he said that rather than being considered "a restricted railroad ticket, good for this day and train only," an earlier decision should be respected by the Court. Hastie questioned the appropriateness of that view regarding *Smith*. The Court should follow precedent "for a considerable period," he said years later, but "a very strong conviction of legal error and consequent injustice" justifies departure from it. "Only the most stupid or the most stubborn men never change their minds."[21]

Hastie and Marshall had avoided a frontal assault on *Grovey*. More than the Fifteenth Amendment, they stressed Article 1 and the Seventeenth Amendment, in keeping with *Classic*. Having established constitutional authority in that fashion, they emphasized Section 31 of Title 8 of the United States Code, which guaranteed qualified electors the right to vote "without distinction of race, color, or previous condition of servitude; any constitution, law, custom, usage, or regulation of any State or Territory, or by or under its authority, to the contrary notwithstanding."[22]

Hastie and Marshall argued that any person who, because of another's color, denied him/her the right to vote in primaries violated federal law. They did not characterize the Democratic party as a state agency, but the Court held that it was. They neither emphasized in their brief nor mentioned in their argument primary elections for state office, but the Court's decision covered state as well as federal offices. By invoking the Fifteenth Amendment, which applies to state and federal elections, rather than Article 1 and the Seventeenth Amendment, which apply only to federal elections, the Court overruled *Grovey* and gave the broadest possible protection to the voting rights of blacks. Indeed, by using the Fifteenth Amendment instead of Article 1 (and skirting the matter of state action), the Court extended its decision beyond voting. Looking back to the oral argument, Hastie thought it significant that justices had interrogated counsel about the state-action concept in other instances in which individual conduct might be regulated by agencies which, although ordinarily considered to be private, functioned under state authority.[23]

At first Chief Justice Harlan F. Stone assigned Justice Frankfurter to write the opinion. Stone then received from Justice Robert H. Jackson a note sent "only with the greatest reluctance and frank fear" that it would be misunderstood. Jackson wrote that Frankfurter "unites in a rare degree factors which may unhappily excite prejudice": He is a Jew, he is from abolitionist New England, and he is thought to be unsympathetic to the Democratic party. "We deny the entire South the right to a white primary, which is one of its most cherished rights," said Jackson. "It seems to me very important that the strength which an all but unanimous decision would have, may be greatly weakened if the voice that utters it is one that may grate on Southern sensibilities." Stone assigned the opinion to Justice Stanley F. Reed of Kentucky.[24]

By an eight-to-one vote the Court held that the Fourteenth and Fifteenth amendments forbade racial discrimination in primaries as well as in general elections. Hastie took Frankfurter's concurrence, in the result only, to mean "that he believes the situation should have been analyzed along the lines of our brief." State officials requested a rehearing. The request, which Hastie thought "merely shows that little things like rules never embarrass the rebels," was denied. At the airport in Miami, en route from the West Indies, he bought a newspaper as he hurried to catch his plane. He was aboard before he saw the headlines that proclaimed the victory in *Smith*. "I am sure the people on the plane thought I was crazy because I just let out one whoop, and had it not been for the seat belt I would have gone straight up in the air."[25]

At Harvard Law School Erwin Griswold welcomed the decision in *Smith*. "I was never able to understand how *Grovey v. Townsend* could have been decided, and I am glad that it is now formally put out of the way," he told Hastie. "Your part in that result was a real public service."[26]

Smith alone made Hastie one of the all-time greats among black jurists, says Darlene Clark Hine, author of the definitive study of that struggle. The victory brought from William Henry Huff, an attorney in Chicago, the comment that Hastie and Marshall deserved "more credit than you will ever receive here below; but, there is always an abundance of satisfaction in knowing that future generations will benefit from the labors and sacrifices you make in . . . breaking down and weakening the gigantic walls of hate."[27]

The significance of *Smith* was also clear to James Marshall, whose father, Louis, alone and with Moorfield Storey had assured the preparation of briefs and made oral argument in the five NAACP cases that the Supreme Court heard between 1913 and 1927. Louis Marshall and Moorfield Storey had donated their services at a time when it was unfashionable for white lawyers to take civil rights cases and impossible for black lawyers to carry the burden. They laid much of the legal foundation from which Hastie, Marshall, and other lawyers later attacked the white primary. Louis Marshall, for example, argued in 1925 that far from being purely private associations, political parties were inherently public entities.[28]

Louis Marshall argued the first case against the Texas white primary. Death prevented his taking on the second case, which his son took to the Supreme Court. Having done that, James Marshall appreciated *Allwright*. "I hope [it] will be a mounting block for active participation by colored people in the elections."[29]

"I only hope that the Texas Primary decision will prove as important as our enemies seem to think it is," Hastie replied to James Marshall.[30] Maybe it would not only enable blacks to vote but also weaken their attachment to the lily-white Republican party. "Since our enemies are so greatly fearful of the consequences of the decision," he told Channing Tobias, "we certainly must do everything possible to prevent them from being disappointed."[31]

Tobias, who thought the decision alone justified the costs of all NAACP operations in 1944, had mentioned one problem in preventing that disappointment. He wrote Thurgood Marshall, "I noted with interest and pleasure your prompt reminder to Attorney General Biddle that we expect him to put the machinery of the Department of Justice to work in the interest of securing compliance with the laws regulating voting in all the states as affected by the Supreme Court decision."[32]

Perhaps Biddle was influenced by Charles Fahy, the solicitor general, who made known his views about helping out in *Smith*. We won *Classic*, Fahy said in effect, and that gave blacks the leverage they needed. Should we do even more for them and offend many Southerners? "I think not." President Roosevelt agreed, and that was reason enough for Biddle not to follow through on the Court's decision. But Biddle's rationalization was larger than that. Convictions would be hard to obtain; election officials would act in good faith; mediation was better than compulsion; successful implementation of the decision rode on public opinion.[33]

Biddle's department "is, to put it mildly, reluctant to act," said Hastie in 1945. "There are powerful leaders of the Democratic party in and from the South who know that they can hold office just so long as the Negro is kept from the polls. The influence of these men is very great in Washington."[34]

Hastie and Marshall knew what to expect from Biddle, who, it was said, was "too strongly influenced by the professional white solvers of race problems."[35] They had asked him to submit a brief in an NAACP case (Marshall thinks it was *Smith*). As they sat in his office, Biddle began to explain the legal limitations on his actions. They listened.[36]

"See, Bill had this marvelous control," says Marshall, adding that Hastie would listen, never interrupt, and nod occasionally. Thinking he had Hastie cornered, the speaker would say, "Well, then you agree," and Hastie would answer, "No." If asked, Hastie would explain—but only if asked.

Hastie listened to Biddle's reading from law books. "This went on for about five minutes," says Marshall. Then Hastie leaned forward, tapped his cigarette over an ashtray, and deliberately grunted, a signal that "he was getting ready to get you." To Biddle, in a tone very polite but very firm, he said, "Mr. Attorney General, you may assume—and rightly so—that Mr. Marshall and I know the basic principles of law." Marshall recalls that Hastie emphasized each word. "I'll never forget that. Incidentally, we didn't get the brief."

Black editors hailed the *Smith* decision and their warriors: Hastie, Marshall, and the other NAACP lawyers. If the South wanted to scream, that was just fine. "Let its leaders yell bloody murder. We licked them fairly and squarely in the courts because, first, we were right and, second, we had better brains than they." The more the South shrilled, said Hastie, the more blacks wanted to vote, and the more they participated in politics, the more likely it was that the South would become the ultimate leader of liberalism in America.[37] In court he had noticed black lawyers who had been sent to Washington by their communities in Alabama, Louisiana, and South Carolina. Their assignment was to prepare for action back home by learning as much as possible from the *Smith* experience. This quickened Hastie's pulse. "Such indications of strategic planning for the future are among the most hopeful indications of our increasing social intelligence."[38]

The significance of *Smith* cannot be overstated. "It broke the back of the white primary", says Judge Robert L. Carter of the United States District Court in New York City. Jack Bass and Walter De Vries, authors of *The Transformation of Southern Politics*, feel that "[it opened] the way for black political participation throughout the South, a central force in the region's political transformation during the next three decades." No other victory so paved the way for the protection of black rights.[39]

Smith was important in another respect. It represented a major step toward legitimizing the entire NAACP legal-redress campaign, for it was, as Judge Constance Baker Motley explains, among the "cases of the first importance" that the Supreme Court heard. "One of the reasons we were able to move the Court was because we were able to demonstrate legal competence and not have the Court feel that it was performing some kind of political function," she explains. "The Court had to have some legal theory on which to proceed to grant some rights to blacks. And when it saw lawyer-like briefs and not briefs of mere orators or political advocates, then it felt secure in taking the action that it took."[40]

Judge Motley thinks that Hastie's "great contribution" was his authorship and critique of briefs. The magnitude of this contribution is impossible to determine. Many briefs that do not bear his name do bear his imprint.[41] *Morgan* bears both. Irene Morgan encountered what Hastie had called "the indignity which every Negro traveler knows may be his own lot tomorrow." It had been his in 1943, on the Chesapeake and Ohio railroad. As the train sped from Detroit to Columbus, he entered the dining car. At most tables sat two or more persons, but in the center of the car a diner sat alone at a table for four. Hastie walked toward that table. From the far end of the car the steward hurried to intercept him.

"Would you like a seat for one?" he asked.

"Yes," said Hastie, and sat down.

The steward pointed to a table for four at the end of the dining car and suggested that Hastie join the three blacks seated at it. "I declined to move," recalled Hastie, "indicating that my table was quite satisfactory." The steward insisted: company regulations. Hastie refused: Ohio laws.

"You're asking for trouble," the steward told him, the same trouble another black passenger got.

"What's your name, anyway?" the steward demanded. Hastie remembers his response: "I replied that I would be glad to give him my name if he would give me his."

Silence. A short interval.

"It being apparent that I would not be cajoled or intimidated, the steward placed a service check on the table and the meal proceeded without further incident."[42]

It was not for food alone that Hastie had entered the dining car. Too few of the handful of blacks who could afford to dine on trains did so, said Hastie. Maybe they were unaware that rights they did not use were rights they stood to lose. "Not a small part of the task of the Negro in America is learning to walk without timidity, neither looking for trouble nor avoiding it, and taking full advantage of the privileges of citizenship wherever he may be."[43]

When Spottswood W. Robinson III, a lawyer in Richmond, told him about Irene Morgan, Hastie recognized her as a kindred spirit. On

16 July 1944 she bought a ticket at Hayes Store in Gloucester County, Virginia, and boarded a Greyhound bus. All the seats were occupied. Another black woman invited Mrs. Morgan, who was returning to Baltimore for postsurgical care, to sit in her lap. In Saluda Mrs. Morgan took a seat vacated by a white passenger. Estelle Fields, a young black carrying an infant, sat down next to her. Immediately, R. P. Kelly, the driver, loomed over them. Twice he demanded that they relinquish their seats to a white couple who boarded the bus at Saluda. Irene Morgan offered to exchange seats with whites who occupied seats behind hers, the third from the rear. She would even move to the rear seat, but she would not surrender her seat and stand, as he insisted.[44]

Kelly left the bus. He returned with the sheriff and deputy sheriff, who had a warrant for Irene Morgan's arrest. "I asked them what the warrant was for," said the twenty-eight-year-old mother of two. "They both grabbed hold of me, each one tugging at one of my arms."[45]

According to the county court record, as the sheriff tried to read the warrant, Mrs. Morgan snatched it from him and threw it behind her seat. Estelle Fields moved out of the seat—she had tried to move when the driver first approached them, but Morgan had restrained her. As the officers forced her from the bus, Morgan tried to hit the deputy sheriff. She missed him but kicked the sheriff's leg three times. They took her across the street to jail, where she was held for six hours until her mother posted a five-hundred-dollar bond.[46]

Three months later, on October 18, Mrs. Morgan was found guilty on two charges: resisting arrest and violating state law requiring racial segregation of passengers on interstate buses. She paid the hundred-dollar fine and costs adjudged in the first charge but appealed her conviction (and ten-dollar fine plus costs) on the second charge, contending that the state law was inapplicable to her as a passenger in interstate commerce.[47] Precedent was against Irene Morgan.[48] The Supreme Court had never invalidated a state law that *required* segregation in transportation and had never validated one that *prohibited* segregation. The rationale of previous cases had never been applied to bus transportation, but there was no reason to think that it would not be applied when *Morgan v. Virginia* reached the Court in June 1946.[49]

What glimmer of hope there was for Irene Morgan was the result of an uncommon stroke of luck under common circumstances. "When Irene Morgan told me that she had been charged with violation of the segregation ordinance, I couldn't believe my ears," says Robinson. The charge had always been disorderly conduct, which meant that the ordinance could not be contested. "I looked at the charge sheet, and there it was: violation of that ordinance."[50]

Robinson, Oliver W. Hill, and Martin A. Martin took the case. They lost. On appeal they did not contest the requirement of separate accommodations, nor did they dispute state regulation of commerce within its borders. Instead they argued that the state could not regulate interstate commerce, and that Morgan was therefore exempt from the ordinance. Abram P. Staples, the state attorney general, countered by arguing that carriers, not persons, were the subject of the commerce clause in the Constitution. The carrier had not complained, he added, and Morgan could not complain in its behalf.[51]

Staples argued that no federal law dealt with segregation in interstate commerce. Three bills prohibiting segregation in interstate travel had died in congressional committee. "This indeed is an eloquent silence," he said. And where Congress was silent the state was free to affect interstate commerce short of regulation. The segregation ordinance at most caused inconvenience but no burden on interstate commerce. Even if such burden resulted, Morgan's challenge was "academic." She had been convicted of violating the carrier's regulations, not state law, which required drivers to reseat passengers who disobeyed the regulations demanding segregation.[52]

Hastie had anticipated that argument. The clue to it had been given in a dispute that was considered an ideal challenge to the Virginia segregation law. Marianne Musgrave, Erma McLemore, Angella Jones, and Ruth Powell had not budged when ordered by the driver to move to the rear of a bus in May 1944. Their conviction in magistrate's court in Vienna was upheld in Fairfax County Circuit Court. The judge, hearing that they would appeal, reduced by half the fine (forty-five dollars) that had been imposed upon each of the Howard University coeds. The state appellate court granted the writ of error sought by their lawyer, Leon A. Ransom. The attorney general dropped charges. "It may well be that the plan is to advise all carriers to issue regulations and to rely upon them in segregating passengers," Hastie told Pauli Murray.[53]

When Staples revealed the plan, Robinson said that state law left carriers no choice but to adopt the regulations and drivers no choice but to enforce them. Persons providing interstate transportation and those purchasing the service were equally engaged in interstate commerce, and even if the state law did not affect Morgan directly, its indirect effect justified her attacking its constitutionality. Again they lost. On 19 November 1945 the Supreme Court agreed to hear their appeal.[54]

Hastie and his colleagues described for the Court the conditions created for passengers and carriers by segregation laws in various states. They asked the Court to consider a journey from Pennsylvania to Mississippi. In Pennsylvania passengers were not segregated, and when they reached the Maryland line the law permitted interstate, but not intrastate, travelers to remain unsegregated. In the District of Columbia all passengers were free to sit where they pleased, but all were required to return to segregated seating in Virginia. Once in Kentucky, interstate passengers were unsegregated, intrastate passengers segregated. In Tennessee interstate travelers were segregated. No one knew what the seating would be in Arkansas. Unaffected by local law in Louisiana, interstate passengers were affected by local law in Mississippi. The logistics of travel were onerous to passengers, the details of compliance were burdensome to carriers, and all this adversely affected interstate commerce.

Hastie and his allies argued that the state appellate court's ruling that the Virginia law was consistent with the commerce clause contradicted the Supreme Court's prohibitions against state action concerning segregation on public interstate carriers. The Supreme Court's decisions had established the principle that interstate passengers must not be hindered by state law "predicated upon provincial notions of social policy." Virginia's location made its law particularly undesirable. Through that state passed all north-south travelers along the eastern seaboard and between the nation's capital and the South and Southwest.[55]

When the Supreme Court declared its readiness to hear *Morgan*, Robinson was ineligible to argue it. Appointed by Hastie to the faculty at Howard Law School in 1939, he had combined teaching with part-time practice in Richmond and had not met the requirement for admission to practice before the Supreme Court, namely, three years' practice in a state's highest court. Hastie and Marshall therefore took on appeal a case they had not handled in the lower courts. That is a hard assignment for a lawyer, says Judge Murray I. Gurfein of the United States Court of Appeals for the Second Circuit. "In gambler's parlance, the cards are stacked against him." Under Robinson's tutelage Hastie and Marshall mastered (says Robinson) "a different order of constitutional mystery."[56]

As always, careful thought went into the selection of the lawyer who would conclude the NAACP's argument before the Supreme Court. It would not be Houston; there was no telling what he would say once he was caught up in argument. In another case Houston had been interrupted by Mr. Justice James C. McReynolds. "I don't understand your point," McReynolds told Houston. Without missing a beat Houston replied, "You've never been a Negro."[57]

The man for *Morgan* was Hastie. "His courtly reserve was emblematic of an ability to approach problems from a perspective divorced from the emotions of a lifetime—emotions felt deeply, but kept completely under control in his personal and professional relationships," Robinson has written. "Calm and dispassionate presentation of ideas was undeviatingly his manner, and clarity of expression his trademark."[58]

So on 27 March 1946 Hastie and Marshall stood before the Court. Marshall presented the facts of the case. Then Hastie, said *The Crisis*, provided "a cool rejection" of Staples's thesis. The distinctive qualities that Robinson admired in Hastie were evident. Among them was "the skill in dialectic" that another distinguished jurist considers "a useful tool" of appellate lawyers. Those who possess it often display their special sparkle in response to questions from the bench. This jurist remembers United States Attorney George Z. Medalie's reaction to his suggestion that Medalie include a point in his opening argument. "No," said Medalie. "The point is so vital that they are bound to ask it. It will be more effective if I give our position in answer to a question."[59]

Eventually Mr. Justice Wiley E. Rutledge asked Hastie whether *Morgan* was subject to the Fourteenth Amendment. Robinson "winced inwardly" because he knew that Hastie "was bursting with arguments against *Plessy*'s separate-but-equal doctrine which he thought were irrefutable."[60]

"I pretended not to hear him," Hastie said. Rutledge repeated the question. "I gave him fifteen minutes of irrelevancies." Rutledge persisted, but Hastie would not nibble at the bait. As Robinson says, *Brown v. Board of Education* would eventually vindicate Hastie's conviction about *Plessy*, but *Morgan* was not the challenge they intended to make. The Court probably would have reaffirmed the separate-but-equal doctrine. "We did not want to make bad law," Robinson says.[61]

So Hastie did not take a chance on losing *Morgan* by digressing into *Plessy*. He kept the Court's attention riveted to the commerce clause. "The law is a peculiar thing," says Secretary of Transportation William T. Coleman, Jr. "A lot of times it is what you *don't* say that counts."[62]

It is, of course, also often what you *do* say that counts, Coleman adds. And it was in the brief for *Morgan* that Hastie had already said so much. Mindful that the Nazis had planned to do what the racists were already doing, he had written that all Americans were "joined in a death struggle against the apostles of racism . . . [and united] to promote universal respect for, and observance of, human rights and fundamental freedoms for all without distinction as to race, sex, language, or religion." He suggested that it must be clearer than ever that interstate commerce should not be marked "by disruptive local practices bred of racial notions alien to our national ideals, and to the solemn undertakings of the community of civilized nations as well."[63]

But what would the Court say? On 3 June 1946 it said that blacks did not have to sit in the "Negro" section of interstate buses. As Judge Louis H. Pollak was to say, Hastie and Irene Morgan's other champions had nudged the Court "the way in which great lawyers help [it] to move, incrementally, to the place where the Constitution will at last be found, and, in this instance, was to be found eight hears hence, when *Brown* v. *Board of Education* gave to our land a new birth of freedom."[64]

Robinson says that *Morgan* illustrates Hastie's indispensability in NAACP litigation. Hastie was unsurpassed in logic and debate, "yet the talent we chiefly sought lay elsewhere" and was found in "the rare and precious qualities that were distinctly his." No matter how unclear or emotionally charged, the words of others took on "a cool but incisive thrust" after conversion by him. "This is not to say merely that he was a great legal composer; it is to say a great deal more." Melding propositions of law and aspirations of blacks, Hastie created "prescriptions for social change" that jurists found legally and morally persuasive.[65]

Robinson's list of cases that bore the Hastie touch and shaped the development of this nation includes *Morgan* v. *Virginia* and *Smith* v. *Allwright*, which advanced liberty in transportation and voting. To these Robinson adds, in particular, the cases that secured for blacks the opportunity to receive professional education equal to that available to whites; guaranteed equal pay as public school teachers; upheld a New York law against discrimination in labor organizations; required nondiscrimination by labor organizations that are the exclusive bargaining representatives of a craft or class; and outlawed racially discriminatory housing covenants. Judge Motley recalls that of the nineteen cases that Thurgood Marshall, representing the NAACP or the NAACP Legal Defense and Educational Fund, Inc., argued before the Supreme Court between 1939 and 1949, Hastie figured as co-counsel or consultant in twelve. Asked to specify the ones in which Hastie was important, Marshall said, "All of them."[66]

But Robinson is quick to add that no list of cases gives a true measure of Hastie's achievement. To appreciate Hastie as "a jurist of the very first rank" we must understand the civil rights alliance that consisted of Howard University Law School, civil rights organizations, and individual attorneys who brought civil rights cases from all over the country to Howard, knowing that help would be available there. "Hastie was a charter member of this informal but closely-knit group," says Robinson, "and one of its most faithful and ardent adherents. . . ." And it was while he was its dean that the law school became, in Robinson's judgment, "headquarters for a legal collective bred by a shared purpose and united by mutual respect."[67]

There were, in the forties, two citadels of civil rights in the United States. One was the NAACP, the other was Howard University Law School; and both were what they were largely because of William Hastie, whose admirers prayed that the gods would favor the champion who came forward, "a fighter for us."[68]

CHAPTER FOURTEEN
Caribbean Outpost

Hastie was still savoring the victory in *Morgan* when he emerged from the Saturday-morning conference at Howard University in June 1946. Mordecai W. Johnson's secretary had not dared to interrupt the president's meeting with administrators, but when it ended she told Hastie that reporters had been telephoning him. He returned a call and learned why they were eager to talk to him. "We'd like to have your comment on your appointment as Governor of the Virgin Islands."[1]

Hastie had no comment. On 27 November 1945 Secretary of the Interior Harold L. Ickes had asked if he would like to succeed Governor Charles Harwood. Hastie had not thought about returning to the islands; he enjoyed his work at the law school. He said that he would not seek the appointment, but consented to Ickes's submitting his name to Roosevelt.[2] Ickes thought Harwood's appointment "a tragic joke."[3] By the time he asked about Hastie's willingness to become governor, Ickes had decided that Harwood spent too little time governing in Charlotte Amalie and too much time "ingratiating himself" with congressmen in Washington.[4]

Roosevelt's death made it necessary for Ickes to persuade President Harry S. Truman to appoint Hastie as governor. Truman sat on Ickes's recommendation and follow-up memorandum because he believed that Hastie had opposed Roosevelt in 1944. Ickes had no knowledge of Hastie's having done that, but he did not think Roosevelt would have been much concerned about it. Besides, having discussed the nomination with Hastie, he thought it inadvisable to withhold it. Truman agreed to send Hastie's name to the Senate.[5]

A subcommittee of the Senate Committee on Territorial and Insular Affairs opened hearings on Hastie's nomination on 20 March. Ickes's successor, Julius A. Krug, endorsed Hastie in a statement read by his solicitor, but Senator Ralph O. Brewster of Maine was unimpressed.[6] Above all else, said Brewster, he wanted to determine the probability of Hastie leading the islands out of their economic distress, but without following Rexford Guy Tugwell's socialistic example in Puerto Rico. No question of Hastie's ability was involved. "I don't think his ability can be challenged; certainly not by a fellow Harvard man," said Brewster.

If Brewster did not make the challenge, Louis Lautier wondered who would. "Certainly not Eastland." Senator James O. Eastland of Mississippi had "attended" three Southern universities and had "studied" law somewhere, sometime, somehow. "By Harvard standards, he is virtually an illiterate." Ellender, then? Senator Allen J. Ellender of Louisiana held master of arts and bachelor of law degrees. But St. Aloysius College and Tulane University were not Amherst College and Harvard Law School.[7]

Brewster had no qualms about Hastie's competence or color, but having "heard or read in the papers quite serious questions raised by your ideological affiliations and views," he asked Hastie, "What have you to say about these?"[8]

Hastie was then vice president of the Washington Committee for the Southern Conference on Human Welfare, the Washington Council for Community Planning, and the National Lawyers Guild. He had been chairman of the NAACP's legal committee and an editor of the *National Bar Journal* and the *Lawyers Guild Review*. He had been on the President's Caribbean Advisory Committee since 1942. Confronting Brewster and three other senators—Abe Murdock of Utah, Ellender, and Eastland —Hastie replied that "politically, I am a voteless resident of the District of Columbia, and I don't have any party affiliation." Clearly, there was room for societal improvement. "But I think that the American Constitution and laws are one of the great landmarks in world progress and in government, and I shall certainly hope and anticipate that we shall continue to be just that."[9]

Hastie's comment satisfied Brewster. Eastland, however, could stand it no longer. Apparently following advice received from the House Un-American Activities Committee, he hurled questions at Hastie. Had Hastie been active in the Abolish Peonage Committee in 1940? Hastie could recall no such role. What about the National Lawyer Guild? Hastie identified himself as a national vice president. The Washington Committee for Democratic Action? Hastie had cooperated with it but was not a member. Had Hastie helped to sponsor a conference on constitutional liberties in 1940? Not that Hastie recalled, but perhaps. Eastland asked whether Hastie had been a member of the National

Negro Congress (yes), the Southern Conference for Human Welfare (yes), and the New Negro Alliance (yes).[10]

Hastie hardly batted an eyelash as he fielded the questions. Eastland persisted. Had Hastie, at 8:30 P.M. on 16 April 1936, at the Metropolitan Baptist Church in Washington, given a speech under the auspices of the Washington-Scottsboro Defense Committee? "I may have . . . I was extremely interested in the Scottsboro case and in the efforts made on behalf of the defendants in that case." Had Hastie supported President Roosevelt's national defense program between 1939 and 1941? "I would not have accepted an appointment in the War Department had I not been wholeheartedly prepared to do so."[11]

Eastland was ill at ease. Shifting in his chair and adjusting his glasses, he asked, "In 1944, did you hold a demonstration against blood bank segregation?"

Turning to squarely face Eastland and looking him dead in the eyes, Hastie answered, "I did, sir."

"Under the auspices of what organization did you appear?"

"Under the auspices of the District of Columbia Branch of the National Association for the Advancement of Colored People."[12]

Hastie had been the president of the local NAACP in 1944 when, along with five other organizations, it sponsored a protest against blood segregation. The demonstrators marched to the blood bank in the Acacia Life Insurance Building in northwest Washington. They donated three hundred pints of blood and protested the blood segregation policy, which the Surgeon General of the Army had characterized as scientifically meritless. Because the demonstrators had no permit, the police forbade them to hold a meeting in the park across the street from the Acacia Building; without a permit no more than four or five persons would be allowed to assemble there, and the demonstrators were two hundred strong. Since Acacia prohibited their assembling on its premises, the police told the inquiring Hastie that the demonstrators had to disperse. He saw to it that they did.[13]

If asked, Hastie would have provided details about this demonstration at the blood bank. But no one asked, and the hearings recessed at 12:25 P.M.[14] For two hours a pair of women had paid keen attention to it. Quiet, her expression unchanging, beautifully attired in a brown and yellow dress, Beryl Hastie studied her husband's interrogators as they grilled him. She thought things would go well. As for the other woman, "They tried to catch my boy napping," said Roberta Hastie, "but they didn't."[15]

They first tried early in the session when Brewster interrupted Hastie's opening statement. Hastie, trying to help them understand his attitudes, began with comments about his antecedents and childhood, for he believed that childhood impressions endured throughout one's life. He told the senators that his paternal great-grandmother, who, after the death of his grandparents, had raised Hastie's father, "instilled in him, as I believe he did in me . . . a great belief in the power of the individual in the American scene through his own industry, determination and courage to achieve a decent life for himself and his own. That impression has never left me; it never will."[16]

The impression, Hastie continued, had been reinforced by his mother's older brothers—one a cobbler, the other the proprietor of a small store—who took up arms to protect their family and friends against terrorists in Reconstruction Alabama. "From them, too," said Hastie, "I have gotten a tradition of the self-reliance, industry and courage which can make for success in the American scene." His mother and his father strengthened the tradition. Hastie then summarized his career since Harvard and offered to discuss the Virgin Islands. But Brewster interjected the question that everyone knew would be asked, because it hit a sore spot with Southern senators. "Before you come to that, under what circumstances did you terminate your services as Civilian Assistant to the Secretary of War?"[17]

Hastie offered to read, in full or in part, the statement he released the day after he resigned from the War Department. "I think we had better have the full statement," said Brewster. Hastie read it. But he sidestepped the attempt by Brewster and Ellender to elicit further criticism of military policies. In fact, he described Stimson, Patterson, and Marshall as men who were understandably preoccupied with the larger concerns of war but nevertheless showed "sympathetic interest" in his area of responsibility. Then Hastie quickly changed the subject of the discussion to the Virgin Islands.[18]

Asked by Ellender whether he had a plan for making the islands self-sustaining, Hastie said he had none. But there were possibilities: expanding handicrafts; having government help private industry to provide jobs and commodities or, as with the Virgin Islands Company, doing so itself when the choice was between having people work on public projects or having them go on public welfare; returning to the Virgin Islands (as to Puerto Rico) federal taxes on rum; and developing the tourist industry.[19]

Tourism is to be a major industry in the islands, Ellender said to Secretary Krug at the session on March 28, "and the question arises as to how much that trade will expand with a colored man as Governor."[20] Irritated, Krug reiterated his opinion of Hastie as the best man to be governor. Ellender's slur did not discombobulate Hastie.[21] Tourists, he explained to Ellender, do not deal with the governor; they deal with "the minor officers, who, regardless of the color of the Governor's skin, happen to be colored."[22]

The subcommittee heard three more witnesses endorse Hastie's appointment: Roy Gordon, chairman of the Municipal Council of St. Thomas and St. John; Ashley L. Totten, president of the American Virgin Islands Civic Association; and Hope R. Stevens, general counsel of the National Negro Congress. Then came Leslie F. Huntt. He claimed to represent a group of "progressive people" who would not proceed with plans to make substantial investments in the islands if Hastie became governor. Given the death struggle between democracy and communism, he said, "we consider the political implications of Mr. Hastie's appointment at least as important as the racial."[23]

Attempting to paint Hastie red or at least pink, Huntt relied on documents supplied by an investigator with the House Un-American Activities Committee, but he was ignorant of the contents of the documents and was unable to produce backup witnesses. He wilted under questioning by Senator Murdock until, glowing with slyness, he offered as evidence an article written by Benjamin J. Davis, Jr., in support of Hastie's nomination. That restored the interest of Eastland and Ellender. Davis had left the Republican party and had become a communist because he was embittered about the mistreatment of blacks, Hastie said in response to questioning about his friendship with the New York City councilman. Ellender wondered if an attitude similar to Davis's accounted for Hastie's having left the War Department. No, said Hastie, for racial problems could be resolved within our constitutional framework.[24]

"If you thought that, why did you walk out?" asked Ellender.

"Because I thought I could do more as a private citizen who could speak publicly than as a Government officer who would not."

"In what respect—by agitating the subject, or what?"

"If the Senator calls public statements agitation, yes."[25]

After Huntt had called Hastie an able legal adviser, Senator Murdock asked if he also considered Hastie an educated, liberal, tolerant man. Huntt did. A grin spread across Murdock's face as he noted that those were Huntt's own criteria for the selection of a governor. Huntt was bewildered; Hastie remained calm; Eastland chewed his pencil.[26]

Huntt had additional reason to squirm when Frederick Dorsch, executive secretary of the Chamber of Commerce of St. Thomas, testified. Dorsch said that Hastie would be decidedly an asset in improving the economy of the islands, that Huntt represented no more than 5 percent of the businessmen there, and that Huntt had lied in saying that he had no role in raising opposition to Hastie's nomination and received no compensation for appearing before the subcommittee. The organization that Huntt claimed to represent, the Committee for Economic Development and Public Cooperative Service, was unheard of in the Virgin Islands. Dorsch described bright prospects of economic development, including the tourist industry, under Hastie.[27]

Hastie made his final comments on April 5. He rebutted personal accusations, giving the greatest attention to the innuendo that he practiced or favored communism. He said it was dysfunctional to label all liberals pro-communist, for that would make it all the more difficult to distinguish subversive from loyal liberals, should that ever be necessary. But something more was on his mind.[28]

"I would like to say this final thing with reference to the participation of Negroes generally, and myself in particular, in specific projects with persons of varying political views." Blacks were not to blame for the proscriptions heaped upon them, regardless of their political affiliations. Their primary goal was to free themselves of the proscriptions. "And I would say that in most cases we do not even know, we certainly do not inquire [about] the political affiliation of our fellows in that effort. Some may be Republicans, some may be Democrats. Some may be socialists and some may be communists.

"But in our minds, if a person sincerely and honestly is willing to join in such a worthwhile effort we say to him, 'Welcome, brother. We are glad to have you; we are glad to work with you.'"[29]

On May 1 the "rather vigorous fight," as Hastie described the dispute over his nomination, culminated in his confirmation by the Senate. He said it had served two valuable purposes: It exposed the red-baiting tactics that victimized thousands, and it unified Virgin Islanders. The outcome electrified the islanders, and they set about arranging a "monster celebration." In Charlotte Amalie, while children searched for flowers, adults searched for bunting to decorate the Triumphal Arch in Market Square, and taxicabs sprouted tiny American flags. Day and night the community band practiced for the parade and reception on inauguration day. Preparing to join policemen and sailors in an honor guard, veterans dug out uniforms and spruced up. On St. Croix preparations were made for the governor's visit on the day after his inauguration.[30]

On Wednesday, May 15, accompanied by Secretary Krug and other officials, the Hasties and their nineteen-month-old daughter, Karen, flew from Washington to San Juan. On Friday they flew to St. Thomas.

Just before noon their C-47 army transport landed and was greeted by the cheers of thousands. Bearing Governor Tugwell, other officials from Puerto Rico, and newsmen, two additional planes emerged from the rain clouds. Acting Governor Morris F. De Castro, United States District Judge Herman E. Moore, and Municipal Council Chairman Roy Gordon led officials in greeting the Hasties, who were the first passengers to alight, and their party. A bouquet from the Charlotte Amalie High School class of 1937, her class, was presented to Mrs. Hastie by Angela Durant. Karen promptly plucked petals from the roses.[31]

Nothing could spoil inauguration day. Not even the drizzle that began as the Hasties and De Castro, in an open car, led the colorful parade westward to Charlotte Amalie, down Main Street, decorated with flags and banners and lined by thousands of festively attired onlookers, under the Triumphal Arch, and into Market Square. The Hasties then went to Government House (the governor's mansion and office) for a brief rest before going to Emancipation Garden for the inauguration. The governor waved "with a confident, debonair air" to the four thousand persons who cheered him and his wife as he and the First Lady moved to their seats on the bandstand.

Hastie took the oath of office from Judge Moore, signed the written version, and kissed his wife. Then, standing within a stone's throw of the place from which the emancipation proclamation had been read ninety-eight years earlier, Governor Hastie gave his inaugural address. As usual, he was optimistic despite the odds against success. "I believe we are ready . . . to make these Islands hum with constructive activity."[32]

He had no program for economic development, but he did have "certain fundamental ideas" to govern its formulation and implementation. If Virgin Islanders were to prosper, they would have to consume and export more of their own products and import fewer goods and services—but not at the sacrifice of such "spiritual values of human life" as dignity, civility, lawfulness, or civil rights. "We shall welcome whatever helps our economy, with the one limitation that our people must not be demoralized, nor their human rights infringed thereby."

That tall order was marvelously received, though not fully grasped, by the crowd in Emancipation Garden. Perhaps that was as it should have been on this festive occasion. After a few more speeches, a basket of gladiolas for the First Lady, a bit of rain, and a lot of cheering and music, the ceremonies in Emancipation Garden ended. Early that evening the Hasties hosted a reception to which the public was invited, an unprecedented action that brought six hundred guests to Government House.[33]

The next day (Saturday) would be spent at festivities on St. Croix; in Frederiksted and Christiansted the gubernatorial party would be warmly and impressively hailed. But the celebration was capped the night of the inauguration (Friday) in the capital, Charlotte Amalie, at Bluebeard Castle Hotel. For then and there, after the state dinner, the inaugural ball was held. The governor, wearing a white dinner jacket and black trousers, and his First Lady, wearing a black evening gown, headed a cast of two hundred celebrants. They danced "an unorthodox CARIOCA with a distinct Howard University flavor," and when the ball ended at one in the morning "they were still going strong."[34]

As the festivities ended, governing began on a disquieting note. Virgin Islanders welcomed the appointment of a black governor. But some would have preferred the deletion of the qualifier "acting" from Governor Louis Shulterbrandt's title. In February 1946 Ickes had elevated him from commissioner of finance during the temporary absence of Morris De Castro, who served as governor during the period between Hastie's nomination and confirmation. De Castro and Shulterbrandt were Virgin Islanders. To many natives Shulterbrandt, a black, was nearer the ideal appointment than was Hastie. "He is one of our boys," Ralph Matthews, a black reporter from the mainland, was told. "Hastie is one of yours."[35]

There were additional bad omens for Hastie. Virgin Islanders were addicted to politics, the more acrimonious the better, and Hastie was not in the best position to do battle with his inevitable opponents. He was subject to congressional whim, and on St. Thomas every governor had to know when and how to push or placate a municipal council that one reporter likened to "a baby's rattle," the prize at stake in local campaigns. Since Virgin Islanders would be proud to have a native as First Lady, Hastie's marriage might work to his political advantage. Not necessarily, however. She was a Lockhart, and for many identified the Lockharts' interests with those of the mulatto group on which darker natives blamed much of their economic trouble. Indeed, color and caste consciousness would be an obstacle to Hastie. For example, his civil rights record would not work much to his advantage because natives did not hold of whites an opinion comparable to that held by blacks on the mainland.[36]

Dire forecasts were plentiful. According to J. Antonio Jarvis, the newspaperman and historian, "the most optimistic well-wishers of the new administration wait to draw an easy breath." Divisiveness among Virgin Islanders flourished, and they were committed to democracy to the extent that it served their personal purposes. Cofounder (with Ariel Melchior, Sr.) of the *Daily News* in 1930, Jarvis also commented, "If the newspapers permit him to succeed, Governor Hastie should have a noteworthy administration." James A. O'Bryan, an activist, predicted sabotage of Hastie's administration by reactionaries.[37]

Similarly, the *Photo News* forecast that out of the ranks of Hastie's allies would spring "the most vicious hatchet men," whose treachery would hasten his resignation in disgust. "Even the most sycophantic, obsequious, hypocritical, deceitful executive never pleases everybody, and Hastie is a straight-shooter, who 'calls 'em as he sees 'em.'" The natives expected Hastie to immediately usher in the millennium, Peter Edson, a mainland columnist, wrote, adding that the governor would have more trouble with blacks, "who expect too much and are probably doomed to disappointment," than with whites, "who expect too little."[38]

On Friday, May 17, Hastie entered a political lion's den. His thoughts about the job ahead were partially encapsulated in his testimony before the Murdock subcommittee. He had no program to solve the islands' economic problems, but their basic cause was that imports exceeded exports. Either alone or in cooperation with other Caribbean communities the Virgin Islands should market their appeal to tourists less interested in "exciting amusements" than in rest in a climatically pleasant place. The Virgin Islands Company, which represented governmental action as a last resort in fending off economic disaster, should be continued even as efforts were being made to entice private industries to the islands. To help the latter become self-sufficient, taxes on rum should be returned to them. Until self-sufficiency was obtained, federal aid would be needed to raise the standard of living to the same level as that existing on the mainland.[39]

The war had ended and things had clearly slackened in the islands, but Hastie wanted to make no comments about corrective action until he saw conditions firsthand. Conditions were somewhat obscured by the dazzling beauty of the islands. They were "a virtual paradise" wrote one columnist. Charlotte Amalie, wrote another, "is a quaint up and down place, overlooking a harbor," with houses painted as though in imitation of "a setting for a Broadway musical." A third saw it as "a lovely tangle of pastel houses and flaming tropical growth, winding roads and stairways—all tumbling down the steep hillside to the vivid harbor below." Bewitching were "the gentle rolling plains," pastures, and sugarcane fields of St. Croix, with "the breathless beauty of sky and sea . . . everywhere," and stunning were the forests, trails, and fantastic beaches on St. John, "an unbelievable haven of peace and quiet."[40]

But the Virgin Islands were a place of poverty as well as beauty. Charlotte Amalie, for example, had not only "pastel houses" but wooden shacks, many of which consisted of one room and rented for three dollars a month. Kerosene lamps were used in seven out of every ten homes. Four types of toilets were in use. Modern flush toilets were in some four hundred homes. In public housing projects were so-called family flush toilets, arranged on a line, flushing automatically and periodically. The "night soil system" required the family to put their can or pan in an outhouse or in the yard, where each night sanitation workers either emptied or replaced it. Some families used the fourth kind of toilet, the Chick Sale-type pit privy.[41]

At the end of his first full year in office Hastie reported that drought, loss of revenue from the sale of rum, and the closing of the submarine base helped plunge the islands into an economic depression. Of the four principal enterprises—shipping, sugarcane, military activities, and the sale of rum—only shipping had improved. But shipping was 2.6 million tons below the volume in 1946 and the dredging and improvement of St. Thomas Harbor, which Congress and the Army Corps of Engineers approved in 1937, had been delayed. Silt clogged the harbor. Its removal being a priority to him, Hastie exercised the privilege he had been given to take matters directly to President Truman. As a result the Bureau of the Budget included in the appropriations bill the dredging project that the Army Corps of Engineers had repeatedly recommended. Hastie thought the Virgin Islands had finally obtained assistance that he considered all-important. But then the congressional ax fell again as the project was scrapped in keeping with pork-barrel politics.[42]

An even keener disappointment for Hastie was the refusal of Congress to return to the Virgin Islands the federal taxes collected on locally produced alcoholic beverages. Assisted by Walter White, he tried to overcome congressional objections, which were (as stated) that the islands should be required to make budgetary requests rather than being assured such automatic funding, and (as unstated) that the consideration sought by Hastie should not be given to a possesion whose inhabitants were 80 percent black. Native legislators did not help matters with their displays of political immaturity.[43]

Unquestionably this immaturity resulted in part from the political inexperience mentioned in the Hibbard Report. As submitted only a few months before Hastie took office, this confidential document to Secretary Ickes had asserted that local legislators had been placed in power without preliminary or continuing instruction in the art of governing. "The effect is just about as bad as the car with a new driver and no instructor. We fix the wrecked car each time it is driven into obstructions but have decided to waste no time in guiding the student."[44]

One obstruction that the student drivers and their teacher, Hastie, tried to remove was bigotry. In 1946 this accounted for the enactment of a civil rights law that forbade discrimination because of race, color, creed, or national origin. Said by Hastie to be the strongest such legislation under American jurisdiction, the law made the suspension of a business license a matter of gubernatorial discretion, which Hastie threatened to exercise in one instance. That situation arose when he learned that a hotel on St. Croix intended to deny accommodations to a representative of a national Jewish organization. Hearing about Hastie's reaction, a white Crucian remarked to her black physician, "Imagine that nigger Hastie trying to tell us how to run our affairs."[45]

This patient spoke for more than one white Crucian. Dr. George D. Cannon, a physician in Harlem, remembers the strategy that called for President Truman, when visiting St. Croix, to be entertained at a private residence. White Crucians were then free to exclude all blacks except one, namely, Governor Hastie. Cannon never discussed this with his friend Bill Hastie, but he could not help but notice that when Hastie hosted a reception for the President on St. Thomas, he invited all the black officials from St. Croix who had been excluded from the luncheon held there at Ward M. Canaday's estate to honor Truman.[46]

It was during Hastie's administration that Arthur S. Fairchild donated Magens Bay Beach and fifty adjoining acres to the people of the Virgin Islands. Having made a fortune on Wall Street while young, Fairchild had begun a search for the perfect place to spend the rest of his life. His sinus trouble made New York unbearable. He divided the world into six areas, lived for at least a year in each, then circled the world twice to visit and revisit places known for their warm, dry climate and for their beauty. One day he came to Charlotte Amalie and knew he had reached the end of his search. For the next seven years a prominent architect designed what would be called Fairchild's Castle high in the hills. Fairchild bought more land, an area one by four miles in size that swept down from a high ridge and included Magens Bay. He traveled the world over collecting all kinds of exotic flora, which he planted in the grove at Magen's Bay. When deciding to donate the beach and land to the people of the Virgin Islands, an important consideration for Fairchild was his belief that Hastie had their best interests at heart. For that reason Fairchild liked dealing with Hastie, and the governor was prominent in the negotiations. Hastie had a big hand in drafting the deed of transfer, served as the first chairman of the Magens Bay Park Authority, and appointed all of its original members.[47]

During Hastie's second year in office economic development continued to lag. A hundred years after the end of chattel slavery, Hastie noted, "economic emancipation" was still a distant goal.[48] Congress had denied his request for the return of taxes derived from the sale of alcoholic beverages and for the dredging of St. Thomas Harbor. He had found it hard to attract new industry, encourage large capital investments, or make exports exceed imports. Although he thought it jeopardized "character, integrity, and human dignity," Hastie emphasized tourism as the islands' brightest economic hope.[49]

Tourists were attracted by the climate, bargain-priced alcoholic beverages, free-port status, and liberal divorce laws of the Virgin Islands.[50] In February 1948 tourism received a boost from President Truman's visit to the Virgin Islands. Although some observers saw the visit as an election year maneuver, Hastie said it indicated the President's "very lively personal interest" in the islands. He and Truman met aboard the presidential yacht *Williamsburg* before exchanging greetings on the West Indian Company dock after the President arrived on 22 February. A lively crowd was on hand, in part because all ministers except one had granted Hastie's request that church services be held in the afternoon rather than in the morning that Sunday. Truman had flown from Washington to Puerto Rico and had boarded the *Williamsburg* there. The Secret Service had planned to bring the President's chauffeur to the Virgin Islands, but a tour over the roller coaster routes there convinced them of the soundness of Hastie's insistence that his own chauffeur, who had served governors since the 1920s, drive the President's car.[51]

A scenic drive with Hastie was one of Truman's many activities. Another was Truman's visit to see his friend Ward M. Canaday, the president of Willy's Overland Corporation, developer of the jeep. Canaday owned vast acreage in the Virgin Islands, raised cattle, and conducted experimental agriculture. Hastie recalled that Canaday liked the Virgin Islands, and had Mrs. Canaday shared this joy, he might have immersed himself even more in the economic development of the Caribbean. Truman appointed Canaday the American chairman of the Caribbean Commission, which Canaday and Hastie served together for some time.[52]

During lunch, which was served on the terrace of Canaday's plantation, the President listened as the leader of a seven-piece calypso band, shaded by a gooseberry tree, beat a lard can while other performers drew music from an exhaust pipe six feet long, two guitars, a gourd, a brass triangle, and a banjo. They sang a song of good wishes to Truman for the coming election. Canaday said the visit was successful largely because Hastie, "with characteristic generosity and thoughtfulness," rescued him from "the anxiety and concern I felt when I learned that my hospitality was involved in the official protocol of the island. . . ."[53]

Virgin Islanders and Truman took to each other. Throngs came to Emancipation Garden to see and hear him. Standing where Governor

Peter von Scholten was said to have issued the emancipation proclamation, Truman spoke in celebration of the hundredth anniversary of that event and unveiled a plaque commemorating it. A banner designating him CHAMPION OF HUMAN RIGHTS flew overhead as he said, "The struggle for freedom is unending and documents alone do not conclude it . . . We must have freedom of the spirit, religious freedom, freedom from want, freedom from fear." He commended Hastie's leadership in that struggle in the islands.[54]

Truman's visit enraged Southern Democrats. They disapproved of both his making the trip and his taking along three black reporters. The three—P. Bernard Young, Jr., Lemuel Graves, and Llewellyn A. Coles—were the only reporters present at the official luncheon given by Hastie for Truman in Charlotte Amalie. Although Truman made some enemies, he also made some friends. He and Hastie spent hours together, often alone. Hastie came to feel "much closer" to him, and his regard for the President, both as a person and as the nation's leader, increased significantly.[55]

As soon as Truman's visit ended, Hastie and the Progressive Guide were again at each other's throat. Hastie was out of his element; he was suited to "gut" politics but not to gutter politics, the name of the game played by his opponents, members of the Progressive Guide. He had seen them play the game when he arrived on St. Thomas in 1946. The one-party system was being rocked then by insurgents who were unwilling to support Progressive Guide candidates for reelection to the Municipal Council of St. Thomas–St. John. The rebels wanted "their turn at the trough," said the *Daily News*. "They represent the same narrow minded intolerant group that [the incumbents] were when they entered office." Why switch to them? No reasons for switching or not switching were given in political platforms. As usual, said the American Veterans Committee, the typical candidate nailed "a few planks (which are full of holes) to his platform" and assassinated character. "Not a single constructive issue has been raised in the whole campaign," said the *Daily News*, "but all the personal and political dirt about each other has been shown in technicolor for the amusement of fans."[56]

A case in point was the Progressive Guide's attack on its chief opponent, Earle B. Ottley. The party said that he wanted to be in the municipal council to improve his chances of obtaining a twenty-thousand-dollar contract to publish its proceedings, while avoiding repayment of the fifteen-hundred-dollar public loan that had enabled him to attend school on the mainland.[57] The Guides did not restrict their attack to the candidate. Don't call us communists, they warned Earle Ottley's brother, manager of the *Photo News*, describing him as "the incongruous Mr. Randolph Ottley, the gentleman who is well known to keep a rendezvous with the moon so very often. . . ." And Carlos Downing added, "I did not make any mistake when I referred to you as a half-wit in response to your 'communist' charge against me." Downing, whose letter to Randolph Ottley was published in the *Daily News*, said it was unfortunate that Randolph Ottley expected to insure his brother's electoral victory on that basis. But Downing was proud of his own life, boasting to Ottley that "when you have lived in like manner you will be at peace with yourself rather than being at odds with your wits."[58]

Hastie deplored such politics, for he believed that self-government was certain to fail wherever judgment was based on personal animosity rather than on the public interest. But the Progressive Guide cared nothing about his views. Roy P. Gordon, chairman of the Municipal Council of St. Thomas-St. John had told the Murdock subcommittee that Virgin Islanders wanted a chance, under Hastie, to show Congress how best to help them become self-supporting. Congress gave them that chance, but as Gordon K. Lewis maintains, they were to show Congress "the peculiar savagery" of their politics, which he attributes partially to the "struggle between colonial status and political power."[59]

The Virgin Islands were somewhat nearer colonial status than self-government. Decisive power was not on St. Thomas but in Washington. Federal departments and agencies functioned as coequals with the governor, who was accountable not to the people but to the secretary of the interior and to Congress, which seemed to be either apathetic or hostile to its subjects. The balance of power between the governor and the legislature (favoring the former) fanned the flames of antagonism, which made life and work difficult for governors.[60] But it did seem at first that things might be different with Hastie residing in Government House.

OFF TO A GOOD START read the *Daily News* headline about Hastie six weeks after he took office. He had appeared before the Senate Appropriations Committee and had recovered most of the funds that had been cut from the budget. He had visited St. John with encouragement for its people. He had urged businessmen to build residential housing. He had coped with a power emergency and suggested a radio station for St. Thomas. He had commended the municipal council for its fiscal responsibility. From eight o'clock in the morning until late in the afternoon he held appointments, taking time out for tennis with staff members. Vindictiveness, pettiness, and compulsion were not characteristic of him; dignity, courtesy, and efficiency were. He welcomed constructive criticism.[61]

"Even skeptics will admit that he is on the ball," the *Daily News* said about Hastie. He had rubbed councilmen on St. Thomas the wrong way by gently criticizing some of their behavior. But the "big-blow-off" would not occur until he took a stand on some volatile issues, the *Photo News* said. When he did, the honeymoon would end.[62]

THE HONEYMOON IS OVER the *Daily News* claimed on September 3. After the councilmen refused to reconsider a pension bill that he thought too expensive for the municipality, Hastie vetoed it. The *Daily News* predicted more clashes because of their philosophical incompatibility. He wanted efficient government, they wanted pork-barrel politics. For instance, he wanted to release twenty-one surplus helpers in the kindergarten program in order to provide better meals for the children, but many councilmen disapproved. He was not the man they remembered as their supporter in the thirties.[63]

At his inauguration Hastie had said that Virgin Islanders could show the world that "so-called dependent areas" were capable of artful and fair self-government. This he tried to facilitate by interpreting the Hatch Act so as to permit governmental employees to participate in political campaigns. He even went so far as to suggest to President Truman that the office of governor be made elective, but voters—led by his opponents—rejected the proposition by more than a three-to-one margin. Hastie favored maximum self-government but with two provisos, namely, that power be wisely and unselfishly exercised and that the people be "honestly critical" of their elected officials.[64] These stipulations, articles of faith with him, were to be a source of friction between him, on the one hand, and the legislators and voters, on the other.

During the 1946 campaign charges of political corruption were hurled by and at the Progressive Guide. Hastie ordered an investigation that led to the trial of Omar Brown, a legislator and member of the Progressive Guide. In February 1947 Brown was charged with defrauding the municipality of St. Thomas–St. John of more than fifty dollars while serving as a committee chairman. He was accused of having represented as expenditures for refreshments payments actually made for oil and gasoline for his own automobile. Tried in March 1947, Brown was acquitted by a jury. For Hastie the acquittal settled the matter—and made a point: "More generally, it should be said that vigilance should never be relaxed. Our best effort should be put forth continually to the end that all officers of the government shall remain above suspicion."[65]

That was Hastie's message to the people. Edward R. Dudley, his executive assistant who had helped District Attorney Croxton Williams in the investigation and prosecution, recalls having received a message in return. "After the trial, the local people said 'If you fellows think that you are coming down here to put our people in jail, you're crazy.'" Hastie and Dudley were not just outsiders; they were continental blacks and therefore were especially not to be considered natives, a racially elastic badge to which even (as with Hastie) marriage into a native family did not entitle them. According to Gordon K. Lewis, it would not have entitled him to this status a quarter of a century later, when every mainland black "finds entry difficult. . . ."[66]

Hastie said that most of the legislators were his friends, "but sometimes we had serious differences." The friendships survived politics, according to Senator Frits Lawetz, largely because Hastie did not throw his weight around. Alphonso A. Christian, Sr., confirms this statement. As the executive secretary of the Legislative Assembly, he was a presidential, not a legislative, appointee. Hastie could have used an item veto to cut Christian's job out of the budget (there was no separate budget for the assembly). "But," says Christian, "he never did anything like that."[67]

Nevertheless, from the moment that Hastie ordered the investigation of charges of corruption, the legislators, sinking to "the nadir of irresponsibility," ignored or opposed his measures without regard to their merit. The Legislative Assembly balked at appropriating funds to effectuate the Merit System Law and Public Health Service programs, and it attacked even more violently in April 1947. Having been authorized by the council to reorganize the government and then report to it, Hastie proposed to discharge fifty old employees and hire eleven new ones at a saving of forty thousand dollars and a predicted increase in efficiency. The *Daily News* hailed him as the first governor brave enough to cut governmental jobs—something long overdue—and saw him ushering in "an inspiring chapter" in the history of the islands. "This is the type of courageous leadership that thinking people want to follow."[68]

Eventually the councilmen realized that they had given Hastie more power than they had intended. He said he would use it. They were horrified. They will block his budget to make him compromise, the *Daily News* said. "A squall is making up." Meanwhile members of the House Subcommittee on Appropriations, before which Hastie had testified, congratulated him on the reorganization, some of which he ordered before the councilmen, who had complained that they had not meant to give him such broad authorization, had a chance to counterattack. They amended the law to make prior legislative approval a condition of reorganization. Hastie's veto was overridden six to zero, with one abstention. Had he vetoed it again, the measure would have gone to Truman for final action. Rather than bothering the President with a measure that was neither of national importance nor of great potential harm to the community, Hastie allowed it to become law without his signature. Actions already ordered, however, would be completed.[69]

Hastie and the legislators were fast becoming less and less to each other's liking. As *Photo News* had said in November 1946, they were not on terms that "one would call sweet," though there had been "no open break." But a break had been in the making during the several months that Hastie had been trying to dissuade legislators from appointing themselves to executive boards and commissions. The particular issue concerned the Police Commission, which was a power base for them because it had the final word about salary, promotion, and disciplinary action concerning policemen. Hastie objected not to the results but to the method of appointment. He argued that the executive should appoint and the legislature approve or reject members. His predecessors disapproved of appointments by legislators but had not contested the practice. Hastie went to court.[70]

The governor complained that councilmanic appointment of Roy P. Gordon and Oswald E. Harris to the Police Commission was invalid because the appointments were the executive's exclusive prerogative. The Ordinance of June 1920, under which the assembly acted, should give way to the Organic Act of 1936, the Constitution, and federal laws. He had given the legislators formal notice to that effect and called upon them to make no more such appointments. But they contended that the separation-of-powers doctrine did not apply to the Virgin Islands because the Organic Act did not expressly provide for it. Assisted by Edward R. Dudley, the governor represented the municipality, while David E. Mass represented the legislature. Judge Herman E. Moore agreed with Hastie's interpretation of the law. Hastie then appointed Gordon and Harris to the Police Commission.[71]

Judge Moore's decision added fuel to the fire, and Hastie's appointment of Gordon and Harris did not mollify the legislators, who were dead set against being either coaxed or coerced into responsible governing. Hastie had made them an unwilling partner in a major development.[72] He had also made them more implacable adversaries. It could not have been otherwise. "To a man of Hastie's strict views and rigorous constitutional training, the doings of the legislators were deplorable," Beverly Smith wrote. "To him their procedures seemed slipshod, their finances loose, their hunger for patronage and privilege excessive."[73]

As Hastie's first year ended, legislators made an issue of another appointment that he wished to make. They rejected his nomination of Amy Joseph Moron as a stenographer-clerk in the personnel director's office, though they admitted that she was qualified for the position. Hastie, in turn, dismissed their suggestion that he appoint instead a former supervisor of the Progressive Guide. In keeping with his unprecedented gubernatorial tendency to fight back, he accused a "willful minority" of having rejected his nominee, placing personal and partisan preferences above the public interest. He vowed to expose those who displayed such irresponsibility if they ran for reelection in 1948.[74]

A venomous attack on Hastie was then made by Carlos A. Downing, Omar Brown, Oswald E. Harris, and Roy P. Gordon—all St. Thomians. Hastie was "rude and out of place" to have blamed his legislative woes on "a wilful minority," shouted Downing. If anyone was "wilful . . . vindictive and malicious" it was Hastie, "an evil force" whose emulation of Russian tactics would lead the islands "straight to hell." And if Hastie wanted something to do, he should go to Washington and try to get some money for the islands; or go back to the mainland and try kicking around the people who had kicked him around before he came to the islands with his "gang of nincompoops," including "dish washers and ambulance chasers" from the NAACP. "Who the heck is Hastie?" Downing asked. Harris answered by calling Hastie a carbon copy of Hitler and a communist, as the House Un-American Activities Committee had said. ("What comfort and delight those who opposed Hastie must be having now," the *Daily News* commented.) To Gordon the governor seemed to be not a reactionary but a man of poor judgment.[75]

The tirade against Hastie went on for almost two hours before legislators from St. Croix moved for adjournment. Brown and Downing had accused Hastie of trying to win over the Crucians. "I do not say you will accept bribes," Downing told them, "but with his gang upon the hill he is playing two ends against the middle." Several councilmen from St. Croix had decided, however, that time and money were being squandered in a session that enacted none of the eighteen items on its agenda. Five of the seven Crucians therefore supported the motion to adjourn, and it passed seven to six.[76]

Hastie's "wilful minority" statement had touched off a "low down, stupid personal attack . . . the most vicious and unwarranted thing that the Assembly has done to date," the *Daily News* asserted. "There was such a stream of epithets that even some hard-boiled, experienced legislators threw up their arms in disgust," *Photo News* added. Both papers thought the legislators had done a great disservice to the islands, weakening "the whole fabric of government" and providing "damaging evidence against a people on trial." Hastie made no public comment on this except to deplore the "political billingsgate" involved."[77]

Hastie's disputes with the Progressive Guide, Earle B. Ottley has written, did not lessen his determination to do all that he could to assure Virgin Islanders the greatest possible measure of self-government.

That was why he advocated the election of the governor and the appointment of a resident commissioner in Congress. He had long held that view, which had been evident in his response to Governor Lawrence W. Cramer's query about Roy A. Anduze, who in 1940 sought an appointment to the medical staff. Vouching for Anduze, Hastie said community morale and welfare "should be and, in my judgment, will be heightened by recognition accorded to native sons who have prepared themselves to render professional service."[78]

Hailed at first, Hastie was soon resented by legislators, who had assumed that their support for him during his confirmation fight would assure them equal standing with him in executive decision making. Hastie's initial popularity prevented their attacking him openly, but Omar Brown's trial marked the end of peaceful coexistence between the governor and the Progressive Guide. From then on the Guide members assailed him throughout St. Thomas and on every imaginable pretext. They said, for example, that he was power crazy, vindictive, and irresponsible.[79]

Resentment against Hastie simmered inside and outside the legislature. It surfaced in the *Daily News*. Hastie's first obligation, the editor said, was to the Virgin Islands. "However, there is evidence that he devotes much of his attention and energy to dealing with racial, educational and political matters on a national scale while his government continues to disintegrate under him." A broader theme was also voiced: One day "all the missionaries who have come to save us" would depart, and Virgin Islanders would have to "put the shambles together" and salvage their own destiny. "That is one reason why we should not submerge our opinions for any consideration—no matter how exalted it might be."[80]

Through it all Hastie was glad to have Government House not only as his office but also as his home. There with him was his daughter Karen, who blended energy, beauty, and mischief. In family photographs she is seen wearing a cap that would have done Bluebeard proud as the badge of a troublemaker. But she was the apple of her father's eye. "She used to do everything with him, everything," says their former butler, Raymond Plaskett. "The apple of his eye" and "a mutual admiration society" are expressions that were coined for the governor and his daughter.[81] Then came Billy.

The fourth member of the family, William H. Hastie, Jr., greeted the world on 5 March 1947, the first child born in Government House.[82] Mercer Cook heard all about the glorious event when Hastie came to Washington.

"Oh, by the way, Mercer, you're his godfather."

"Damn. That's a great honor. Thank you."

"Yeah, you're the only friend I have who goes to church."

The Hasties were a "very close, very caring" family. For a while there was a fifth member, Mother Hastie. Upon leaving and returning home, Hastie would kiss her. Plaskett found that surprising, "but it made you feel happy." By then age and arteriosclerosis had enfeebled her, and she was not much seen. But a photograph of her walking in Charlotte Amalie with Billy in tow revives memories of the strong Roberta Hastie whom Roy Anduze never forgot. As a student at Howard University in the 1930s, he could have saved money by going to People's Drug Store to buy three bars of soap for the quarter that purchased one bar at a nearer store.[83]

"I would walk these seven blocks," says Anduze. "There would be Mother Hastie with this sandwich thing [i.e., a sign]. And you dared not go in and get your three bars of soap because of this old lady walking there in the snow, up and down, for hours." The memory made Anduze chuckle. "She was a strong person. Had to be a strong person to be out there in the snow in February with a damn sandwich sign to keep you from buying three bars of soap for a quarter."

When Anduze returned to the Virgin Islands, he told James A. Bough and others about her. Years later, when he told Bough he would not join a march on Government House to demand that the governor's office be made elective—"I am interested in medicine, not in politics," he said— Bough countered, "Remember Mother Hastie." The next night there was Anduze, marching. "She wasn't *my* mother," he says, "but I was sufficiently impressed with this old woman, walking up there in the damn snow, and here I am, a young student, saying, 'Why in the hell doesn't she get away from there so I can get these three bars of soap?'"

Both Anduze and Bough saw much of the mother in the son, and thus understood why some legislators could not possibly deal with Hastie. As James A. Bough put it, "He would not compromise on principle."[84]

What the governor could not do—namely, assure responsible behavior by the legislators—Hastie thought the voters could and should do. In Virgin Islands politics he detected too much "advantage," the habitual exploitation of others for one's own benefit, and too little inclination by bystanders to do anything about it. "I think we will grow and mature as a worthy free community as we not only cease taking advantage but prevent others from doing so." This meant that more people had to become better informed about what was going on in politics, government, business, and personal relations. "And learning, we must care. Caring, we must exert ourselves. And exerting ourselves, we will make 'advantage' a memory rather than an accepted practice." But politics made Hastie's an impossible mission in the Virgin Islands.[85]

Hastie was a man of missions, all of them trying, some of them impossible, and several of them historic. This was especially true in 1948, when Harry S. Truman faced almost certain defeat in the presidential campaign.

CHAPTER FIFTEEN
Truman's Rescuer

A button was missing from the soldier's sleeve. The MP stopped him. They argued and, their companions standing back, they eventually fought. Things took a turn for the worse for the MP, so he yelled for help from the sheriff.

The sheriff fired. "Any more niggers you want killed?"

This murder in Centerville, Mississippi, said Hastie, was part of the "bloody notice" to the Roosevelt administration that its "attempts to muddle through cannot succeed . . . The Government must take the issue to the people boldly, giving America a measure of understanding and of leadership which is now lacking."[1]

Presidential leadership was still lacking in 1948. "Truman was . . . having trouble with the blacks," Robert J. Donovan writes. "It was over discrimination in the armed forces . . . [He] was vulnerable." Truman's vulnerability, however, was not solely attributable to military racism. It was he, not the military, who had said in 1940 that his endorsement of equal rights for blacks should not be misconstrued. "I wish to make it clear that I am not appealing for social equality of the Negro," he told the National Colored Democratic Association in Chicago. "The Negro himself knows better than that, and the highest type of Negro leaders say quite frankly that they prefer the society of their own people. Negroes want justice, not social relations." In 1944 Truman's selection as the Democratic vice presidential candidate angered blacks, who resented Roosevelt's dumping of Vice President Henry A. Wallace in favor of Truman, a move that the Pittsburgh *Courier* considered "an appeasement of the South."[2]

When Truman became President after Roosevelt's death in 1945, Hastie had regarded him as "just another midwestern Senator." That opinion changed after Truman's visit to the Virgin Islands in February 1948. Hastie said that his regard for Truman, "both as a human being and as the head of our government, grew tremendously." He believed that not politics but the President's "very lively personal interest" in the islands accounted for the visit.[3] Perhaps—and perhaps not, for Truman's visit and other actions with civil rights overtones should be viewed in light of a memorandum that he had received several months earlier.

The memorandum was to become a truly historic document, for its prescription of strategy made a close winner out of a president who seemed to be a clear loser. Clark M. Clifford, special counsel to the President, signed the memorandum, which James Rowe, a presidential assistant, had drafted. Submitted to Truman on 19 November 1947, it held that he probably could win the 1948 presidential election if he held on to the South and the West, above all else the West. Its authors thought the Republicans would nominate New York Governor Thomas E. Dewey, "a highly dangerous candidate," but they gave much more attention to Henry Wallace. Backed by communists, Wallace would not have to beat Truman in order for the Reds to be the real winners. If he siphoned off independent and labor votes, enabling Dewey to capture the White House, Wallace would advance the communist cause. All that Wallace had to do in a close election was to win 5 or 10 percent of the vote in a few states.[4]

Clifford and Rowe said that Wallace's strength should not be underestimated. The communists were but part of his large national following. If he could not be dissuaded from running, then he had to be identified as their candidate, and "prominent liberals and progressives" had to be persuaded to help do that publicly. To undercut Wallace, however, the President had only to move to the left, as Wallace's own advisers admitted. The move could be more apparent than real, said Rowe and Clifford: Since the GOP Congress would give Truman nothing, thus making attempts at bargaining pointless, he could take whatever stand he wanted on whatever domestic issues he chose and cast himself as a man of fine principle by refusing to compromise. His messages to Congress and to any other audience should be prepared with this in mind and tailored to voters, not to congressmen. Clifford and Rowe expected the crucial voters to be the independents and progressives, not the party faithful, for parties had been supplanted by pressure groups, including blacks, who were quite likely to support Dewey.

This grand strategy, which should have brought Hastie to its designers' minds, did not. He remained in the Virgin Islands as the Democrats made a number of efforts to neutralize Wallace's appeal to black voters. Truman told the NAACP that the national government must defend "the rights and equalities of all Americans. And . . . I mean all Americans." He appointed a Committee on Civil Rights and endorsed its report. He urged Congress to abolish poll taxes, create a Fair Employment Practices Commission and a permanent civil rights commission, and outlaw lynching. He promised to deal with discrimination in the civil service and the armed forces.[5] But when he called for Universal Military Training (UMT) and for Selective Service renewal legislation, he made no mention of segregation, though his advisory Commission on Military Training had recommended a ban on discrimination under UMT. He issued no executive order against discrimination in the military, though his Committee on Civil Rights had been particularly revolted by that brand of racism, and his civil rights message to Congress had contained his pledge to discharge Jim Crow.[6] The president rejected A. Philip Randolph's plea for a ban on discrimination in the Selective Service. Randolph then founded the League for Non-Violent Disobedience Against Military Segregation in an attempt to persuade him to integrate the armed forces, vowing a lawsuit to promote nonregistration by blacks and whites unless Truman issued an appropriate executive order.[7] On 26 July 1948 Truman ordered an end to discrimination in the civil service and armed forces. The orders would have been more impressive had their issuance preceded rather than followed the Democratic National Convention that adopted a civil rights plank much stronger than the one Truman wanted.[8]

To his credit, Truman had refused to knuckle under to Governor Strom Thurmond of South Carolina and others who resented his overtures to blacks, but even that had not sufficiently allayed black voters' misgivings about him. For all we know, said one black newspaper, he would "ruthlessly trade away" our interests if that would satisfy a more important group.[9] Clifford and Rowe had told Truman that blacks might be the swing vote in such states as Ohio, Illinois, Michigan, Pennsylvania, and New York, all of which could go to Dewey by default.[10] According to Henry Lee Moon, the black vote had become important in the 1944 presidential election and would be crucial in 1948. Politicians conceded this, he wrote, though they usually did not realize —and certainly did not concede—that the outcome of presidential elections was more likely to ride on black votes than on Solid South votes. Unlike the Solid South vote, the black vote was not wedded to any party, including Roosevelt's. Furthermore, it was strategically diffused through sixteen states whose electoral votes were generally considered decisive and in which close presidential contests could be decided by blacks. In twelve of the twenty-five states where a change as small as

5 percent would have reversed the outcome in 1944, the potential black vote was larger than that margin. Roosevelt had lost two of those twelve states, namely, Indiana and Ohio.[11]

But the President and his men made civil rights "the most silent and unmentioned issue in [the] campaign."[12] The election rode as much on the black vote as on any other, and it, in turn, rode on that issue. But neither Truman nor any other celebrated strategist knew how to capitalize on it. In their own feeble and frenetic way, they more or less tried.[13] Donald R. McCoy and Richard Reutten assert that "the most significant tactic" was the summoning of Governor William Hastie of the Virgin Islands, whose activities were designed almost entirely to undermine Wallace."[14]

Hastie, who was not "summoned," would have been important even if he had not been an effective campaigner. "Just being around with faith in Truman was a contribution," says G. James Fleming.[15]

Harry E. Harman, a friend of Hastie's, tried to persuade him to support Dewey rather than Truman. "Do that and you might keep your job," Harman said. "You haven't got a chance with this guy Truman. Throw him down the drain and come out immediately for Dewey." To which Hastie replied, "Job or no job, I'm going for Truman." He thought that blacks should stand by their friends.[16]

In September Hastie visited Truman. Telling the President that he had never been active in party politics, Hastie offered to take leave without pay and campaign for him. Truman asked him to get in touch with Senator J. Howard McGrath, the Democratic national chairman. "So sometime after the middle of September, I came to Washington with the thought of remaining on the mainland until the election to do whatever I could. I had a very disillusioning experience the first couple of weeks I was in Washington."[17]

Party leaders made no effort to press Hastie into service. "Well, I don't think you're being discriminated against," Julius Krug said when he mentioned this, "because I've had the same experience." Hastie thought that party leaders were "going through the motions of organizing a campaign." They seemed to have resigned themselves to Truman's defeat. Moreover, Congressman William L. Dawson, the black from Chicago who coordinated black campaigners, was unenthusiastic about Hastie's coming aboard.[18] Dawson had organized a national committee seeking Truman's reelection. Hastie declined his invitation to join. Walter White had come into the Democratic fold. The hiring of a black assistant director of publicity and the integration of its black division into the staff of the Democratic National Committee indicated that the party was making some attempt to increase Truman's appeal to blacks.

But as a presidential assistant was later to observe, Hastie was its "primary weapon against the Wallace move to suck in the uninformed Negro vote."[19]

Hastie struck at once. Wallace's party was just a puppet for the communists, he said. That was why it opposed the Marshall Plan and called honest critics of communism red-baiters.[20] Beginning with a New York City press conference held at Democratic headquarters in the Biltmore Hotel, he was to make a nineteen-day speaking tour of nineteen cities in nine states.[21] "But the discreet Negro Governor does not plan to visit South Carolina," Drew Pearson wrote. He had been invited to do so by Thurmond. Belatedly informed of Hastie's personal features, Thurmond said, "I would not have written him if I knew he was a Negro."[22]

Hastie regretted his inability to accept Thurmond's invitation. His Administration and his children were young, making extensive travel difficult. He looked forward to meeting Thurmond at the next governors' conference. Meanwhile he reciprocated by saying that the Hasties would be pleased to have the Thurmonds as their houseguests.[23]

By the time he hit the campaign trail, Hastie had removed his tongue from his cheek. It would be no joking matter to bring blacks into the camp of the candidate whom the black editor said would "ruthlessly trade away" their rights if politics demanded it.[24] But in city after city Hastie rallied blacks to Truman. Their choice, he told them, had to be made among a vigorous civil rights program under Truman, a reactionary administration under Dewey, and a Mother-Russia-First policy under that "will-o'-the-wisp," Wallace. Having sounded this theme in New York on October 13, Hastie left late that night for Ohio, where he spoke in Toledo, Cincinnati, Dayton, and Cleveland on successive days. At Friendship Baptist Church in Cleveland on Sunday, October 17, he told eight hundred persons the basic issues in the campaign.[25]

He asked if the wings of labor should be clipped. If you think so, vote Republican. Should the federal government spend less to provide "ordinary people" with social security, housing, and other assistance? Vote Republican if you want that. "I get plenty mad when the Republicans appeal to the Negroes and attempt to make you and me think of everything but these fundamental issues. How stupid do they think we are?"

In Hastie's opinion, Wallace took blacks to be even bigger fools. "What he says now is very good, but his record in earlier years was not good at all." Later Hastie said that he might have resented Wallace because the former secretary had been Ickes's rival, and "I was perhaps an Ickes man." But Wallace had not been an effective New Deal liberal and his Department of Agriculture had not been a significant force for improving race relations.[26]

Hastie could have told how Wallace enthused about local control under the Agricultural Adjustment Administration (AAA) when such control was inimical to blacks. Not one black sat on the AAA county committees that made policy binding on all farmers. Under the AAA many blacks were pushed down the economic ladder: from farm owners to tenants to sharecroppers. County committees assured them neither fairness in receipt of payment benefits nor right of appeal from unfair decisions. The AAA outstripped all New Deal programs in being crude while assuring poverty among blacks on farms, declared John P. Davis of the Joint Committee on National Recovery.[27]

"What kills us here," one tenant lamented, "is that we just can't make it cause they pay us nothing for what we give them, and they charge us double price when they sell it back to us." Some expected improvement, but not before they died.[28]

"When Mr. Wallace was in positions of great power and influence he was not interested in the Negro's problems," Hastie said. He could have added that in barring blacks as members, AAA county committees could take their cue from Wallace, whose department trailed all others in percentage of black employees and in appointing a black adviser. "You didn't dare take a Negro to lunch at Agriculture," said Will Alexander, one of those rare New Dealers who supported civil rights. Wallace would not risk offending Southerners in Congress and at Agriculture. While a black tenant farmer grieved, "Ain't make nothing, don't speck nothing more till I die," Wallace complained to Alexander, "Will, don't you think the New Deal is undertaking to do too much for Negroes."[29]

Think of Wallace now as he thought about us then, Hastie told blacks in St. Louis. "I'm glad to see he's talking differently now, and I hope he's changed, but he's talking civil rights against a background of non-action." Wallace would draw black votes from Truman, benefiting Dewey, who was being "all things to all people." Hastie acknowledged that he could lose his job if Dewey won. But the election would determine whether the nation advanced or reversed the New Deal. "As a Negro, I am bound to be influenced by the additional fact that never before in the history of the United States has a President had the courage and vision to take a completely forthright position on civil rights issues."[30]

The governor spent the night in St. Louis at the home of friends—none of the finest hotels would allow him to rent a room—and went by taxi to East St. Louis, Illinois, the next day. He spoke there and in Chicago, Gary, and Pittsburgh, becoming all the more confident that

Truman would win. "Even though I've been optimistic that the people would rally in the last days of the campaign," he said upon his arrival at the airport in Harrisburg on October 22, "the extent of the growing enthusiasm amazes me."[31]

From Harrisburg Hastie flew to Charleston, West Virginia, where George L-P Weaver joined him. "He had an extremely high sense of integrity, which flowed from his character, honed by his legal training," Weaver remarked about his former dean. "He said what he meant and meant what he said succinctly and with precision." Curious about Hastie paying his own expenses ("My experience with politicians was just the opposite") Weaver asked him about that "extremely odd" practice.

Truman measured up on civil rights, Weaver was told. "It was personal support of the President, not the Democratic party, and the hope that he represented," says Weaver. Hastie's effectiveness as a campaigner even won over Congressman Dawson, the veteran who had doubted the neophyte.[32]

Hastie carried Truman's banner by plane from Charleston through Pittsburgh to Newark, then by train to Connecticut, where he stumped in Bridgeport, Ansonia, and Hartford. Next he flew to Indianapolis by way of New York City, returning to the latter the next day. Although he stressed the economic reasons for preferring Truman to Wallace or Dewey, Hastie continued to emphasize the broader humanitarian argument for Truman's election. Part of the argument was the views expressed by Truman to the NAACP on 29 June 1947—the speech about which Truman wrote to his sister, "Mama won't like what I have to say because I wind up by quoting Old Abe. But I believe what I say and I am hopeful we may implement it."[33]

Hastie did not know what Truman had written to his sister, but he did know what the President had said to the NAACP. He carried a copy of Truman's address. Certain passages Hastie had underlined: "[W]e can no longer afford the luxury of a leisurely attack upon prejudice and discrimination. There is much that state and local governments can do in providing positive safeguards for civil rights. But we cannot, any longer, await the growth of a will to action in the slowest state or the most backward community. Our National Government must show the way."[34]

In the final days of a tour that took him to Philadelphia, Newark, and Buffalo, Hastie continued to criticize Wallace but increasingly hammered Dewey. Why not take civil rights out of the campaign, Mr. Dewey, by saying that the President's program is a good one and that you would carry it out? Why not at least be forthright? Why try to agree with both sides on this issue, as on all others? The GOP strategy on civil rights, Hastie said, is clear: Dewey would "make vague statements to us about what his policy will be and at [the] same time do business with [the] Dixiecrats."

To entice voters from the Dixiecrats, Hastie said in Cleveland and repeated in Buffalo, Senator Robert A. Taft was trying to win Dixiecrat votes by quietly playing John Alden to Governor Dewey's Miles Standish. "For every vote President Truman loses through a Dixiecrat, it's up to us to see that he gets two." More than two thousand persons at the rally in Buffalo heard him say that the Republicans, having "saddled themselves with a candidate they didn't want," were trying to create a bandwagon effect with their propaganda about Dewey's impending election. But they were deluding themselves. Dewey was "the warmed-over candidate who is not any better today than he was in 1944." Truman was the candidate creating great enthusiasm among voters.[35] If so, the effect was undetected by Howard E. Bell. In the week preceding the election he attended two rallies in Madison Square Garden. Half filled for Truman, it overflowed for Dewey. Bell's conclusion: "Dewey seemed unstoppable."[36]

Truman stopped him. Among the reasons was Hastie's success in thwarting Wallace's attempt to win black voters over.[37] But voters themselves indicated that it was not solely Hastie's exploitation of the communist issue that accounted for that success. In Cincinnati an elderly black lady was accosted by Republican and Democratic poll workers. The former cited Lincoln's greatness, the latter Truman's civil rights stand. As she left the polling place, the lady was asked by the Republican whether she had voted "the right way." She answered, "Abraham Lincoln ain't running, and I want some of them civil rights now."[38]

The literature is silent about Hastie's contribution to Truman's victory. Truman's daughter and biographer, Margaret Truman Daniel, has no recollection of it. Hastie himself called it "relatively unimportant" in a victory that he attributed to campaigning by Truman and organized labor.[39] "But he did one heck of a good job," says Mr. Justice Marshall. "He got the votes."[40] Politicians who were on the campaign trail with Hastie attested to his effectiveness.[41]

It is certain that Truman did not win on black votes alone. It is equally certain that he would not have won without them. In California, Illinois, and Ohio blacks gave him majorities large enough to clinch victory. Had he lost and had Dewey won in Ohio and Illinois, the Southern-dominated House of Representatives would have decided the election. For example, his margin in Ohio (7,107) was less than half his margin among blacks in Cleveland (14,713).[42]

"Those four meetings—Toledo, Cincinnati, Dayton [and] Cleveland did the trick," Hastie was told by John O. Holly of the Democratic State Executive Committee. Governor Frank J. Lausche believed that Hastie had "undoubtedly contributed" to his and Truman's victories. Holly said that Hastie had been "largely responsible not only for the huge Negro vote in this state, but for actually putting Ohio in the [D]emocratic column."[43]

It was a hard campaign. "I don't think I've ever been as tired physically as I was after five weeks on the road," Hastie would say, "because it involved daytime activities around in the community and going to places where people gather and talking with people." These were not political leaders, but the man in the street. "Usually a meeting large or small [was held] in the evening with spirits, helping to build up the spirits of the faithful. . . ."

"Well, how do you think it's going to come out?" Hastie would ask.

"Well, I guess Mr. Dewey's going to win, but I'm going to vote for Mr. Truman."

Hastie received that response with "amazing consistency." For example, in Ohio he heard this not only from persons at Democratic rallies but also from workers in factories, restaurants, and bars. And in Connecticut, where local Democratic leaders were telling people, "Even though you're going to vote for Dewey, split your ticket and vote for Chester Bowles for Governor." They had thrown in the towel. Hastie thought about that while listening to the radio back in Government House on election night. He brightened at reports that Truman was easing ahead of Dewey in Connecticut. Until that moment he had thought that Truman would lose two states: New York (because of Wallace) and Connecticut (because of local Democrats). But when he heard the returns from Connecticut, he said to the people who were with him, "You know, Truman's re-elected." Someone questioned that assessment. But, knowing the situation in Connecticut, Hastie took Truman's performance there as a good omen even for the campaign in New York, which Hastie had expected Dewey to win—and which, thanks to Wallace's taking black votes from Truman, Dewey did win.[44]

Truman, however, triumphed, and in an exchange of postelection letters he promptly acknowledged Hastie's "strenuous and loyal efforts in the great contest we have just won." The nation, he said, had caught the "wonderful enthusiasm" first shown by Virgin Islanders during his visit there and later injected into the campaign by Hastie.[45] The zest for battle that impressed Hastie was typical of the President and was evident at the final rally. "Oh, it was really tremendous," said Hastie about that event in Madison Square Garden. Truman's ebullience had surprised him, "because I, a person considerably his junior, could barely drag myself to the meeting, and I was way down, I mean, physically and emotionally, just exhausted from the campaign. . . ." Truman enlivened the crowd. Photographers snapped away, people moved around the platform, the hall buzzed with excitement. Hastie was surprised when the President came over to him. "Governor, I haven't seen you for several months now," he told Hastie, "but I know where you've been and I know what you've been doing. I just want to say thank you."[46]

Immediately thereafter Hastie left for the Virgin Islands. He had vowed to oppose irresponsible legislators. Some friends and advisers urged him to stay out of the campaign. They thought he would be no verbal match for the legislators and would alienate voters, who would resent being told whom to elect—especially by Hastie, an appointed official, an outside, a continental black. Nevertheless Hastie flew into St. Thomas late in the afternoon on Monday, November 1, scheduled to speak at eight thirty that evening in Roosevelt Park. "The place was jam-packed," says Alphonso A. Christian, Sr. "Maybe ten to twelve thousand people, a lot for this little community. You couldn't stick a pin in the crowd." At home Vivian Anduze made the drinks a little stronger than usual that night—with reason. "I knew it was probably going to be a little strained there." Among the friends who had joined the Anduzes on their porch overlooking Roosevelt Park were Beryl Hastie and the wives of some of the legislators. Caught in the middle, the Anduzes were all the more certain that Hastie should have taken the advice against speaking.[47]

"He didn't have the chance of a snowball in hell," says Dr. Roy A. Anduze. But he took it. Ariel A. Melchior, Sr., editor and publisher of the St. Thomas *Daily News*, remembers Hastie saying that some legislators snoozed in session, some were incapable of self-expression, and others were drunkards. He also said that they made good, progressive government impossible. He urged voters to "Throw the rascals out!" Melchior says that Hastie had a point. "But in that emotion-packed group, truth was the worst weapon he could have used."[48]

Hastie's opponents heard him out—or, rather, they heard whatever comments were not drowned out by the bands they had instructed to play as he spoke. When he left the platform, they mounted it and urged the crowd to come to Emancipation Garden (the usual place for political speechmaking). "We told them we were going to emancipate them down there," Christian says. "They loved that. They wanted to see what in the hell we were going to do to him."[49]

From 11 P.M. until 4 A.M. the crowd heard twelve speakers and listened to a good orchestra. "When we saw the mood of the people, we played to keep that mood up," says Christian. The game was not gut politics but gutter politics, and in it anything goes. In those days before television, local politicians got "very personal," says Roy Anduze. They played the dozens, that is, they spoke disparagingly of their opponent's ancestry. Their calypso talk confused Hastie. He could not understand most of what they said. But the crowd understood and cheered. And Beryl Hastie understood—and cried.[50]

Unable to discredit Hastie as a person, the rascals tried to discredit him as a politician. Omar Brown was in great form. "Just what does this man want to do? Throw us all into the *sea*? And when he has thrown us into the *sea*, he gets in the boat and he's gone!"[51]

What Hastie said should have been said—but not by him. He saw "the sharp tone of things, not the gray areas," in matters of honesty, morality, and responsibility, says Isidor Paiewonsky, a local businessman. The electorate, knowing the excesses of "the gang of bandits," would undoubtedly have voted some out of office. Hastie thought that he could assure that by speaking out against them. Instead "he provided a cloak to cover up all their sins."[52]

Later Alphonso Christian, vice chairman of the board of elections, sat in the small anteroom reserved for judges at Fort Christian. Across the railing was the area in which the audience sat. "I'm calling out the votes, and Governor Hastie was right in front of me. And when he heard that there was a landslide in favor of the party that he opposed, he just walked out of the place in a very dignified way." His political effectiveness had ended.[53]

For some persons this showdown between Hastie and the Progressive Guide is the most vivid of recollections about his administration. But others also recall the eradication of elephantitis on St. Croix, the acquisition of Magens Bay, the enactment of civil rights legislation, the birth of the tourism industry, and the inclusion of the Virgin Islands under coverage of such federal legislation as the National Social Security Act. The Building Code and Merit System Law passed by local legislators were additional boosts to the general welfare. So progress was made under Hastie, though not nearly as much as he wanted.[54]

Hastie had accepted the post of governor because he considered it an honor to his race. He had been content at Howard University: His students were bright and eager; his law school had been a beehive of social engineering; and he had been active in Gideon's band. His move to the islands had taken him out of the thick of battle over civil rights, though he remained involved in strategic cases by returning (at his own expense) to the mainland with some frequency.[55] He would not have characterized himself as such, but he was a pathfinder. He had played the part in the Virgin Islands in the thirties and would have played it again in the forties, but Virgin Islanders were less than willing to be honestly critical of their legislators and more than willing to be ethnocentrically censorious of their governor. He wished that it had been otherwise. In the end, however, it was less a matter of what he wished them to be than what he continued to be that accounted for greater progress than even he might have hoped for. "Hastie was such an example of the proper use of political power," notes Isidor Paiewonsky, "that he facilitated its transfer from whites to blacks in the Virgin Islands. What more can you say for anybody [except] that in a period of history when there was need for a figure to help with the transition, he was that figure?"[56]

CHAPTER SIXTEEN

The Greatest Need

After the 1948 campaign President Truman asked Hastie what he could do to show his appreciation for the black support he had received. Hastie suggested the appointment of Ralph J. Bunche as assistant secretary of state. Bunche declined the post, saying that he did not want to subject his family to racism in Washington and that he would be of greater service remaining at the United Nations. Hastie did not discuss with the President a reward for himself. So it might well have been news to him to learn that Truman had asked Senator Francis J. Meyers of Pennsylvania, the Democratic leader, if he would support Hastie's nomination to a seat on the United States Court of Appeals for the Third Circuit. Myers agreed rather than insisting that first call go to a Pennsylvanian.[1]

In the third circuit white lawyers objected—but not as much as black lawyers in Philadelphia, the "City of Brotherly Love." Still prevalent was the attitude that one black lawyer had revealed to Walter White in 1945: "As I see it, our good friend Bill Hastie has been quite signally honored by the Democratic Party during the last number of years . . . For that reason alone, if for no other, it would seem that these few honors which the Negro enjoys, especially in the legal profession, should be passed around . . ."[2]

Hastie's opponents denied that they were against him. It was just that they preferred appointment for one of their own, namely, Raymond Pace Alexander, Walter E. Gay, Maceo W. Hubbard, or Carlyle Tucker. They contended that Truman was trying to offset the opposition he encountered for having refused to nominate Hastie to the bench in Washington in 1945, a refusal that had spared him the wrath of Southern senators, who would not have wanted a black federal judge to enter the battle over segregation in the District of Columbia. Hastie's appointment, they argued, was a "political reward"; it reflected the view that only one black was qualified to hold important posts; and it was an affront to black lawyers in Philadelphia. Hastie had been backed by the NAACP, the National Lawyers Guild, the National Bar Association, the American Federation of Labor, and the Congress of Industrial Organizations, but Truman had not appointed him to the United States Court of Appeals for the District of Columbia. He had been caught between Representative William L. Dawson of Illinois and the NAACP in the congressman's struggle to wrest from the association and other social-welfare organizations control over whatever political crumbs were available for distribution by blacks. Dawson won the fight, so Hastie was out and Irvin C. Mollison of Chicago was in as Truman's first black appointee (to the United States Custom Court).[3]

The NAACP explained Hastie's rejection by saying that he had not labored in the Democratic party's vineyards. By 1949, of course, that was no longer the case. Judge Henry W. Edgerton, the dissenter in *Mays v. Burgess,* a restrictive covenant case in the District of Columbia, told Hastie that he would have welcomed him to the appellate court there, but that was not to be. For Hastie the appointment would not be in Washington but in Philadelphia, where he was in fact welcomed by some black lawyers. Herbert R. Cain, Jr., for example, told him that most members of the Langston Law Club felt no resentment toward him. And Arthur C. Thomas said that the club's protest, "the act of little, selfish men with axes to grind," was unofficial because it had been agreed upon at a meeting about which some members had not been notified. Thomas said that he and many other members "heartily endorse one so representative as you have proved to be in the past."[4]

Nevertheless, the opposition to Hastie was described as "vehement" and of "unprecedented proportions." Top-level black Democrats vowed to desert their party in city elections in 1949 and state elections in 1950 to teach Myers and other power brokers a lesson. In September the senator's reply to one query was that he was not recommending Hastie for the job. Within a month ranking Democrats, black and white, and all Democratic congressmen made clear their displeasure about Hastie's probable nomination. A black newspaper told Myers to "stop this funny business" of selecting a Virgin Islander for a job that rightfully belonged to a Pennsylvanian.[5]

Black lawyers in Philadelphia could not block Hastie's appointment, said Lem Graves, who considered their gripes "incarnate stupidity." Residence in the circuit was not required of an appointee, Louis Lautier added. And if it were, then residents of the Virgin Islands, New Jersey, or Delaware had a more valid claim to the judgeship than Pennsylvanians. The court's jurisdiction and the residential distribution of sitting judges included four from Pennsylvania, one from New Jersey, one from Delaware, and none from the Virgin Islands. Two seats were added to the federal district court in Philadelphia at the time that the circuit judgeship was created, and Hastie's appointment did not preclude black lawyers in Philadelphia from trying to obtain one or both of the seats. "We can keep yelling 'politics' like it was a nasty cuss word, if we want to," Graves worte. "But the sooner we learn that politics is a respectable and significant aspect of U.S. life, the more gains we will make."[6]

Hastie avoided the controversy. "Quite apart from any personal factor, such contention within the ranks of Negro lawyers cannot help in our efforts to have as many competent Negro lawyers as possible appointed to the bench." On October 15 he was appointed.[7]

"I am getting a little tired of sending you letters of congratulation," John Aubrey Davis wrote to Hastie. "As a matter of fact, I think we of the younger set have a legitimate gripe against you and Ralph [J. Bunche], Percy Julian, Allison [Davis], and Charles Drew. The pace is too damn fast." Of course, the appointment thrilled Davis and Hastie's other friends. Harold L. Ickes, whom Hastie credited "as much [as] or more than any one person" for his career in public service, thought that Truman had made the best possible appointment. "This is simply swell," opined Governor Ernest Gruening of Alaska. George Slaff of Samuel Goldwyn Publications told Hastie that "your pervasive humanity will have real scope." Recalling Hastie having said in 1939 that blacks could go far under the New Deal, Cleveland attorney William T. McKnight wrote, "You are now leading the pack. . . ." To Pauli Murray, who predicted that he would lead it to a seat on the Supreme Court, Hastie broached another topic: "While I am sorry that our efforts a few years ago were not successful immediately, we can at least be happy that this barrier has been now removed." Harvard Law School was open to women.[8]

Several letters came out of Hastie's past at Harvard. "You have far outstripped most of your classmates, but I know that they all, like myself, have greeted [your several appointments] with no other sentiment than admiration and friendship," Stuart N. Scott wrote on 24 October 1949. Five days earlier Paul A. Freund had written, "I look forward to the pleasure of reading—and indeed teaching—your opinions." And the next month, on the fifth, Dean Erwin N. Griswold wrote to express his belief that Hastie would provide a nationwide example in his new office, "and I am very glad that, in a sense, you have come back to the country again."

This, however, was not the universal sentiment, for in 1949 Hastie was one of three prominent and rebellious black men who were on the red baiters' hit list. Another was Benjamin J. Davis Jr., who had been tried in federal court on charges of conspiring to teach and preach subversion. Although not accused of having taken any steps to overthrow the government of the United States, Davis and ten fellow communists were convicted and sentenced in October 1949 in Judge Harold R. Medina's court at Foley Square in Manhattan. Davis was sentenced to five years in prison and fined ten thousand dollars.[9]

The day after the trial Truman nominated nineteen judges. Two of them were familiar with Ben Davis's experience, namely, John F. X. McGohey, his prosecutor, and William Hastie, his friend. The latter would never have condoned subversion, but he never forgot the circumstances that propelled Davis toward Foley Square. Davis had stepped into the legal profession ahead of Hastie. The son of the Republican National Committeeman from Georgia, he tried to practice in Atlanta. "But he was not allowed to succeed," Hastie said. "Even on his way to an upstairs courtroom, he was required to take a service elevator rather than the regular passenger elevator. In the courtroom he was 'boy' or that 'nigger lawyer.' He was insulted and humiliated at every turn."[10]

Hastie knew full well "the bitterness and desperation" that drove Davis from the Republicans to the communists. And when Davis wrote, on 1 October 1943, to ask his endorsement for election as a New York City councilman on the communist ticket, he told Hastie that the endorsement could be given in "careful disassociation" with the Communist party. But, surely, he added, "two friends, like ourselves, can clasp hands over Party lines" to advance shared ideals. "More than glad" to do so, Hastie replied on 6 October 1943, sending the endorsement that day to Dr. George D. Cannon. "If I were a citizen of New York, I would most certainly join in your splendid effort." He wrote this not just because he admired Davis but because blacks would increase their political power "when we show the nation that we are determined to disregard party labels, to have fair Negro representation in elective office, and to support candidates on the basis of their individual fitness." When Medina gave Davis the maximum sentence, the latter wrote, "One thought crowded everything else out of my mind—in the whole history of the United States, with more than 5,000 brutal and monstrous lynchings of Negroes, not one perpetrator had received a sentence of five months—to say nothing of five years."[11]

Just as lynchings were on Ben Davis's mind in 1949 when he faced Medina, these horrors had been on Paul Robeson's mind in 1947 when he faced Truman. Robeson was in Washington representing the Council on African Affairs at a national conference against lynching. At his request Truman received a delegation. The President reddened when a member of the delegation asked how the United States could press the war trials in Nuremberg while condoning injustice against blacks? "Domestic problems are domestic problems and foreign problems are foreign problems," the President snapped. Robeson said that blacks, particularly veterans, did not see it that way, especially where lynching was concerned. "If they can't have protection [locally] they want to know they will have federal protection." To the delegation Truman had just said, "All right-thinking Americans abhor lynching." Despite Robeson's apology ("I'm afraid I was carried away"), Truman's farewell matched the warmth of the Arctic.[12]

Two years later Truman retaliated against Robeson. His message to the NAACP convention in 1949 attacked the actor and singer for having backed Henry A. Wallace in the 1948 presidential campaign. Roy Wilkins, the NAACP's acting secretary, read to the convention Truman's greetings, which typified those politicians drag out of a fusty chest and send to civil rights gatherings. The message was well received. And although the NAACP was disappointed by the President's civil rights program, Wilkins was as much agitated by Robeson as by Truman. Racism is not our only concern, said Wilkins. In demanding our civil rights, "we do not cry out bitterly that we love another land better than our own, or another people better than ours."[13]

Wilkins lashed out at Robeson because he thought that the civil rights cause was endangered by the combination of colors given to Robeson's skin by the Lord Almighty and his politics by the House Un-American Activities Committee (HUAC). A number of black spokesmen accused Robeson of "adding the burden of being 'red' to that of being black." HUAC was investigating subversive activity among minorities, and Robeson was its favorite target.[14]

Robeson had said that blacks would not take up arms against the Soviet Union. One of his black critics, Jackie Robinson, second baseman for the Brooklyn Dodgers, told HUAC that he and others had too great an investment in America "for any of us to throw it away for a siren song in bass." Another critic, Ralph Bunche, gave this witty riposte when asked about Robeson's statement: "I have always admired Mr. Robeson's singing more than his social philosophy . . . I do not believe in dignifying his remarks with a comment." But delegates to the NAACP convention at which Bunche made this remark wanted the fight against racism to be waged vigorously and the one against communism to be waged sensibly. They wanted Truman to fulfill his party's civil rights promises and to revoke his loyalty order. They saw Strom Thurmond as a greater menace to freedom than Paul Robeson. To their way of thinking HUAC was uniquely abhorrent.[15]

Hounding persons whose ideology it disapproved of, HUAC inflicted punishment through innuendo that could not have been imposed through litigation, because its victims were persecuted not for crimes they were accused of having committed but for those they were "suspected of desiring to commit." Constitutional safeguards were available against criminal charges of espionage, sabotage, and treason, but not against indiscriminate accusations of disloyalty. Moreover, any congressional committee or administrative tribunal pressing such accusations was, in effect, grand jury, trial jury, prosecutor, and judge: Mere accusation was itself proof of guilt; and punishment was assured because the press invariably carried out the sentence: stigmatization through public exposure. Hastie was vulnerable to this type of persecution, which reached the stars—those in Hollywood—as HUAC struck the movie industry with the help of the president of the Screen Actors Guild, Ronald Reagan.[16]

Some of the stars voiced alarm.[17] "Before every free conscience in America is subpoenaed, please speak up!" Judy Garland pleaded. Frank Sinatra asked, "If you make a pitch on a nationwide radio network for a square deal for the underdog, will they call you a Commie? . . . Are they going to scare us into silence?" And Frederick March wondered who the targets really were and would be. Ministers? Teachers? Students? Parents? "They're after more than Hollywood. This reaches into every American city and town!"

But why reach into black communities? "Negroes . . . who have never produced a traitor . . . who have had only a minute proportion of conscientious objectors . . . and who have no other national background, resent any implication of disloyalty," Lem Graves wrote. So why did HUAC—which had never seen fit to investigate klansmen—want to investigate blacks? It knew that communism had not taken root in the black population. But Robeson had spoken, and polls in seven cities disclosed that 52 percent of the one thousand white respondents believed that he spoke for blacks in general. Many respondents assumed that in racist America self-respect would drive blacks into the arms of Reds.[18]

Of course, not every self-respecting black was Red. But many, as Benjamin Quarles says, were individualistic; and individualism was, as Alan Barth notes, taboo, particularly if it was to the Left. Thousands of Americans, claimed Alger Hiss, "entered into a nightmare world of inquisition"—by local, state, and federal agencies. Persons accused of subversion were condemned, prosecuted, and tormented if they failed the litmus test of loyalty: guilt determined not by personal conduct but by association, particularly with persons in organizations on the list maintained by Attorney General Tom Clark.[19]

After Truman issued his executive order on 22 March 1947, the government was swamped with loyalty boards that heard and recommended the disposition of cases. Department and agency heads were personally responsible for assuring that disloyal employees were fired. Every federal employee was investigated. Entitled to hearings before loyalty boards, accused employees were allowed to bring their lawyers, to testify, and to present evidence through witnesses and affidavits; but they had no right to confront, or even to know the identities of, their accusers. If recommended for dismissal, they could appeal to their departments and, if turned down, to the Loyalty Review Board (in the Civil Service Commission) that was to establish rules, coordinate departmental policies, maintain a file on every person subjected to a loyalty investigation, and receive the attorney general's list.[20]

Membership in an organization on that list could figure in the disposition of a case or in the decision to conduct an investigation. Tom Clark's predecessor, Francis Biddle, kept the list private; Clark published it. Regardless of his or her job, every civilian worker was checked against FBI fingerprint files. Unfavorable reports would spark an investigation, and persons in sensitive positions could be summarily fired. Truman said that civil rights were to be protected. "Nevertheless," writes Robert J. Donovan, "a basic fault of the program . . . was that it placed a perceived need for national security ahead of the traditional rights of individuals." Roosevelt's men had been forbidden to act as though radical politics and espionage were one and the same thing, whereas Truman's men were encouraged to do so. The influence of congressional zealots increased.[21]

Hastie had faced zealots during the hearings on his gubernatorial nomination in 1946. He told them that he was an anticommunist, but insisted that he "[had] no intention of engaging in red-baiting." In his opinion, red-baiting was unfair—and dysfunctional, since it complicated the task of distinguishing between the liberal-but-loyal and the disloyal. "I would hope, therefore, in the interest of our future security, that there might be an abandonment of this blunderbuss and dragnet method. . . ."[22]

At the 1949 NAACP convention at which Wilkins attacked and Bunche belittled Robeson, Hastie dealt with the Red Scare.[23] "I think a great educator offered the most devastating answer to this business of ascribing guilt by association when he reminded us that it was said of Christ 'that he consorted with publicans and sinners, and therefore must be guilty.'" Many persons were associates in such "proper causes" as the elimination of job discrimination, Hastie added. This association sufficed to bring the anticommunist wrath down upon the heads of many blacks. As a result, "timid persons" were frightened out of any kind of association with blacks. Asserting that a person's own actions, not the identities of his or her associates, were the only proper basis for judging the individual's loyalty, Hastie wanted to make certain that government's "misguided efforts" in ferreting out communists did not destroy the liberty that it claimed to protect. "Under the guise of seeking the disloyal, government power can be abused to force compliance with popular or generally accepted modes of thought and action."

Loyalty investigations, however, were but one aspect of governmental behavior that concerned Hastie. "The manner in which [g]overnment itself behaves in the civil rights area has a large educational effect upon all people everywhere . . . But beyond mere examples, government directly organizes and sets the pattern of much more of our lives than we realize." Schools, housing, military service, recreational facilities—its operations in these and other areas shape as well as regulate social conduct, and "until it becomes color blind, government itself will be guilty of the worst and most inexcusable form of 'Un-American Activity.'"

On all levels, said Hastie, officials claimed that they should be neutral in the conflict between those seeking and those denying justice. Such neutrality was forbidden by the nature of their offices; their responsibilities were not to be discharged according to whim, convenience, expediency, or prejudice. "He who will not use his office to fight for the ideals written into our basic law is false to his oath to support that law. He is the true subversive and deserves to be branded as such."

Hastie declared his support of "radicals" and his criticism of "social standpatters" to be harmonious with the American principle of equality. But with Walter White absent from the NAACP convention; rumors flying that he would seek employment elsewhere upon his return from a leave of absence in Europe; reports circulating that Wilkins's allies had dissuaded White from attending the convention so that Wilkins could prepare for his own "succession" by planning the 1950 program; Henry Wallace's supporters and communists accusing Wilkins of red-baiting and pulling off a power grab by misrepresenting the aims and actions of NAACP "progressives and militants"; and Wilkins citing editorials from *The Crisis* to the effect that the communists were devoted solely to their party line—with all this palace intrigue at the convention, Hastie's speech created quite a stir.[24]

Wilkins's backers accused Hastie of wooing leftists by inviting them into the NAACP if they agreed to fight racism. Hastie, they said, was also trying to align leftists on the NAACP board of directors against Wilkins. Walter White's reliance on Hastie's counsel, according to George Streator, was heavy, and Hastie's friends wanted him "closer" to the NAACP "when so much is in flux." Hastie's basic attitude was that racists were no more capable of being good Americans than were communists. To him it was inconceivable that one could love and believe in America while accepting separate standards of justice for blacks and whites, or that one could be a patriot and still regard race or religion as disqualifiers for residence in a given neighborhood or elevation to high public office. Expressed quite a while after the 1949 NAACP convention, these views were as much a part of him as those he had voiced at the convention.[25]

It was, however, the latter views that were the topic of Streator's article in the New York *Times* on 14 July 1949. Entitled "Governor Hastie Defends Radicals," the article would be used against Hastie almost four years later, during his fight for confirmation of his appointment to the United States Court of Appeals for the Third Circuit.

"I had thought the appointment was coming," Judge Herbert F. Goodrich had written in welcoming Hastie to the court, "but then heard that there were lions in the path."[26] These were not lions but jackals. "I think one of the difficulties was the atmosphere of fear and distrust which Senator [Joseph R.] McCarthy and others have so effectively created," Hastie would tell another friend months later.[27]

It was, however, hard to tell where the Red Scare ended and the Redneck Scare began as the drama of his confirmation unfolded.[28]

    1949—October 15: Truman nominates Hastie.

        October 18: Hearings on all judicial nominations are postponed.

    1950—January 5: Hastie's nomination is referred to the Committee on the Judiciary.

        January 12: A subcommittee consisting of Chairman Patrick A. McCarran (Dem., Nev.), Harlan M. Kilgore (Dem., W. Va.), and William Langer (Rep., N.D.) is named.

        April 1: McCarran and Langer hold hearings. American Bar Association and an assistant to the attorney general endorse Hastie.

April 3: Subcommittee files a favorable report. Eastland's motion, seconded by Forrest C. Donnell (Rep., Mo.), to return the nomination to the subcommittee fails; Judiciary Committee (on Kilgore's motion) votes to hear Hastie in executive session and instructs its staff to investigate the nominee.

Harold Ickes took the stalling as a bad omen, telling Hastie that McCarran was "out to get you," as he had gotten Nathan R. Margold. McCarran had prevented Margold's nomination to the appellate bench in Washington from getting out before Congress adjourned. Margold was a Jew.[29] Hastie was a black, and time was running out on him just as it had run out on Margold. Congress would adjourn at the end of July. Hastie asked McCarran about the accuracy of news accounts that the committee had postponed action pending further hearings. He offered to provide whatever additional information the committee wanted and sent to Kilgore and Langer copies of his letter. "Your situation . . . is much worse than your short letter would indicate, much as I regret to advise you," Langer replied. To improve it Hastie's supporters concentrated on Senator Donnell, who had sided with Eastland on April 3. Knowing that Donnell would not risk the ABA's anger by voting for a former officer of the National Lawyers Guild, Chief Judge John Biggs of Hastie's court had Pennsylvania's representative on the ABA Special Committee on the Judiciary take the matter up with Howard Burns, its chairman. Burns, delighted by his school ties (viz. Amherst College and Harvard Law School) with Hastie, sent a letter to all senators on the Judiciary Committee. To Donnell he sent a personal letter on April 22 (a copy of which Biggs gave to Hastie).[30]

"I can say to you what I perhaps could not say to . . . any other member of your Committee," Burns told Donnell. "I understand there is grave danger that if Judge Hastie's nomination is not confirmed the President may nominate some other colored candidate from Philadelphia and that the ones who are most prominently mentioned have no qualifications, either legal or otherwise, for such a position. . . ." Burns added that he had talked with Herbert F. Goodrich several times, once when the judge called long distance to say that Hastie enjoyed the full confidence of his court.

By then Goodrich had said as much to Donnell. "We have a pretty close association in this Court, much more so than in some other groups elsewhere. We always lunch together and see a great deal of each other outside the regular court hours." In the manner of good friends, they discussed all sorts of topics. Never had Hastie displayed thoughts or sentiments different from those of any solid American. "This is true not only as to what has been said but as to what might be left unsaid . . . I can hardly say more except that on the professional side everything is really first-rate. He knows his law and he applies it like a good Judge."[31]

Another colleague, Judge Albert B. Maris, summarized what he had told Donnell about Hastie. "He combines legal scholarship with a practical approach . . . I believe him to be ideally fitted for the appellate bench and I think that in this respect he stands in the front rank of the bar of our circuit regardless of race." Hastie disloyal? "To me this is incredible."[32]

Still, Francis Biddle told Biggs, "I think there is going to be real difficulty in getting Hastie confirmed." Not for lack of their trying, though. Both wrote to McCarran. Biddle asked Chief Justice Arthur T. Vanderbilt of the Supreme Court of New Jersey to vouch for Hastie. To Biggs he passed along Senator Kilgore's tip about having some prominent ABA members intercede with Donnell (and McCarran). Biggs arranged Burns's endorsement of Hastie, but Donnell joined Eastland in bombarding colleagues with anti-Hastie allegations whose documentary basis they refused to disclose.[33]

Donnell was told by Judge Francis E. Rivers of New York City that Hastie's confirmation was crucial to the GOP's bid for black votes. The senator "seems less hostile," Hastie told Rivers about a week later, but was "still an uncertain quantity." Asking Rivers to urge two other Republicans, Senators Alexander Wiley of Wisconsin and Homer Ferguson of Michigan, to support him, Hastie said the problem was more indifference than hostility. Although McCarran and the Democrats should take the lead in bringing the nomination to the floor, it would be proper for the minority party to do so.[34]

Judge John J. Parker considered Hastie "thoroughly loyal to our country," and told McCarran that he should not be "defeated on grounds which many would regard as a mere pretext." Regarding the opposition to Hastie in that light, Leslie Perry, the NAACP's man in Washington, tried in vain to see Senator Wiley. He had not contacted Senator Ferguson "because I regard him as the possible key on the Republican side and careful thought should be given to the first approach."[35]

Perry had talked to someone whom he considered to be an effective intermediary with Ferguson. As chairman of the Senate Campaign Committee, Ralph O. Brewster was shopping for votes. Perry was tempted to point to political advantages available among blacks if the GOP denounced the Democrats' delaying tactics or broke the logjam. But he was unsure about how or whether to do so. Not to worry. "[Brewster] solved that problem by bringing up the 'old school tie,'" Perry told Hastie. The senator said that he would inform Republicans on the

Judiciary Committee that in 1946 his subcommittee had cleared Hastie of the charges that were being made and would urge them to insist that the committee act. Perry advised Hastie to ask Brewster to talk to Ferguson. As for McCarran, who "simply does not respond to pressure," Perry got Msgr. John O'Grady to approach him. McCarran knew that no one could hurt him in Nevada.[36] Nevertheless, Roy Wilkins asked black newspapers to send "a flood of strong protests" against McCarran. Sixteen district court nominees and all circuit court nominees (except Hastie) had been approved, Wilkins noted. "The delay on Judge Hastie is clearly based on race." If Hastie were not confirmed during the current session, he would be ejected from the bench.[37]

It was touch and go on Hastie's nomination. He had said all along that congratulations should be withheld until he had been confirmed. Hearing that only Senator James O. Eastland seemed to be dead set against Hastie, Walter White said, "I'd begin to get suspicious of you if Eastland also went over to your side." Hastie had identified the Mississippian as perhaps his only outspoken opponent, but added that "there seems to be no real drive within the committee to act." It was said that Hastie would be called before the senators, but as of 1 June he had not been. "The delay does give me concern since the end of the session is approaching."[38]

On 27 June Hastie made the first of three appearances before McCarran's committee. McCarran tried at once to discredit the ABA's endorsement of him by saying that Howard F. Burns belonged to the National Lawyers Guild. J. Edgar Hoover, the FBI director, would testify that the NLG was Red, he said. Donnell objected. And after John J. Carmody, a member of the ABA's Standing Committee on the Federal Judiciary, vouched for Hastie, Senator Wiley said that no one was impugning the ABA, but that one just had to wonder how much time a man as busy as Burns could have spent evaluating Hastie.[39]

"Are you satisfied [Hastie] is a good, loyal American?" Wiley asked Carmody.

"I am."

"A lot of people think they are loyal Americans, but they don't believe the American system is worth a damn."[40]

McCarran then asked Hastie whether he had ever belonged to, or associated with, a red organization?[41]

"My answer to that, Mr. Chairman, has to be 'I don't know.' I have been associated with organizations which various persons have felt were in various positions on the left."[42]

What about the NLG? asked Richard Arens, a staff member of the committee.

Hastie said that he had joined the NLG sometime between 1935 and 1940, had been a vice president, but had departed in 1947. The NLG, he said, definitely was not Red. It had received approbation from patriots such as Supreme Court Associate Justice Stanley F. Reed. Dean Lloyd K. Garrison had been a vice president, and Governor Harold Stassen of Minnesota had been a member. Hastie would let his affiliation be judged on the basis of their esteem for the NLG.[43]

"How did you happen to become connected with the [National] Lawyers Guild and not also be connected with the American Bar Association?" Senator Ferguson asked.

"Well, at the time, Mr. Chairman, it is my understanding that the American Bar Association did not welcome the membership of Negroes. That is my reason."[44]

Then what about the New York *Times* article? Hastie had not mentioned the FBI by name when talking about investigators who equated white-black friendship with disloyalty on the part of whites, but the reporter attributed the accusation to him. Donnell called it an irresponsible accusation, especially since Hastie admitted it was based on discussions with Walter White and not on FBI reports he himself had actually seen. Hastie's handwritten notes disclose that, having been asked about this, he consulted the Department of Justice and learned that persons subjected to loyalty hearings were given transcripts, some of which included information about FBI investigations. He believed the NAACP materials on which he relied were of that nature. At the time of Donnell's criticism Hastie said that the FBI had served the nation well. He also said the FBI had not denied the charge that it rendered unfavorable reports on whites solely because they had black friends.[45]

But was it not conceivable that a black might fraternize with a white communist organizer, making the association pertinent? Hastie was asked. "That is correct," he answered, "but in such circumstances there would be no relevancy in referring to race . . . The same thing would be true if he were associating with a Negro communist organizer."[46]

What had Hastie meant in calling "subversive" some investigators and officials? "I mean by that tending to the destruction of our institutions, tending toward causing Americans to lose faith and confidence in the law and the enforcement of the law."[47]

Hastie had characterized as "more than subversive" government officials "who pretend to be 'neutral' in matters between aggrieved minorities and those who seek to maintain the inequality of our society." He was asked whether he had said that disparagingly.

"Yes."

"And is it your definition of the connotation of that word [viz. subversive] as it is used in this sense that it was to describe people who are undertaking to undermine our form of government?"

"Yes, because I think very definitely that people who do not undertake to the best of their ability to carry out the guarantees of our Constitution, of the underlying tenets and premises of our society, are tending to break down the society itself by not making real the guarantees of the protections that our law attempts to give us."[48]

Arens had inserted into the record a portion of Hastie's testimony during the hearings on his nomination as governor. He wanted the committee to know that Hastie had said that communists were welcome to the civil rights struggle. Hastie wanted it to know what Arens had omitted, namely, Hastie's explanation that blacks had to take allies where they could find them and to work with persons whose basic ideology and allegiance they disapproved of.[49]

Did that include the Americans for Democratic Action (ADA)? Hastie had worked with it. Was the ADA pro-Russia?

No, said Hastie. "[The ADA] resulted from a division of persons who had actually worked together in various liberal causes, at a time when it was felt the cold war was beginning and it was necessary to slough off, if you will, any persons who apparently were sympathetic with the cause of Russia."[50]

The ADA was the only organization Hastie had joined since 1940. He had joined the NAACP in the 1920s. The National Negro Congress? It had not been Red in its earliest days. He had left it in the late 1930s. About Hastie's affiliation with the National Negro Congress (NNC), Senator Langer said that he should be complimented for joining it to help his race and congratulated for leaving it when it became pro-communist. "What is the use of wasting our time on a lot of rot like this?"[51]

It was 1946 all over again. Eastland and his allies accused Hastie of aiding organizations that were on the House Un-American Activities Committee's list. As in the hearings on his gubernatorial nomination in 1946, Hastie countered the accusations. So did the Kansas City *Call*, which deplored the practice of accusing liberal whites and blacks of being Red. Hastie fought Jim Crow openly, constantly, vigorously, said the *Call*. If fighting for equality made a person a communist, then every black was a Red.[52]

So Hastie had been active in the NLG, which enlisted black lawyers, condemned lynching, and supported liberalism—all in striking contrast to what one United States senator called "the stupid anti-social behavior of the American Bar Association." The seeds of the NLG, whose Washington chapter Hastie helped to organize, germinated in the ABA's racism.[53]

"We belonged to all those different organizations," says Thurgood Marshall. He, as well as Hastie, held a special opinion about the House Un-American Activities Committee's list. "If a Negro wasn't on it, he shouldn't be alive. We went through all of that and I don't think we had any problem . . . The commies had no effect on us at all."[54]

Some senators brandished the New York *Times* article that contrasted Roy Wilkins's castigation of communists with a few left-wingers' enthusiasm about Hastie's remarks at the NAACP convention. That had been much ado about nothing. The reporter who wrote the article, George Streator, told Wilkins that he had used that slant because it made "a good story." For the same reason another article highlighted what Wilkins called the "Fiction of Factionalism in the NAACP," based on Walter White's absence from the convention, some fellow travelers' prattle, and discussions common to conventions. Two things were well known, said Wilkins: The NAACP's criticism of the way the loyalty order was enforced was not the same thing as left-wingers' condemnation of any and all such orders, and there was no Wilkins-Hastie split in the NAACP. No right wing led by Roy and no left wing led by you, Hastie was told by Walter White, who offered to ask Streator to write Hastie a letter explaining that the reference to him as the left-wing faction's leader "meant that you favored a more militant program and emphatically did not mean a communist wing."[55]

To learn Hastie's views about communism, the senators did not have to rely on a Streator's newspaper article. They could have relied on Hastie's own newspaper column. "I have been meditating upon the subject of stupidity," he had written in 1934, phrasing an issue for debate: "Resolved that the rabid anti-communist agitator in America is more stupid than the rabid communist agitator." Both sides should win. The rabid anticommunist, smug in ignorance about communist theory, tarred everyone—communist, socialist, radical, anyone else who advocates change—with the same brush. He was against all change, except the kind he jingled in his pocket. Ours is the burden of proving that our system is suited to human needs, Hastie added. "This is much too big a job to permit wasting time whipping heads and denouncing communists. If we prove that capitalist institutions can cure a sick world, the communist is licked; if we don't—Yeah!"[56]

Hastie was not one to complain about personal victimization by racist slurs. But he must have shared the view expressed by Bernard G. Segal, the Philadelphia lawyer whose assistance Judge Biggs obtained for him, that in this instance opposition to segregated blood banks was a more important reason for protest than was his membership in the National Lawyers Guild. And Hastie did tell Walter White that he found helpful "recent publicity of the type represented by your column. . . ."[57] White had asserted that Hastie was being pilloried because certain senators found him guilty of "an unforgivable crime—being born with a dark skin."[58]

Eastland's behavior aroused the suspicion of the New York *Herald Tribune*, which on June 15 asserted that Hastie's color "has more than a little bearing with some members of the committee." A Philadelphia reporter felt that Hastie's difficulties resulted in part from his fidelity to the NAACP. Marquis Childs pronounced Hastie fit for the bench, saying, "It is unfair to discriminate against him for any reason whatsoever and especially because he happens to have been born a Negro."[59]

Hastie had been cleared by the FBI before being nominated by Truman. He had been confirmed by the Senate on two prior occasions. He had reminded McCarran's committee of his fight against Henry Wallace in 1948. Why was Truman's own congressional party blocking the nomination? When Truman took McCarran to task for this, the senator declared that no one could force the nomination on his committee. On July 13 Hastie again appeared before McCarran's committee. Donnell asked most of the questions during the five-hour secret meeting. Eastland was absent. The committee called no witnesses. Although the black Barristers Club of Philadelphia did not send its endorsement of Hastie until the next day, the committee had heard from other supporters, particularly Philip Murray, president of the Congress of Industrial Organizations.[60]

Murray had asked Majority Leader Scott Lucas to help pry Hastie's nomination out of the committee. Lucas was sure that the Senate would confirm Hastie and had done "everything I can" to get a favorable committee report. He asked Murray to appeal to McCarran.[61] On June 13 Murray telegraphed all committee members for "a prompt, favorable report," saying that failure to provide it "can only be attributed to the fact that [Hastie] is a Negro. . . ."[62]

On July 17 the committee reported favorably on Hastie's nomination. The vote was nine to one (Eastland). Without holding hearings and by unanimous vote the Senate confirmed Hastie on 19 July. Nine months had passed since Truman made the nomination, over six since he had resubmitted it to the Senate.[63] Finally, in the words of one observer, Hastie was off "the anxious seat."[64] His second "inquisition" in four years, said another, was at an end.[65] He had considered his nomination "literally one of those prospects that seems almost too good to be true."[66] And now his confirmation would enable him to render the public service that, in his opinion, gave a lawyer the deepest personal satisfaction.[67]

Hastie's confirmation punctuated the kind of story that Langston Hughes found rare in the literature about blacks. Books and plays emphasized their plight, Hughes wrote. Those accounts had their place. "But for ourselves who are colored and know our plight too well, for ourselves there is a need, more than anything else, of great patterns to guide us, great lives to inspire us, strong men and women to lift us up and give us confidence in the powers we, too, possess. . . ."[68]

This, of course, is more than a need of black people. It is a need felt by all people and satisfied by the truly great, "those who in their lives fought for life," as Stephen Spender described them.[69] At times defeated but never defeatist, these fighters take their stand at the point, beyond the cutting edge of battle. Greatness is therefore one part locality—and one part performance: The great turn stumbling blocks into stepping-stones on their way to singular achievement. Greatness is one part character and is measured by its possessors not so much by what they achieve for themselves as by what they achieve for others.

William Hastie thought of himself as just another soldier in the bittersweet war against bigotry. But was he? He enlisted, he excelled, he inspired. And the beneficialness of his greatness transcends race and place and time.

EPILOGUE

On 5 April 1976, in the city of the African princess's enslavement, scholars and public officials assembled to celebrate the birth of freedom in the United States. The speaker at the dedicatory ceremony had been selected because he was one of them, not because he was one of her descendants. However, at Independence Hall in Philadelphia on that day, Hastie might well have had her in mind when he told the celebrants that there was reason to salute the nation on its bicentennial. "But a nation's beginning is a proper source of reflective pride only to the extent that the subsequent and continuing process of its becoming deserves celebration."[1]

On 14 April Hastie sat at his desk, elaborating on that theme in the keynote speech he was preparing for delivery when the Washington Bar Association awarded him and five other black lawyers its Charles Hamilton Houston Medallion of Merit. Hastie planned to pay tribute to Houston and other black lawyers who had fought to make America's "becoming" worthy of celebration. He intended to tell his audience about the bitter and the sweet in their struggle.[2] He would not have talked about the role he had played. It would have been out of character for him to have done so.

"He never blew his own trumpet," Judge Charles E. Wyzanski, Jr., has said. "But that is no reason for us not to make sure that his music becomes part of the heritage of the ages."[3]

Hastie would not have told his audience an anecdote that reveals the leitmotiv of his life and work. But that is no reason for us not to make sure that it is known.[4]

One day on St. Thomas an old man was ushered into Hastie's chambers carrying a tin box. The man seemed almost speechless, and Hastie had no idea as to why he had come. But soon the ice was broken.

242

EPILOGUE

"This aged gentleman, hair grizzled, face deeply lined, bent almost double from age and infirmities, asked to consult me about 'putting my personal affairs in order.' I suggested that he should see an attorney, but he explained that he only wanted some 'advice.'"

Just as Hastie's curiosity about the tin box peaked, the old man opened it and removed a batch of frayed papers. He handed some of the papers to Hastie, then leaned back as far as infirmity permitted, his eyes never leaving Hastie's face as the judge examined the stock certificates issued by Marcus Garvey's Black Star Line.

Several minutes passed while Hastie tried to decide how best to tell his visitor what was on his mind.

And then the knotted body eased forward. "I know what you're going to say, Your Honor. You're going to tell me they aren't worth a penny." A frail hand gripped Hastie's knee, and a sudden, surprising firmness entered the old man's voice. "But I want you to know that I don't regret any of the hard-earned dollars I spent for them."

Sharing the anecdote with a reporter who had requested a vignette about his judicial experience, Hastie added, "That's the end of the story. A simple, little thing, perhaps. But I hope that at the end of my life I can be as certain that I have done the right things, that I have invested in the truly worthy causes, as was the old man."

NOTES

CHAPTER 1. BLACK DIAMOND

1. William Hastie's remarks at the University of Rochester, Rochester, N.Y., 13 April 1967; Horace Mann Bond, "American Missionary Association Colleges and the Great Society," Inauguration of Herman Hodge Long, B.A., M.A., Ph.D., as President of Talladega College (New York: United Church Board of Homeland Ministries, n.d.), pp. 8–9, 11–17, 28–30; E. Franklin Frazier, *Black Bourgeoisie* (New York: Collier Books, 1962), pp. 56–59, 65–71; "Obituary of Roberta Child [sic] Hastie"; interviews: Bertha Childs, William Hastie's cousin, Marion, Ala., 23 September 1977; Ella Childs Walker, William Hastie's cousin, Hartford, Conn., 25 September 1977; and Louise Russell Hall, William Hastie's cousin, Chicago, Ill., 2 July 1976. Hereafter, reference to Hastie will usually be made through the use of his initials (WH). Unless otherwise indicated, incomplete references to materials are to those in the Hastie Papers at Harvard Law School. References to materials from the Howard University Law School are acknowledged by the acronym HULS.

2. WH's remarks at the University of Rochester; Beverly Smith, "The First Negro Governor," *Saturday Evening Post* (17 April 1948), 15, 17, 151; Don W. Berg to the author, 7 December 1978; Newton M. Roemer, "Judge William Henry Hastie of the United States Court of Appeals, Third Circuit," *New Jersey State Bar Journal* 7 (Spring 1964), 1113; "References to W. H. Hastie," original document in the Amherst College Library; WH's remarks, Negro Newspaper Week Broadcast, NBC, 2 March 1946; Herman C. Walker to WH, 19 December 1949; "Obituary of Roberta Child [sic] Hastie"; Bond, "American Missionary Association," pp. 16–17; Herman H. Long to WH, 21 January 1966; WH to James R. Lawson, 28 April 1975; interviews: Bertha Childs and WH's friends in Knoxville, Tenn., including Leon S. Nance (23 September 1977), Margaret Gaiter (23 September 1977), Margaret Carson (23 September 1977), and Mrs. J. Herman Daves (25 September 1977).

3. WH's remarks at the University of Rochester; interview with Leon S. Nance.

NOTES

4. WH's remarks at the University of Rochester; Smith, "The First Negro Governor," p. 151; "America's First Negro Governor," an article of unspecified source and date in the Moorland-Spingarn Research Center (MSRC); Virginia Gardner, "Meet Governor Hastie," *New Masses*, 12 February 1946, p. 12.

5. Interviews with Etta Childs Walker and Louise Russell Hall.

6. Interview with Etta Childs Walker; Robert C. Weaver, "William Henry Hastie, 1904–1976," *The Crisis* (October 1976), 267.

7. Smith, "The First Negro Governor," p. 151; interview with Mrs. J. Herman Daves.

8. Interviews: Leon S. Nance, Etta Childs Walker, Mrs. J. Herman Daves; Alma Scurlock, WH's first wife, Washington, D.C., 2 September 1977; Dr. Robert C. Weaver, Distinguished Professor of Urban Affairs, Hunter College, interviewed in Philadelphia, Pa., 19 November 1976; Dr. W. Mercer Cook, educator and ambassador, Washington, D.C., 26 August 1976; and Margaret Smith, friend of Hastie's parents, Knoxville, Tenn., 24 September 1977.

9. Interviews: John J. Johnson, attorney, Knoxville, Tenn., 23 September 1977; Mrs. J. Herman Daves; Dr. Robert C. Weaver; and Carl A. Cowan, attorney, Knoxville, Tenn., 23 September 1977.

10. Emmett D. Preston, Jr., "The Development of Negro Education in the District of Columbia," *Journal of Negro Education* 9 (October 1940), 595–603; Howard H. Long, "The Support and Control of Public Education in the District of Columbia," *Journal of Negro Education* 7 (July 1938), 390–99; Mary Church Terrell, "History of the High School for Negroes in Washington," *Journal of Negro History* 2 (July 1917), 252–58; Mary G. Hundley, *The Dunbar Story (1870–1955)* (New York: Vantage Press, 1965), pp. 15–17, 66; Jervis Anderson, "Our Far-Flung Correspondents: A Very Special Monument," *The New Yorker* (20 March 1978), 101; Robert N. Mattingly, *Autobiographic Memories . . . 1897–1954: M Street-Dunbar High School* (Washington, D.C., May 1974), apparently privately printed, and enclosed with a letter to the author from Mattingly's daughter, Mrs. Mary R. M. Turner, 8 June 1977; Constance M. Green, *The Secret City* (Princeton: Princeton University Press, 1967), p. 102; interviews with Mary G. Hundley and Dutton Ferguson, WH's friend, Washington, D.C., 6 December 1978.

11. Interviews: Annette Hawkins Eaton (by Robert Grayson McGuire III), Washington, D.C., n.d. (courtesy of Mrs. Georgianna McGuire); Dr. W. Montague Cobb, physician and educator, Washington, D.C., 13 November 1976; and Dr. W. Mercer Cook.

12. Interviews with Dr. W. Montague Cobb and Dr. W. Mercer Cook.

13. Ibid.

14. Anderson, "Our Far-Flung Correspondents: A Very Special Monument," 107–8; Adelaide Cromwell Hill, "Black Education in the Seventies: A Lesson from the Past," in *The Black Seventies*, ed. Floyd B. Barbour (Boston: Porter Sargent, 1970), p. 63; Hundley, *The Dunbar Story*, pp. 14, 31–32; interviews: Annette Hawkins Eaton (by McGuire); Dr. Mercer Cook; Dutton Ferguson; and Dr. Robert C. Weaver, held at New York City, 17 June 1980.

15. Interview with Annette Hawkins Eaton (by McGuire); Carter G. Woodson and Charles H. Wesley, *The Negro in Our History*, 12th ed. (Washington, D.C.: The Associated Publishers, 1972), pp. 528–29.

16. Interview with Dr. Robert C. Weaver, held at New York City, 17 June 1980.

17. Ibid.; interview with Dr. Michael R. Winston, Director, Moorland-Spingarn Research Center, Washington, D.C., 11 September 1975.

18. WH, "Persons and Affairs," *New Negro Opinion*, 3 March 1934.

19. Edith Menard to WN, 23 November 1963, and WH's reply, 2 December 1964; interviews with Margaret Smith and Dr. W. Montague Cobb; Ivan C. Brandon, "Kisses, Hugs in Style as Class of '21 Meets," Washington *Post*, 21 June 1971.

20. David McCullough, "Mama's Boys," *Psychology Today* (March 1983), 32–37, 38.

21. Ibid., p. 38.

22. Interview with William H. Hastie, Jr., attorney, held at New York City, 21 April 1980.

23. WH's remarks at the University of Rochester.

CHAPTER 2. AMHERST COLLEGE

1. Quoted in Derrick A. Bell, Jr., "Black Students in White Law Schools: The Ordeal and the Opportunity," 3 *University of Toledo Law Review*, 539 (1970); Harold Wade, *Black Men of Amherst* (Amherst, Mass.: Amherst College Press, 1976), pp. 5, 28, 41.

2. Interview with Dr. W. Montague Cobb, physician and educator, Washington, D.C., 22 July 1980.

3. Interviews: Dr. W. Mercer Cook, ambassador and educator, Washington, D.C., 26 August 1976, and Clotill M. Houston, friend of the Hastie family and aunt of Charles Hamilton Houston, Washington, D.C., 24 September 1976; WH's father's letter to him, 3 April 1913.

4. Interview with Dr. Robert C. Weaver, Distinguished Professor of Urban Affairs, Hunter College, New York, City, 17 June 1980.

5. WH, Foreword, *Black Men of Amherst*, p. xvi; interview with Dr. W. Mercer Cook; Wade, *Black Men of Amherst*, p. 44.

6. Wade, *Black Men of Amherst*, p. 43.

7. Ibid., p. 44; W. Montague Cobb's remarks at the testimonial to Dr. Ralph Johnson Bunche, Hotel Pierre, New York City, 22 October 1949, attached to the copy of WH's letter to Cobb, 8 August 1950; W. H. Hastie, "Alpha Psi Chapter," *The Oracle* (Winter 1976/Spring 1977), 85; Dr. Merton L. Griswold to the author, 7 January 1978.

8. Interview with Robert J. McKean, Jr., attorney, New York City, 11 January 1978; WH to Alexander Meiklejohn, 12 November 1922; Winthrop H. Root, "Amherst Writing," *Amherst Student*, 11 May 1923; Winthrop Tilley, "Check!" *Amherst Student*, 11 May 1923.

9. WH to Alexander Meiklejohn, 12 November 1922.

10. Alexander Meiklejohn to WH, 14 November 1922.

11. WH to Alexander Meiklejohn, 12 November 1922; "Liberal Policies Have Been Favored by Amherst Throughout Its History," *Amherst Student*, 11 May 1923; interviews: Dr. W. Mercer Cook; Professor Walter Gellhorn, Columbia University School of Law, interviewed in Philadelphia, Pa., 20 April 1978; Oliver B. Merrill, attorney, New York City, 9 January 1978; Stephen H. Millard to the author, 12 May 1978.

12. Interview with Dr. W. Mercer Cook; *The 1925 Olio*, published by the junior class of Amherst College, pp. 199, 207.

13. *The 1925 Olio*, pp. 199–200; interviews: Oliver B. Merrill, Dr. W. Mercer Cook, and Clotill M. Houston.

14. WH, "The Colored Man in the White Man's College," an address of unspecified date, place, and occasion; WH's remarks at Jackson State University, Jackson, Miss., 28 April 1970; interview with Dr. W. Montague Cobb.

15. Interview with Sterling A. Brown, poet and educator, Washington, D.C., 27 October 1978.

16. WH, "The Colored Man . . ."

17. Interviews with Dr. W. Mercer Cook and Oliver B. Merrill; Rev. Wendell Phillips to WH, 29 January 1964; Lawrence K. Blair to the author, 27 March 1978; Newton F. McKeon to the author, 1 May 1978.

18. "References to W.H. Hastie '25," a document in the Amherst College Library; "Trackmen End Season with Middlebury Meet," *Amherst Student*, 14 May 1925; Jonathan Jay Rusch, "William M. [sic] Hastie and the Vindication of Civil Rights" (M.A. thesis, University of Virginia, 1974), p. 10.

19. Interview with Dr. W. Mercer Cook; Richard Hardwick, *Charles Richard Drew* (New York: Scribner, 1967), pp. 34–37; Wade, *Black Men of Amherst*, pp. 47–49; E. King Graves's memorandum ("Charles R. Drew, Amherst College, 1926") to Bill Brown, 6 May 1967, enclosed in Graves's letter to the author, 16 March 1978.

20. Graves's memorandum . . . to Bill Brown.

21. Hardwick, *Charles Richard Drew*, p. 41; "Brown Captures First Meet from Amherst," *Amherst Student*, 27 April 1925.

22. "Track Team to Oppose Brown at Providence," *Amherst Student*, 23 April 1925.

23. Interview with Dr. W. Mercer Cook; "America's First Negro Governor," an article in the Moorland-Spingarn Research Center (MSRC); Newton F. McKeon to the author, 1 May 1978; Newton F. McKeon to Beverly Smith, 27 January 1948, enclosing the information from the yearbook (Amherst College Library).

24. Interview with Dr. W. Mercer Cook; WH to Professor Stewart Lee Garrison, 28 March 1960; Horace W. Hewlett, ed., *In Other Words: Amherst in Prose and Verse* (Amherst: Amherst College Press, 1964), p. 203.

25. Interview with Dr. W. Mercer Cook.

26. "Gibney and Hornbeck Win Hygiene Prizes," *Amherst Student*, 8 May 1922; "Drew Takes Two Places in Intercollegiates . . . ," *Amherst Student*, 26 May 1924; "Ten Men Elected to Phi Beta Kappa," *Amherst Student*, 21 April 1924; "Honorable Mention Awarded Forty-four," *Amherst Student*, 8 October 1923; "Announce Prize Winners After Kellogg Speaking," *Amherst Student*, 17 June 1924; "References to W.H. Hastie '25"; Stephen H. Millard to the author, 30 May 1978; WH, "Free School for a Free People," an address of unspecified occasion, place, and date; Beverly Smith, "The First Negro Governor," *Saturday Evening Post* (17 April 1948) 16; interview with Dr. W. Mercer Cook; Gerald M. Mager to the author, 29 July 1980, enclosing a copy of WH's transcript.

27. *The 1925 Olio*, p. 207; interview with Sterling A. Brown.

28. Interviews: Dr. W. Mercer Cook, Dr. W. Montague Cobb, Dr. Robert C. Weaver, and Sterling A. Brown.

29. Interviews with Drs. W. Montague Cobb and W. Mercer Cook.
30. Interview with Robert J. McKean, Jr.

CHAPTER 3. OLD IRONSIDES

1. Interview with Dr. W. Mercer Cook, ambassador and educator, Washington, D.C., 26 August 1976.
2. "Pictorial Bulletin, State of New Jersey Manual Training School, Bordentown, New Jersey" (1940); "Bulletin of Information, State of New Jersey Manual Training School for Colored Youth" (1943); "Through the Years with the Echo," *The Ironside Echo*, June 1955 (all three items are housed in the New Jersey State Library at Trenton); Marion M. Thompson Wright, *The Education of Negroes in New Jersey* (New York: Arno Press, 1971), pp. 178–79.
3. Wright, *Education of Negroes*, pp. 179–80; "Pictorial Bulletin, State of New Jersey Manual Training School."
4. "Biographical Statement Concerning William H. Hastie," 26 August 1935; "Chronology" and "Through the Years with the Echo, 1927," both in *The Ironsides Echo*, June 1955; Thomas Yenser, ed., *Who's Who in Colored America*, 3rd ed. (Brooklyn: Thomas Yenser, 1933), pp. 433–34; "Pictorial Bulletin, State of New Jersey Manual Training School"; "Bulletin of Information, State of New Jersey Manual Training School for Colored Youth"; interview with Charles B. Ray, WH's successor at the Bordentown Manual Training School (BMTS), Trenton, N.J., 6 April 1977 and 23 June 1977.
5. "Pictorial Bulletin, State of New Jersey Manual Training School"; "Bulletin of Information, State of New Jersey Manual Training School"; interview with Charles B. Ray.
6. Yenser, *Who's Who in Colored America*, p. 433; interview with Vivian Anduze, WH's friend, St. Thomas, V.I., 21 July 1977.
7. Interviews: Mr. and Mrs. Boyd Eatmon, WH's colleagues at BMTS, Burlington, N.J., 12 April 1977; Charles B. Ray; Dr. Roy A. and Vivian Anduze, physician and wife, St. Thomas, V.I., 21 July 1977.
8. Interviews: Mr. and Mrs. Boyd Eatmon; Charles B. Ray; and Frances O. Grant, WH's colleague at BMTS, New York City, 16 April 1977; Vivian Anduze and Helen Reid, Valentine's former daughter-in-law, St. Thomas, V.I., both interviewed on 3 February 1978; Ivory Buck, administrator, Glassboro State College, Glassboro, N.J., 5 August 1980; Catherine E. Scott, WH's colleague at BMTS, Boston, Mass., 5 April 1977.
9. Interviews: Mr. and Mrs. Boyd Eatmon, Dr. Roy A. and Vivian Anduze, and Frances O. Grant.
10. William H. Jones, "Some Theories Regarding the Education of the Negro," *Journal of Negro Education* 9 (January 1940), 39; W. E. B. Du Bois, *The Education of Black People: Ten Critiques, 1906–1960*, ed. Herbert Aptheker (Amherst: University of Massachusetts Press, 1973), pp. 61, 65–66.
11. Henry Lee Moon, *The Emerging Thought of W. E. B. Du Bois* (New York: Simon and Schuster, 1972), p. 27.
12. W. E. B. Du Bois, "The Talented Tenth," in *The Negro Since Emancipation*, ed. Harvey Wish (Englewood Cliffs, N.J.: Prentice-Hall, 1964), p. 62.

13. W. E. B. Du Bois, *The Souls of Black Folk* (Greenwich, Conn.: Fawcett Publications, 1968), pp. 76–77.
14. Ibid., pp. 77–78.
15. John Hope Franklin, *From Slavery to Freedom*, 3rd ed. (New York: Vintage Books, 1969), pp. 390–93.
16. Ibid., p. 395.
17. Du Bois, "The Talented Tenth," p. 70.
18. "Pictorial Bulletin, State of New Jersey Manual Training School"; interview with Giles R. Wright, director, Afro-American History Program, New Jersey Historical Commission, in Willingboro, N.J., 29 February 1984; Jones, "Some Theories Regarding the Education of the Negro," p. 39.
19. Robert D. Bole and Laurence B. Johnson, *The New Jersey High School: A History* (Princeton, N.J.: Van Nostrand, 1964), p. 11; Wright, *Education of Negroes*, pp. 167–68, 173.
20. "State of New Jersey Manual Training School"; "Through the Years with the Echo"; The Interracial Committee of the New Jersey Conference of Social Work in Cooperation with the State Department of Institutions and Agencies, *The Negro in New Jersey* (New York: Negro Universities Press, 1932; repr. 1969), pp. 37–40.
21. WH's handwritten notes for unspecified occasion, place, and date; interview with Sterling A. Brown, poet and educator, Washington, D.C., 27 October 1978; Gerald F. Norman, "Colored Tennis Championships," *The Crisis* (November 1925), 18.
22. Interview with Sterling A. Brown.
23. Ibid.
24. Ibid.
25. Interview with Frances O. Grant.
26. Interview with Helen Reid.
27. Interview with Frances O. Grant; Catherine E. Scott to the author, 27 April 1977.
28. Walter J. Leonard, *Black Lawyers* (Boston: Senna & Shih, 1977), p. 9.
29. Alonzo Moron to WH, 15 November 1944 (HULS).
30. Jonathan J. Rusch, "William M. [sic] Hastie and the Vindication of Civil Rights" (M.A. thesis, University of Virginia, 1978), p. 19.
31. WH to Alonzo Moron, 20 November 1944 (HULS).

CHAPTER 4. HARVARD LAW SCHOOL

1. Charles H. Houston, "The Need for Negro Lawyers," *Journal of Negro Education* 4 (January 1935), 49.
2. Gilbert Ware, "A Word to and About Black Lawyers," *The Crisis* (August-September 1978), 247–48; Carter G. Woodson, *The Negro Professional Man and the Community, with Special Emphasis on the Physician and the Lawyer* (New York: Negro Universities Press, 1969), pp. 193–97.
3. Woodson, *Negro Professional Man*, pp. 199–245; "The Founders," *50th Annual Convention Program of the NBA* (Washington, D.C.: National Bar Association, 1975), pp. 4–6; J. Clay Smith, Jr., "The Black Association and Civil

Rights," 15 *Creighton Law Review*, 652–53 (1982); interview with James B. Morris, Sr., an NBA founder, Washington, D.C., 22 August 1975.

4. Houston, "Need for Negro Lawyers," p. 51; Jerold S. Auerbach, *Unequal Justice* (New York: Oxford University Press, 1976), pp. 65–66.

5. Charles H. Houston, "Findings on the Negro Lawyer," (3 May 1928), p. 15, a report on file at the Rockefeller Archive Center, Hillcrest, Pocantico Hills, North Tarrytown, New York; Houston, "Need for Negro Lawyers," p. 49; interview with Oliver W. Hill, attorney, Richmond, Va., 8 April 1977.

6. Walter J. Leonard, *Black Lawyers* (Boston: Senna & Shih, 1977), pp. 1–2, 9, 142–43.

7. Ibid., pp. 142–43. (Erroneously, Leonard gives 1932 as the year that Hastie received an S.J.D. degree); "Hastie on Harvard Law Review Staff," Pittsburgh *Courier*, 29 September 1928.

8. Interviews: Louis L. Redding, attorney, Cheyney, Pa., 3 October 1976; Mr. Justice William J. Brennan, Jr., Supreme Court of the United States, Washington, D.C., 24 September 1976.

9. Interview with Elwood H. Chisholm, associate general counsel, Howard University, Washington, D.C., 20 September 1976.

10. Confidential interview.

11. Ibid; Joseph P. Lash, *From the Diaries of Felix Frankfurter* (New York: Norton, 1975), p. 35; Scott Turow, *One L* (New York: Penguin Books, 1978), p. 194; Harold Wade, Jr., *Black Men of Amherst* (Amherst: Amherst College Press, 1976), p. 69.

12. Mrs. W. H. Wahlen to the author, 23 July 1980, enclosing Hastie's transcript; U.S., Congress, Senate, Committee on Territories and Insular Possessions, "Nomination of William H. Hastie for Appointment as Governor of the Virgin Islands," Hearings before a subcommittee of the Senate Committee on Territories and Insular Possessions, 21 March 1946, Vol. 2, pp. 44; Harvard Law School Annual Examinations, June 1928; a printout, "Registration of Negro Students in American Law Schools for the year 1927–1928," taken from the Annual Review of Legal Education, 1926 and 1927, the Carnegie Foundation for the Advancement of Teaching, n.d., pp. 22–23 (HULS); confidential interview; Wendy S. Yang to the author, 20 June 1977; interviews with Professors Paul A. Freund and Louis Loss, Harvard Law School, Cambridge, Mass., 7 June 1977 and 8 June 1977, respectively; Bruce A. Murphy, *The Brandeis/Frankfurter Connection* (New York: Oxford University Press, 1982), p. 201.

13. WH to Dean Erwin N. Griswold, 21 November 1949.

14. Lloyd K. Garrison to Hon. Henry F. Ashurst, 10 March 1937 (National Archives).

15. Confidential interview; Maynard J. Toll to the author, 16 September 1976.

16. Maynard J. Toll to the author, 16 September 1976; Lloyd K. Garrison to Senator Henry F. Ashurst, 10 March 1937 (National Archives).

17. Alger Hiss to the author, 9 July 1977.

18. Leonard, *Black Lawyers*, p. 45.

19. Ibid., p. 2; Genna Rae McNeil, *Groundwork: Charles Hamilton Houston and the Struggle for Civil Rights* (Philadelphia: University of Pennsylvania Press, 1983); Richard Kluger, *Simple Justice* (New York: Alfred A. Knopf, 1976). Interviews: Mr. Justice Thurgood Marshall, Supreme Court of the United States, Washington, D.C., 17 February 1977; Hon. William T. Coleman, Jr.,

secretary of transportation, Washington, D.C., 26 August 1976; Louis L. Redding; Hon. A. Leon Higginbotham, Jr., United States Court of Appeals for the Third Circuit, Philadelphia, Pa., 21 May 1979; Charles T. Duncan, dean, Howard University Law School, Washington, D.C., 10 September 1975; Professor Herbert O. Reid, Howard University Law School, Washington, D.C., 20 September 1976; Professor Ralph R. Smith, University of Pennsylvania Law School, Philadelphia, Pa., 17 May 1981; Hon. Constance Baker Motley, United States district court, New York City, 12 September 1978; Hon. Robert L. Carter, United States district court, New York City, 9 January 1978; Dean Louis L. Pollak, University of Pennsylvania Law School, Philadelphia, Pa., 10 November 1976.

20. Leonard, *Black Lawyers*, p. 14; Derrick A. Bell, Jr., to the author, 18 July 1978.

21. Interview with Professor Walter Gellhorn, Columbia University School of Law, held at Philadelphia, Pa., 20 April 1978.

22. Wade, *Black Men of Amherst*, pp. 68–69.

23. Judge Charles E. Wyzanski, Jr., to Judge A. Leon Higginbotham, Jr., 21 March 1977; interviews with Hon. William T. Coleman, Jr., and Hon. Charles E. Wyzanski, Jr., senior district judge, United States district court, Cambridge, Mass., 7 June 1977.

24. Interview with Hon. William T. Coleman, Jr.

25. Confidential interview.

26. McNeil, *Groundwork*, p. 66.

27. WH to Charles H. Woolford, 20 November 1950.

28. Interviews: Mr. Justice William J. Brennan, Jr., Elwood H. Chisholm, and Louis L. Redding; Peggy Mann, *Ralph Bunche* (New York: Coward, McCann & Geoghegan, 1975), pp. 48, 51–52.

29. Interview with Dr. Robert C. Weaver, Distinguished Professor of Urban Affairs, Hunter College, held at Philadelphia, Pa., 19 November 1976; Beverly Smith, "The First Negro Governor," *Saturday Evening Post* (17 April 1948), 153.

30. "Variety Features Graduates' Plans," *Amherst Student*, 20 April 1925; Frank I. Wheeler, "Ex-Engineer Burns His Bridges" (Moorland-Spingarn Research Center).

CHAPTER 5. GIDEON'S BAND

1. WH, "A Look at the NAACP," *The Crisis* (September 1938), 263.

2. B. Joyce Ross, *J. E. Spingarn and the Rise of the NAACP 1911–1939* (New York: Atheneum, 1972), pp. 65, 107–8, 130–31, 144–46, 162–63, 168–85; Raymond Wolters, *Negroes and the Great Depression* (Westport, Conn.: Greenwood Press, 1970), pp. 219–22.

3. Ross, *J. E. Spingarn*, pp. 46–48, 153; Wolters, *Negroes and the Great Depression*, pp. 110–11, 224–27; Gunnar Myrdal, *An American Dilemma* (New York: Harper & Brothers, 1944), p. 817; Roy Wilkins to WH, 22 August 1933.

4. Wolters, *Negroes and the Great Depression*, pp. 110–11; Roy Wilkins to WH, 22 August 1933.

5. WH to Roy Wilkins, 2 September 1933; "Minutes of the Joint Committee on National Recovery at the Whitelaw Hotel," Wednesday, 11 October 1933; "Two Years with the Joint Committee on National Recovery, 1933–1935," n.d.; Joint Committee On National Recovery, "Negro Workers under the NRA," n.d.;

"Report of the Executive Secretary, Joint Committee on National Recovery," 1 June 1935.

6. Harvard Sitkoff, *A New Deal for Blacks* (New York: Oxford University Press, 1978), pp. 47-48; WH's remarks at Dunbar High School, Washington, D.C., 1934; WH to Stephen R. Shapiro, 2 January 1963 (actually written in 1964), enclosing Hastie's response to a questionnaire; John P. Davis, "A Brief Note on the Negro and the New Deal," n.d.; "Report of the Executive Secretary, Joint Committee on National Recovery," 9 February 1935; Joint Committee on National Recovery, "Negro Workers Under the NRA."

7. Wolters, *Negroes and the Great Depression*, pp. 331-34.

8. Ibid., pp. 304-6; Sitkoff, *A New Deal for Blacks*, pp. 173-89; "Federal Judge W. H. Hastie Addresses Large Audience," Cleveland *Call and Post*, 5 August 1937.

9. Ross, *J. E. Spingarn*, pp. 52, 217-26.

10. Wolters, *Negroes and the Great Depression*, pp. 322-27.

11. WH, "A Look at the NAACP," p. 274.

12. Wolters, *Negroes and the Great Depression*, pp. 314, 332-33; Ross, *J. E. Spingarn*, pp. 239-45; Sitkoff, *A New Deal for Blacks*, pp. 254, 256.

13. Wolters, *Negroes and the Great Depression*, pp. 304-6, 314, 331-34; Ross, *J. E. Spingarn*, pp. 239-45; Sitkoff, *A New Deal for Blacks*, pp. 254, 256; "Dear Friend" letter from John P. Davis, 15 April 1935, giving notice of the conference and enclosing the "Program for National Conference"; "Two Years with the Joint Committee on National Recovery, 1933-1935."

14. Sitkoff, *A New Deal for Blacks*, p. 57; "In the Matter of a Conference Held at Howard University 18, 19 and 20 May"; Statement of William H. Hastie, 20 July 1935.

15. "Dear Friend" letter from John P. Davis; Wolters, *Negroes and the Great Depression*, pp. 354-58; Davis, "A Brief Note on the Negro and the New Deal," including italics; "Report of the Executive Secretary, Joint Committee on National Recovery," 9 February 1935.

16. Wolter, *Negroes and the Great Depression*, pp. 358-74; Myrdal, *An American Dilemma*, pp. 817-18; Jervis Anderson, *A. Philip Randolph* (New York: Harcourt Brace Jovanovich, 1972), p. 230.

17. WH to Walter White, 4 February 1936; WH to Charlie [Charles Hamilton Houston], 4 February 1936; "Whither the Negro Congress?"; "Randolph Never Opposed Red Money, Davis Says" and "Candid Camera Catches Varying Reactions During Sessions of the National Negro Congress," Washington *Afro-American*, 4 May 1940; Myrdal, *An American Dilemma*, p. 818; Wolters, *Negroes and the Great Depression*, pp. 364-66; Sitkoff, *A New Deal for Blacks*, p. 259.

18. Wolters, *Negroes and the Great Depression*, pp. 358-74; Myrdal, *An American Dilemma*, pp. 817-18; Anderson, *A. Philip Randolph*, p. 230; "Rep. Mitchell Blasts National Negro Congress," Washington *Afro-American*, 11 May 1940; "Whither the Negro Congress?"; "Randolph Never Opposed Red Money, Davis Says"; "Candid Camera Catches Varying Reactions During Sessions of the National Negro Congress"; "Bar Association Splits; Forms New Organization with Dobbins as President," Washington *Tribune*, 17 March 1936; Wolters, *Negroes and the Great Depression*, pp. 372-76; Myrdal, *An American Dilemma*, p. 818

19. Editorial, *New Negro Opinion*, 10 March 1934.

20. WH, "Persons and Affairs," *New Negro Opinion*, 24 February 1934.

21. WH, "Toward an Equalitarian Legal Order, 1930–1950," *Annals of the American Academy of Political and Social Science* 407 (May 1973), 21.
22. Ibid., pp. 19–21.
23. Ibid., p. 21.
24. Richard Kluger, *Simple Justice* (New York: Alfred A. Knopf, 1976), p. 73.
25. Ibid., pp. 74, 79–80.
26. Ibid., pp. 132, 132n.
27. Ibid., pp. 132–33.
28. Ibid., pp. 132–33.
29. Ibid., pp. 133–34; Sitkoff, *A New Deal for Blacks*, p. 219.
30. Kluger, *Simple Justice*, pp. 136–39; Sitkoff, *A New Deal for Blacks*, pp. 220–21.
31. WH, "Toward an Equalitarian Legal Order: 1930–1950," p. 21.
32. Ibid.; Gilbert Ware, ed., *From the Black Bar* (New York: Putnam, 1976), p. xxvii; interview with Oliver W. Hill, attorney, Richmond, Va., 7 April 1977.
33. Interview with Oliver W. Hill.
34. WH, "Toward an Equalitarian Legal Order: 1930–1950," p. 26.
35. Ibid., p. 24.
36. Ibid., p. 25; Sitkoff, *A New Deal for Blacks*, p. 222.
37. Sitkoff, *A New Deal for Blacks*, pp. 222–23.
38. Ibid.; Kluger, *Simple Justice*, p. 155; interview with Nell Hocutt, widow of Thomas R. Hocutt, Bronx, N.Y., 16 April 1977.
39. Interview with Conrad O. Pearson, attorney, Durham, N.C., 8 April 1977; Walter White, *Rope and Faggot* (New York: Arno Press, 1969), pp. 234–36.
40. Interview with Conrad O. Pearson; Walter B. Weare, *Black Business in the New South* (Urbana: University of Illinois Press, 1973), pp. 232–35.
41. Interviews: Conrad O. Pearson; Elwood H. Chisholm, associate general counsel, Howard University, Washington, D.C., 20 September 1976; August M. Burns III, "Graduate Education for Blacks in North Carolina," *Journal of Southern History* 46 (May 1980), 196–98.
42. Interview with Conrad O. Pearson.
43. Cecil A. McCoy to Walter White, 22 February 1933, 19 March 1933, and 21 March 1933 (NAACP Papers).
44. Cecil A. McCoy to Walter White, 21 March 1933; Walter White's wire to Cecil A. McCoy, 21 March 1933; Walter White to Charles H. Houston, 20 March 1933 (NAACP Papers).
45. Walter White, *A Man Called White* (New York: Viking Press, 1948), p. 156.
46. Mrs. W. H. Whalen to the author, 23 July 1980, enclosing a copy of Hastie's transcript; WH to Edwin R. Embree, 24 June 1930; WH to Roscoe Pound, 5 December 1931 and 24 March 1932; WH to the Registrar, Yale Law School, 30 December 1931; A. B. Hadley to WH, 31 December 1931 and 8 January 1932; Guy H. Holliday to WH, 8 January 1932; Zachariah Chafee, Jr., to WH, 9 January 1931; Joseph H. Beale to WH, 7 January 1932; WH to Joseph H. Beale, 11 February 1932; Roscoe Pound to WH, 24 December 1931; WH to A. B. Hudley, 13 April 1932; WH to Mordecai W. Johnson, 13 April 1932.
47. Memorandum from [Walter] White, 22 March 1933 (NAACP Papers); WH's remarks at Temple University, Philadelphia, Pa., 27 October 1965; interview with Karen Hastie Williams, attorney, Washington, D.C., 4 March 1978.
48. Kluger, *Simple Justice*, p. 131; interview with Conrad O. Pearson.

49. Interview with Conrad O. Pearson; "Mandamus in North Carolina 'U' Case Denied. To Appeal Decision" (NAACP press release, 31 March 1935).

50. Interview with Conrad O. Pearson; "Mandamus in North Carolina 'U' Case Denied"; WH's wire to Walter White, 25 March 1933 (NAACP Papers).

51. Kluger, *Simple Justice*, p. 157.

52. Interview with Conrad O. Pearson.

53. "Durham Youth Given Pardon by Governor," Greensboro (N.C.) *Daily News*, 17 March 1933.

54. Interview with Conrad O. Pearson; *Thomas R. Hocutt v. Thomas J. Wilson, Jr., Dean of Admissions and Registrar, and The University of North Carolina*, In the Superior Court, Durham County, Complaint, 16 March 1933; Burns, "Graduate Education," p. 196; "Memorandum Of Authorities For Defendants in Hocutt v. Wilson and University of North Carolina," n.d., pp. 14-15.

55. White, *A Man Called White*, p. 158; Kluger, *Simple Justice*, pp. 157-58; *Thomas R. Hocutt v. Thomas J. Wilson, Jr., Dean of Admissions and Registrar, and The University of North Carolina*, In the Superior Court, Durham County, North Carolina, Judgment, 28 March 1933 (NAACP Papers), M. V. Barnhill to Roy Wilkins, 6 April 1933; Florine Swaggert to the author, 5 May 1977.

56. *Thomas R. Hocutt v. Thomas J. Wilson, Jr., Dean of Admissions and Registrar, and The University of North Carolina*; M. V. Barnhill to Roy Wilkins, 6 April 1933; The case was not officially reported. Florine Swaggert to the author, 5 May 1977.

57. WH to Walter White, 31 May 1933, and Hastie's "Dear Fellows" letter of 19 June 1933, which was evidently addressed to Pearson and McCoy; interview with Conrad O. Pearson.

58. Interview with Conrad O. Pearson. No bill to provide the tuition grants was introduced into the legislature. Carolyn B. Farr to the author, 20 May 1977.

59. Interview with Conrad O. Pearson; "Barnhill Denies Writ to Hocutt," Raleigh (N.C.) *News and Observer*, 29 March 1933; White, *A Man Called White*, pp. 157-58; Kluger, *Simple Justice*, p. 158; Order submitted by plaintiff but not issued by Court in Thomas R. Hocutt v. Thomas J. Wilson, Jr.

60. WH to Walter White, 31 May 1933; WH's "Dear Fellows" letter; WH to J. M. Tinsley, 31 May 1933; interview with Conrad O. Pearson.

61. White, *A Man Called White*, p. 158; Walter White to WH, 26 May 1933.

62. Walter White to WH, 26 May 1933.

63. WH to Walter White, 31 May 1933.

64. Walter White to WH, 26 May 1933; WH's "Dear Fellows" letter; interview with Conrad O. Pearson, 17 August 1983.

65. Cecil A. McCoy to WH, 16 November 1949.

66. WH's address at a convocation, Virginia Union University, Richmond, Va., 13 November 1957.

67. Interview with Dean Charles T. Duncan, Howard University Law School, Washington, D.C., 10 September 1975.

68. 347 U.S. 483 (1954); Albert P. Blaustein and Clarence Clyde Ferguson, Jr., *Desegregation and the Law*, 2nd rev. ed. (New York: Vintage Books, 1962); interview with Professor Clarence Clyde Ferguson, Harvard Law School, Cambridge, Mass., 7 June 1977; Kluger, *Simple Justice*.

69. Mr. Justice Thurgood Marshall's remarks were made at Howard University Law School, 18 November 1978.

CHAPTER 6. TEACHERS' PAY

1. Walter White to WH, 8 September 1933.
2. William B. Gibbs, Jr., to Thurgood Marshall, with enclosure, 10 December 1936 (NAACP Papers).
3. "Editorial," *Journal of Negro Education* 9 (January 1940), 1.
4. Thurgood Marshall to Dr. J. M. Tinsley, 30 August 1937 (NAACP Papers).
5. Richard Kluger, *Simple Justice* (New York: Alfred A. Knopf, 1976), p. 214.
6. WH, "Toward an Equalitarian Legal Order," *Annals of the American Academy of Political and Social Science*" 407 (May 1973), 24.
7. Ibid., p. 26.
8. Walter White to WH, 26 May 1933.
9. George W. Streator's wire to Walter White, 14 September 1933, and Walter White to WH, 14 September 1933 (NAACP Papers).
10. Walter White, *A Man Called White* (New York: Viking Press, 1948), pp. 125–30; Harvard Sitkoff, *A New Deal for Blacks* (New York: Oxford University Press, 1978), pp. 144–48; Francis Butler Simkins, *A History of the South*, 3rd ed. (New York: Knopf, 1967), pp. 517–18; Langston Hughes, *Fight for Freedom: The Story of the NAACP* (New York: Norton/A Berkley Medallion Book 1962), p. 87; Kluger, *Simple Justice*, pp. 144–46, 153–54, 160–61; WH, Introduction, *In Any Fight Some Fall*, by Geraldine R. Segal (Rockville, Md: Mercury Press, 1975), pp. 8–9; Loren Miller, *The Petitioners* (Cleveland: Meridian Books, 1966), pp. 265–76.
11. George W. Streator's wire to Walter White, 14 September 1933, and Walter White's wire to WH, 16 September 1933 (NAACP Papers): WH to Walter White, 11 September 1933; WH to Conrad O. Pearson, 11 September 1933.
12. "W. H. Hastie, Homer Brown on N.A.A.C.P. Legal Body," *Washington Tribune*, 7 December 1933.
13. WH to Walter White, 11 September 1933; WH to Conrad O. Pearson, 11 September 1933.
14. WH to Walter White, 22 September 1933.
15. "Federal Court May Get Negroes' Case," newspaper article attached to a letter to Walter White and evidently from WH, 11 September 1933; WH to Walter White, 22 September 1933.
16. WH to Walter White, 22 September 1933.
17. J. N. Mills to Walter White, 28 September 1933 (NAACP Papers).
18. Ibid.
19. WH to Walter White, 22 September 1933.
20. "Federal Court May Get Negroes' Case."
21. WH to Walter White, 22 September 1933.
22. WH to Walter White, 13 October 1933.
23. WH to Walter White, 22 September 1933.
24. WH to Walter White, 8 September 1933.
25. WH to Walter White, 22 September 1933.
26. Interview with Brenda A. Spears, attorney, New York City, 9 October 1983.
27. WH to Walter White, 22 September 1933.
28. *Alston v. School Board of City of Norfolk*, 112 F. 2d 992 (1940).

29. Walter White to "Dear Cecil [A. McCoy] and Conrad [O. Pearson]," 25 September 1933 (NAACP Papers).

30. WH to Conrad O. Pearson, 2 October 1933.

31. Walter White to "Dear Cecil [A. McCoy] and Conrad [O. Pearson]," 25 September 1933 (NAACP Papers).

32. WH to Walter White, 22 September 1933.

33. Thurgood Marshall, "STATEMENT to the Joint Committee on Teachers' Salaries in Virginia concerning cases of Aline E. Black and Melvin O. Alston," 30 October 1939; and Thurgood Marshall to Dr. J. M. Tinsley, 30 August 1937 (NAACP Papers).

34. WH, "Toward an Egalitarian Legal Order," p. 26.

35. Walter White to "Dear Cecil [A. McCoy] and Conrad [O. Pearson]," 25 September 1933 (NAACP Papers); William B. Gibbs, Jr., to Thurgood Marshall, October 1936 (day unspecified; NAACP Papers); Jill M. Singer, "Black Educator Honored for Equal Pay Struggle," Washington *Post*, 2 August 1979, (the "Maryland Weekly" section); and "Equal Pay for Colored Teachers in Maryland" (NAACP Papers); William B. Gibbs, Jr., to Thurgood Marshall, 9 February 1937 and 6 July 1937; Thurgood Marshall to William B. Gibbs, Jr., 5 April 1937 (NAACP Papers); NAACP, "Equalization of Teachers' Salaries" (New York: NAACP, 1 March 1938), pp. 1-3; Kluger, *Simple Justice*, p. 214.

36. "Equalization of Teachers' Salaries," pp. 3-4; "Federal Court Hears First Case for Equalization of Salaries for Teachers," NAACP press release, n.d.; Thurgood Marshall to Professor Ralph Harlow, 8 January 1938 (NAACP Papers); "Editorial," *Journal of Negro Education* 9 (January 1940), 4.

37. *Mills* v. *Lowndes*, 26 F. Supp. 796-806 (1939); "Equalization of Teachers' Salaries," pp. 1-3; Kluger, *Simple Justice*, p. 214.

38. *Mills* v. *Board of Education of Anne Arundel County*, 30 F. Supp. 245-49 (1939).

39. Ibid., p. 249; "Federal Court Hears First Case for Equalization of Salaries for Teachers"; *Mills* v. *Lowndes* 26 F. Supp. 792 (1939) and 30 F. Supp. 245 (1939); "Federal Court Orders Equal Salaries for Teachers," *The Crisis* (January 1940), 10-12, 29; Dewey A. Stokes, Jr., "Negro Education and Federal Courts, 1877-1955" (M.A. thesis, University of North Carolina, 1955), p. 51.

40. "The Nation's Honor Roll for 1939," *The Nation* (6 January 1940), 6; Morris L. Ernst to WH, 28 November 1939, and WH's reply, 29 November 1939; Stokes, "Negro Education and Federal Courts," p. 5.

41. Walter White to Alfred Baker Lewis, 21 December 1933.

42. "Federal Court Orders Equal Teachers' Salaries In Maryland County," NAACP press release, n.d.; "Equalization of Teachers' Salaries," p. 4; Stokes, "Negro Education and Federal Courts," pp. 51-52; Thurgood Marshall to P. B. Young, Jr., 29 March 1938; Confidential Memorandum to Members of the Joint Committee on Teachers' Salaries in Virginia, from the Legal Staff (signed by Marshall), 13 May 1938; Melvin O. Alston to Thurgood Marshall, 6 October 1938; and Thurgood Marshall to Mevlin Alston, J. Thomas Hewin, and P. B. Young, all on 19 October 1938; "Melvin O. Alston Files Petition in Second Salary Suit," *Virginia Teachers Bulletin* (November 1939), 24; Aline Black to Thurgood Marshall, 9 November 1938; P. B. Young to Thurgood Marshall, 4 November 1938; Memorandum to Roy Wilkins and George Murphy (from

Thurgood Marshall), 2 November 1938 (all in NAACP Papers); Stokes, "Negro Education and Federal Courts," p. 52; interview with Oliver W. Hill, attorney, Richmond, Va., 7 April 1977.

43. J. Thomas Hewin, Jr., to Thurgood Marshall, 12 October 1938; Melvin O. Alston to Thurgood Marshall, 28 July 1939 and 21 August 1939; Thurgood Marshall to Melvin O. Alston, 3 and 29 August 1939 (all in NAACP Papers); interview with Oliver W. Hill; *Alston v. School Board of City of Norfolk.*

44. Thurgood Marshall to P. B. Young, 19 October 1938 (NAACP Papers).

45. Thurgood Marshall to Melvin O. Alston, P. B. Young, and J. Thomas Hewin, all on 19 October 1938; Melvin O. Alston to Thurgood Marshall, 28 July 1939, 21 August 1939, and 8 October 1939; Thurgood Marshall to Melvin O. Alston, 29 August 1939, and to W. L. Davis, 20 September 1939; Thurgood Marshall to P. B. Young, Sr., 11 October 1939; P. B. Young, Sr., and P. B. Young, Jr., to Thurgood Marshall, both letters dated 13 October 1939; Melvin O. Alston to Thurgood Marshall, 8 October 1939; memorandum to Andy Ransom and Bill Hastie from Thurgood Marshall, 11 October 1939; WH to Thurgood Marshall, undated but stamped 13 October 1939 (all in NAACP Papers).

46. Thurgood Marshall to P. B. Young, Sr., 11 October 1939; P. B. Young, Jr., to Thurgood Marshall, 13 October 1939; memorandum to WW [Walter White] and RW [Roy Wilkins] from Andy and Thurgood [Leon A. Ransom and Thurgood Marshall], 18 October 1939 (all in NAACP Papers).

47. "Melvin O. Alston Files Petition in Second Salary Suit," *Virginia Teachers Bulletin* (November 1939), 24 (NAACP Papers); *Alston v. School Board of City of Norfolk*; interview with Oliver W. Hill.

48. White, *A Man Called White*, pp. 104–10.

49. Interview with Oliver W. Hill.

50. Ibid.

51. WH to Samuel B. Horovitz and to Addison T. Cutler, both letters dated 10 June 1940; WH to Dr. Garnet C. Wilkinson, 3 May 1940; White, *A Man Called White*, p. 114.

52. *Alston v. School Board of City of Norfolk.*

53. Ibid.

54. White, *A Man Called White*, p. 114; Stokes, "Negro Education and Federal Courts, p. 53; WH, "Toward 1950," an address of unspecified date, place, and occasion.

CHAPTER 7. NEW NEGRO ALLIANCE

1. John Hope Franklin and the Editors of Time-Life Books, *An Illustrated History of Black Americans* (New York: Time-Life Books, 1970), p. 129; WH, "The Why of the Alliance," *New Negro Alliance Year Book*, 1st ed., 1939 (from the files of H. Naylor Fitzhugh), p. 14; Bernard Braxton, "The Negro in the History of Washington," *New Negro Alliance Year Book*, pp. 4–6; Constance M. Green, *The Secret City* (Princeton: Princeton University Press, 1967), p. 232; *New Negro Alliance, A Corporation, et al., vs. Sanitary Grocery Co., Inc., A Corporation,* Brief for the Petitioners, Supreme Court of the United States, October Term 1937, pp. 31–39; interviews with John Aubrey Davis, professor of political science, The City College of the City University of New York, held at

New Rochelle, N.Y., 21 April 1977, and H. Naylor Fitzhugh, Pepsi Cola Company executive, Purchase, N.Y., 29 November 1977.

2. Interviews with John Aubrey Davis and H. Naylor Fitzhugh; Michele F. Pacifico, "A History of the New Negro Alliance of Washington, D.C., 1933–1941" (M.A. thesis, The George Washington University, 1983), p. 35.

3. John A. Davis, "We Win the Right to Fight for Jobs," *Opportunity* (August 1938), 231; interview with Dr. John Aubrey Davis.

4. "Facts About the Alliance," *New Negro Alliance Year Book*, p. 48; "The Programme of the New Negro Alliance," *New Negro Opinion*, 16 December 1933; interview with Dr. John Aubrey Davis.

5. "Come In! The Alliance Needs You," *New Negro Opinion*, 6 January 1934; Ralph J. Bunche, "The New Negro Alliance," in *Black Nationalism in America*, ed. John H. Bracey, Jr., August Meier, and Elliott Rudwick (Indianapolis: Bobbs-Merril, 1970), pp. 384–86; "Facts About the Alliance," p. 48; "Howard Professor Scores Racialism in N.N.A. Program," *New Negro Opinion*, 16 March 1935; "Bunche Calls N.N.A. Program 'Vicious' at Sunday Forum," *New Negro Opinion*, 23 March 1935; "Alliance Leaders Answer Critics," *New Negro Opinion*, 23 March 1935; Editorial, *New Negro Opinion*, 23 March 1935.

6. WH, "The Why of the Alliance," pp. 14, 19.

7. Interview with H. Naylor Fitzhugh, 30 November 1977; WH, "Toward 1950," remarks of an unspecified date and occasion.

8. Hastie might well have held additional offices in 1940 and 1941. The Alliance's records do not enable us to determine all the officers for those years. Pacifico, "History of the New Negro Alliance," pp. 223–27.

9. This section about the civil rights bill is based on the following sources: untitled and undated flyer; Roger N. Baldwin to Arthur E. Spingarn, 15 May 1935; Walter White to WH, 20 May 1935 and 29 July 1935; WH to Roger N. Baldwin, 22 October 1935; Charles Hamilton Houston to WH, 14 February 1936, and Houston's memorandum for Roger N. Baldwin, 15 February 1936; Roger N. Baldwin to WH, 17 February 1936; "Civil Rights Bill for D.C. Before Congress Would Bar Discrimination in Restaurants and Theatres," a flyer, n.d.; "Mass Meeting for a Civil Rights Bill in Washington," a flyer, n.d.; Pacifico, "History of the New Negro Alliance," pp. 87–89.

10. This account of the Scottsboro case is based on the following sources: Walter White, *A Man Called White* (New York: Viking Press, 1948), pp. 125–30; Harvard Sitkoff, *A New Deal for Blacks* (New York: Oxford University Press, 1978), pp. 144–48; Richard Kluger, *Simple Justice* (New York: Knopf, 1976), pp. 145–46; Francis Butler Simkins, *A History of the South*, 3rd ed. (New York: Knopf, 1967), pp. 517–18; Langston Hughes, *Fight for Freedom: The Story of the NAACP* (New York: Norton/A Berkley Medallion Book, 1962), p. 87; WH, "Introduction," *In Any Fight Some Fall*, by Geraldine R. Segal (Rockville, Md.: Mercury Press, 1975), pp. 8–9; WH, "Persons and Affairs," *New Negro Opinion*, 12 January 1935 and 6 April 1935.

11. This account of the controversy is based on the following sources: editorials that appeared in *The Crisis* in January, February, April, May, and June of 1934; WH, "Persons and Affairs," *New Negro Opinion*, 25 January 1934, 3 February 1934, and 18 August 1934.

12. This account of the dispute is based on the following sources: WH, "Persons and Affairs," *New Negro Opinion*, 1 September 1934, 1 December 1934,

8 December 1934, and 19 January 1935; "School Ban on Crisis Magazine," Washington *Afro-American*, 22 February 1936; "NAACP Raps School Board for Crisis Ban," Washington *Afro-American*, 7 March 1936; Walter C. Daniel, "*The Crisis* and *Opportunity* vs. Washington, D.C., Board of Education," *The Crisis*, (June/July 1978), 205-6; Walter C. Daniel to the author, 17 July 1978; WH, "In the Matter of *The Crisis* and *Opportunity*: Memorandum for the Board of Education," 2 December 1936; Garnet C. Wilkinson, memorandum in re The Crisis Magazine, 4 March 1936; Frank W. Ballou to Roy Wilkins, 20 March 1936; Adeline (Mrs. Frank W.) Ballou to the author, 12 January 1978; "School Board Is Flayed for Crisis Stand," Washington *Tribune*, 25 February 1936; "Crisis Ban to Be Issue at NAACP Executive Meet," Washington *Tribune*, 10 April 1936; WH to the Board of Education, 3 March 1936 and 20 June 1936; Lewis N. Walker, Jr., "The Struggles and Attempts to Establish Branch Autonomy and Hegemony: A History of the District of Columbia Branch National Association for the Advancement of Colored People, 1912-1942" (Ph.D. dissertation, University of Delaware, 1979), pp. 191-93, 204; Pacifico, "History of the New Negro Alliance," pp. 99, 226; WH, "In the matter of *The Crisis* and *Opportunity*: Supplementary Memorandum to the Board of Education," 15 December 1936; "Board of Education Is Praised," a newspaper article; Roy Wilkins to Mrs. Henry Grattan Doyle, 3 April 1936; Marion Wade (Mrs. Henry Grattan Doyle) to the author, 16 January 1977; Garnet C. Wilkinson to Elmer A. Carter, 11 March 1936; Elmer A. Carter to Garnet C. Wilkinson, 17 March 1936.

13. Interview with H. Naylor Fitzhugh, 30 November 1977.

14. M. F. [M. Franklin] Thorne to Aubrey G. Russell, 13 September 1933; J. Aubrey Davis to John Hartford, 14 September 1933; "NNA History Highlights," *New Negro Opinion*, 13 October 1934; WH, "Persons and Affairs," *New Negro Opinion*, 7 April 1934; interviews with H. Naylor Fitzhugh, 29 November 1977, and Dutton Ferguson, WH's friend, Washington, D.C., 6 December 1978.

15. [M.] Franklin Thorne to Aubrey G. Russell, 29 September 1933; "NNA History Highlights"; "Court Denies Picket Right to Alliance," Washington *Tribune*, 31 July 1937; "The Public Swings to A & P Stores," *New Negro Opinion*, 16 December 1933; interviews with H. Naylor Fitzhugh and Dutton Ferguson.

16. Dutton Ferguson, "Thoughts While Picketing," *New Negro Year Book*, p. 20.

17. Interview with H. Naylor Fitzhugh, 29 November 1977; "NNA History Highlights"; "Alliance vs. High's Ice Cream," *New Negro Opinion*, 16 December 1933.

18. By 1936 the campaign had produced seventy-five thousand dollars in income for blacks as salespersons. Pacifico, "History of the New Negro Alliance," p. 71.

19. Carl B. Swisher, *The Theory and Practice of American National Government* (Boston: Houghton Mifflin, 1951), p. 671; Arthur M. Schlesinger, Jr., *The Crisis of the Old Order 1919-1933* (Boston: Houghton Mifflin, 1957), pp. 238-39; ibid., *The Coming of the New Deal* (Boston: Houghton Mifflin, 1959), p. 136.

20. Carl B. Swisher, *American Constitutional Development* (Boston: Houghton Mifflin, 1943), pp. 806-12.

21. *Transcript of Record*, Court of Appeals of the District of Columbia, January Term, 1934, no. 6187, Special Calendar; "The New Negro Alliance, A Corporation; J. Aubrey Davis and [M. Franklin] Thorne, Intercollegiate League

of Industrial Democracy, An Unincorporated Body, Appellants, v. Harry Kaufman, Inc., A Corporation," Filed 13 March 1934 (Printed 11 June 1934), pp. 7–11; "The Kaufman Case: A Statement," *New Negro Opinion*, 6 January 1934; "Kaufman Files Motion to Dismiss Alliance Case," *New Negro Opinion*, 31 March 1934; "Alliance Wins Second Round in Kaufman Fight," *New Negro Opinion*, 12 May 1934; "Appeals Court Hears Picket Rights Argued," *New Negro Opinion*, 8 December 1934; "Preliminary Injunction Granted," *New Negro Opinion*, 6 January 1934; "Alliance Asks Court Examination of Kaufman Facts," *New Negro Opinion*, 7 April 1934; "Alliance Gets Extension in Kaufman Case," *New Negro Opinion*, 28 April 1934.

22. "Editorial," *New Negro Opinion*, 11 January 1934.

23. *Transcript of Record*, Court of Appeals of the District of Columbia, No. 6187, pp. 3–5.

24. Howard Fitzhugh, "Kaufman Case Explained," *New Negro Opinion* 11 January 1934; "Appeals Filed in Injunction Fight," *New Negro Opinion*, 18 January 1934.

25. "Ice Cream Co. Refuses to Hire Colored," *New Negro Opinion*, 28 July 1934; "Pickets Halt As High Co. Hears N.N.A.," *New Negro Opinion*, 18 August 1934; "Postponement Asked by High," *New Negro Opinion*, 25 August 1934; "Alliance Attorney Argues for Right to Picket, After Trick 'Dismissal' Fails," *New Negro Opinion*, 15 September 1934; "Alliance Wins First Round in High Jobs Fight," *New Negro Opinion*, 8 September 1934; Return to Rule Show Cause, "In the Supreme Court of the District of Columbia Holding An Equity Court, Equity #57545, *L. W. High v. the New Negro Alliance*, A Corporation, and Howard Fitzhugh, individually and as administrator of the New Negro Alliance [and] Frank [M. Franklin] Thorne, individually and as a member and representative of the New Negro Alliance," n.d., pp. 3–4; Pacifico, "History of the New Negro Alliance," pp. 63–66.

26. "High Ice Cream Co. Gets Verdict in Picket Case," *New Negro Opinion*, 29 September 1934; "Alliance Attorney Argues for Right to Picket, After Trick 'Dismissal' Fails"; Philip Rosenfeld to Frank [M. Franklin] Thorne, 18 September 1934, enclosing a copy of the injunction.

27. Leon A. Ransom, "The Supreme Court Speaks . . ." *New Negro Alliance Year Book*, p. 17; Pacifico, "History of the New Negro Alliance," pp. 69–71.

28. *Transcript of Record*, The New Negro Alliance, A Corporation, And William H. Hastie, Individually and As Administrator and As A Member of the New Negro Alliance, and Harry A. Honesty, Individually and As Deputy Administrator and As A Member of The New Negro Alliance, A Corporation, Appellants, vs. Sanitary Grocery Company, Inc., A Corporation, United States Court of Appeals for the District of Columbia, No. 6836 (April Term, 1936), pp. 1–9, 16.

29. *The New Negro Alliance, A Corporation, et al. Appellants vs. Sanitary Grocery Co., Inc., Appellee*, United States Court of Appeals for the District of Columbia, No. 6836 (April Term, 1937), *Brief for Appellants*, pp. 1–13.

30. "Court Denies Picket Right to Alliance," Washington *Tribune*, 31 July 1937; "Court Bars Job Picketing," Washington *Afro-American*, 31 July 1937; Ransom, "The Supreme Court Speaks," p. 18; Pacifico, "History of the New Negro Alliance," pp. 141–46; *The New Negro Alliance, A Corporation, et al., vs. Sanitary Grocery Co., Inc., A Corporation, Petition for A Writ of Certiorari to*

the United States Court of Appeals for the District of Columbia and Brief In Support Thereof, Supreme Court of the United States.

31. W. F. P., "Recent Decisions," 13 St. John's Law Review, 172-73 (1938).

32. Interview with Dr. John A. Davis; Pacifico, "History of the New Negro Alliance," pp. 147-48, 218; Petitioners' Brief at 20, New Negro Alliance vs. Sanitary Grocery Co., Inc., 303 U.S. 552 (1938).

33. Petitioners' Brief at 9, 29-30, 37, New Negro Alliance vs. Sanitary Grocery Co., Inc., 303 U.S. 552 (1938).

34. New Negro Alliance vs. Sanitary Grocery Co., Inc., 303 U.S. 552 (1938).

35. Ibid.

36. Interview with Dr. John A. Davis.

37. Interview with H. Naylor Fitzhugh.

38. "N.N.A. Bi-Monthly Bulletin," 13 January 1937.

CHAPTER 8. A FOUNDING FATHER

1. Harold L. Ickes, The Secret Diary of Harold L. Ickes, Vol. 1 (New York: Simon and Schuster, 1953), p. 416; Harold L. Ickes, "My Twelve Years with FDR," Saturday Evening Post (26 June 1948), 79; WH to Clark Foreman, 18 October 1933, and Foreman's reply, 19 October 1933; Newton M. Roemer, "Judge William Henry Hastie of the United States Court of Appeals, Third Circuit," New Jersey State Bar Journal 7 (Spring 1964), 1130; "Biographical Statement Concerning Nathan R. Margold"; WH's interview (by Jerry N. Hess) for the Harry S. Truman Library, 5 January 1972.

2. WH "America from My Vantage Point," an address of unspecified date and occasion; John H. Hollands to the author, 20 May 1978 and 4 June 1978.

3. Interviews with Dr. Robert C. Weaver, Distinguished Professor of Urban Affairs, Hunter College, held at Philadelphia, Pa., 19 November 1976, and New York City, 17 June 1980.

4. Ibid.; WH, "America from My Vantage Point"; "Judge Hastie Highly Lauded at Banquet," Pittsburgh Courier, 10 April 1937.

5. Frederick S. Weaver, "As It Seems," Washington Tribune, 17 April 1936; Beverly Smith, "The First Negro Governor," Saturday Evening Post (17 April 1948), 153-54; John H. Thompson, "Forty-Five Years: Highlights from the Daily News," St. Thomas Daily News, 1 August 1975 (see also the articles by Aubrey C. Ottley, Ivan Brandon, Carter Hague, James O'Bryan, and WH in this issue of the Daily News, as well as its editorial of 2 February 1937); Darwin D. Creque, The U.S. Virgins and the Eastern Caribbean (Philadelphia: Whitmore, 1968), pp. 90-96; John F. Grede, "The New Deal in the Virgin Islands, 1931-1941" (Ph.D. dissertation, University of Chicago, 1962, pp. 68-81; interviews on St. Thomas, V.I.: James A. Bough, attorney (15 July 1977), Aubrey C. Ottley, postmaster (16 July 1977), and Darwin D. Creque, author (19 July 1977).

6. Gordon K. Lewis, The Virgin Islands (Evanston, Ill.: Northwestern University Press, 1972), pp. 42-67, 72-89; Valdemar A. Hill, Sr., Rise to Recognition (privately printed, 1971), pp. 78-81; interview with James A. Bough.

7. Lewis, The Virgin Islands, pp. 43-46, 51-52.

8. Ibid., pp. 52-53; Hill, Rise to Recognition, pp. 78-81; WH's interview (by Hess) for the Truman Library.

9. WH's interview (by Hess) for the Truman Library; interview with Roy Bornn, businessman, St. Thomas, V.I., 18 July 1977.

10. James A. Bough, "Is Commonwealth Status the Next Step?" *Virgin Islands Bar Journal* 5 (1974), 2–3.

11. Ibid., pp. 4–5.

12. Ibid., James A. Bough and Roy C. Macridis, eds., *Virgin Islands* (Wakefield, Mass.: Walter F. Williams, 1970), pp. 40, 48.

13. WH's interview (by Hess) for the Truman Library.

14. Ickes, *Secret Diary*, p. 416.

15. Ickes, "My Twelve Years with FDR," pp. 79, 81; ibid., *The Secret Diary of Harold L. Ickes*, Vol. 2 (New York: Simon and Schuster, 1954), p. 94; "Raps Hastie as U.S. Judge," New York *Amsterdam News*, 6 March 1937; "King Gives In on Hastie Nomination," Washington *Afro-American*, 13 March 1937; "Random Shots," *Flash*, 20 March 1937 (black magazine, published in Washington, D.C.); "W.H. Hastie Is Quietly Married," an article in the Beck Cultural Exchange Center (BCEC).

16. Ickes, *Secret Diary*, Vol. 2, p. 94; ibid., "My Twelve Years with FDR," p. 81.

17. Interview with Alphonso A. Christian, Sr., attorney, St. Thomas, V.I., 23 July 1977.

18. Creque, *The U.S. Virgins and the Eastern Caribbean*, pp. 100–1; Ivan Brandon, "St. Thomas Women Challenged Law in 1935," St. Thomas *Daily News*, 1 August 1975; Benita Cannon, ". . . And in St. Croix Women Filed Suit to Vote in 1936," St. Thomas *Daily News*, 1 August 1975; Cyril Michael's memorandum to Judge William H. Hastie, 8 April 1938; interview with Judge Cyril Michael, retired, St. Thomas, V.I., 3 February 1978.

19. Creque, *The U.S. Virgins and the Eastern Caribbean*, p. 108.

20. Ibid., pp. 108, 112; Hill, *Rise to Recognition*, pp. 91–92; interview with James A. Bough.

21. "Women Must Vote," *Progressive Guide*, 2 July 1938; editorials, *Progressive Guide*, 23 April 1938, 11 June 1938, and 30 July 1938; "Attention Everybody," *Progressive Guide*, 7 May 1938; "Excerpts from 'Standard of Living and Wages' by Valdemar Hill," *Progressive Guide*, 13 August 1938; interviews: Omar Brown, businessman, St. Thomas, V.I., 21 July 1977; Alma Scurlock, WH's first wife (who remarried after their divorce), Washington, D.C., 2 September 1977.

22. Aubrey C. Ottley, "Wages," *Progressive Guide*, 30 July 1938; ibid., "A Serious Problem," *Progressive Guide*, 25 June 1938.

23. "A Word on Business," *Progressive Guide*, 2 July 1938.

24. Editorials, *Progressive Guide*, 2 July 1938, 9 April 1938, and 7 January 1939.

25. "Attention Everybody," *Progressive Guide*, 30 April 1938; "Things I Never Knew Till Now," *Progressive Guide*, 30 April 1938; "Adult Education Program Needed," *Progressive Guide*, 25 June 1938; "The Class of 1938 Graduates," St. Thomas *Daily News*, 28 June 1938; printed program for the graduation exercises, Friday, 24 June 1928, 8:30 P.M.

26. "The Class of 1938 Graduates"; "Civilian Conservation Corps 1933–1940," *Progressive Guide*, 6 April 1940.

27. Alvaro de Lugo, William H. Hastie, and J. S. Moorhead to Harold Hubler,

14 March 1939, enclosing a "Memorandum of Interview" on Monday, 13 March 1939; WH to William Trent, 15 March 1939, and Trent's reply, 24 March 1939; Harold A. Hubler to Alvaro de Lugo, William H. Hastie, and J. S. Moorhead, 15 March 1939 (all from Howard University Law School [HULS] collection); interview with William Trent, former adviser to Secretary Ickes, Greensboro, N.C., 27 November 1977.

28. WH to William Trent, 15 March 1939 (HULS); Creque, *The U.S. Virgins and the Eastern Caribbean*, pp. 112–13.

29. Interview with Alphonso A. Christian, Sr.; articles in St. Thomas *Daily News*, 20 January 1937 and 19 March 1937.

30. "Hastie Confirmed Fed. Judge for V.I." St. Thomas *Daily News*, 20 March 1937; "Mr. Hastie's Appointment," St. Thomas *Daily News*, 15 March 1937; editorials, St. Thomas *Daily News*, 19 March 1937 and 28 April 1937; "The Judge Hastie Dinner" and an editorial, St. Thomas *Daily News*, 20 May 1937; Adolph Gereau, "Judge Hastie Arrives at Post," Pittsburgh *Courier*, 8 May 1937; interviews with Adolph Gereau, newspaper reporter, St. Croix, V.I., 22 July 1977, and Roy Bornn.

31. Interview with Roy Bornn.

32. WH to the Editor, *Color Magazine*, 14 October 1953, enclosing corrected pages of Dorothy Anderson's account of her interview with him for an article; interviews on St. Thomas, V.I.: Aubrey C. Ottley; Geraldo Guirty, newspaper reporter, 19 July 1977; Enid Baa, librarian and historian, 22 July 1977; Dr. Roy A. Anduze, physician, 21 July 1977.

33. Interview with Alphonso A. Christian, Sr.

34. "Observation Tower," St. Thomas *Daily News*, 7 and 9 May 1938; "Here & There," St. Thomas *Daily News*, 6 May 1938; interview with Dr. Roy A. Anduze.

35. "Round A' Round," *Progressive Guide*, 14 May 1938; interview with Alma Scurlock.

36. "Negro Judge Resents Bid to Reich Officers," New York *Times*, 17 May 1938; "Judge Hastie Resents Dance Bids to Reich," "Mrs. Hastie, Wife of Judge Hastie, Challenges Color Barrier in Virgin Islands," "He Said 'No' to a Nazi Ship Party," and Adolph Gereau, "Hasties Flay Color Line in Virgin Islands" (all in Moorland-Spingarn Research Center [MSRC]); interviews: Adolph Gereau, Alma Scurlock, and Roy Bornn.

37. Maud Proudfoot and Elsa A. Lindqvist, "The Other Side of the Story About the High School Benefit Dance," St. Thomas *Mail Notes*, 19 May 1938.

38. "Round A' Round"; Gereau, "Hasties Flay Color Line"; editorial, St. Thomas *West End News*, 8 June 1938, reprinted in *Progressive Guide*, 18 June 1938.

39. WH to the Editor, St. Thomas *Mail Notes* (published 20 May 1938); interview with Alma Scurlock.

40. Lelia A. Pendleton, "Our New Possessions—The Danish West Indies," *Journal of Negro Education* 2 (July 1917), 278–79; Ambassador Henrik de Kaufman's address, 3 July 1948, published in St. Thomas *Daily News*, 8 July 1948; WH to William W. Boyer, Jr., 2 June 1949 (courtesy of Boyer); William W. Boyer, Jr., *America's Virgin Islands* (Durham: Carolina Academic Press, 1983), pp. 29–31, 55–58; interviews: Darwin D. Creque, Enid Baa, James A. Bough, and Professor William W. Boyer, Jr., Department of Political Science, University of Delaware, on St. Thomas, V.I., 21 July 1977.

41. WH's interview (by Hess) for the Truman Library; "Hastie Quits Virgin Isle Judgeship," *Washington Tribune*, 4 February 1939; President Franklin D. Roosevelt to WH, 28 January 1939.

43. Interview with Alphonso A. Christian, Sr.; WH's interview (by Hess) for the Truman Library.

CHAPTER 9. STIMSON'S STABLES

1. Lloyd K. Garrison to Mordecai Johnson, 10 October 1938 (Howard University Law School [HULS]); WH, "Legal Aspects of Racial Discrimination in the Armed Forces," *National Bar Journal* 2 (June 1944), 17; Walter White, *A Man Called White* (New York: Viking, 1948), p. 186.

2. Ulysses Lee, *The Employment of Negro Troops* (Washington, D.C.: U.S. Government Printing Office, 1969), p. 31; C. K. Albertson to WH, 18 October 1941, enclosing the thirtieth chapter of Bullard's memoirs.

3. Lee, *The Employment of Negro Troops*, pp. 44–45; Major General H. Ely's memoranda for the Chief of Staff, one each dated 10 November 1925, and 30 October 1925; WH's handwritten notes.

4. Lee, *The Employment of Negro Troops*, pp. 39, 49–50.

5. Ibid., p. 68; Phillip McGuire, "Black Civilian Aides and Problems of Racism and Segregation in the United States Armed Forces: 1940–1950" (Ph.D. dissertation, Howard University, 1975), pp. 24–28.

6. Correspondence in HULS.

7. Warren A. Seavey to WH, 18 July 1940, and WH's reply, 22 July 1940 (HULS).

8. McGuire, "Black Civilian Aides," p. 43.

9. Lee, *The Employment of Negro Troops*, pp. 73–74.

10. Ibid., pp. 74–76; White, *A Man Called White*, p. 186.

11. McGuire, "Black Civilian Aides," pp. 51–58; Lee, *The Employment of Negro Troops*, pp. 74–76; White, *A Man Called White*, pp. 186–87.

12. White, *A Man Called White*, pp. 187–88; Richard Bardolph, *The Negro Vanguard* (New York: Vantage Press, 1961), p. 353; "Steve Early Knees Cop," Washington *Afro-American*, 2 November 1940.

13. Bardolph, *The Negro Vanguard*, pp. 353–54.

14. WH's interview (by Jerry N. Hess) for the Harry S. Truman Library, 5 January 1972; McGuire, "Black Civilian Aides," pp. 60–61.

15. McGuire, "Black Civilian Aides," pp. 60–61; James Rowe to the author, 21 April 1977.

16. McGuire, "Black Civilian Aides," pp. 62–68.

17. WH to Major Alan M. Osur, 13 November 1974, enclosing a completed questionnaire.

18. The Diaries of Henry Lewis Stimson, 52 vols., Manuscript and Archives, Yale University Library, New Haven, Conn., 28 October 1942 (hereafter Stimson Diaries).

19. McGuire, "Black Civilian Aides," p. 44; Joseph P. Lash, *Roosevelt and Churchill 1939–1941* (New York: Norton, 1976), pp. 235–36; Liva Baker, *Felix Frankfurter* (New York: Coward-McCann, 1969), pp. 28–29.

20. WH's handwritten notes without title or date.

21. Lee, *The Employment of Negro Troops*, pp. 140-41, 148; "Hastie for End of Army Jim Crow," Washington *Afro-American*, 8 November 1941.
22. Richard M. Dalfiume, *Desegregation of the Armed Forces* (Columbia: University of Missouri Press, 1969), pp. 45-47.
23. WH's address at the NAACP Emergency Conference, Detroit, Mich., 6 June 1943.
24. WH's handwritten notes without title or date.
25. "Army Takes Back 'Negro Warning,'" *PM* (New York City) 6 January 1942; "Army Officials Create Problems by Kowtowing to Segregation," Philadelphia *Tribune*, 3 January 1942.
26. Lena Horne's comments in her Broadway show *Lena* on 25 July 1981.
27. Gordon Parks, *A Choice of Weapons* (New York: Harper & Row/Perennial Library, 1966), pp. 202, 212, 214, 219.
28. Lee, *The Employment of Negro Troops*, pp. 141-42; Brigadier General Alexander D. Surles's memorandum to WH, 24 October 1941; WH to Howard C. Petersen, 27 October 1941; WH's memorandum to the secretary of the general staff, 1 December 1941, enclosing a copy of a letter; WH to the publisher of the Philadelphia *Independent*, 27 November 1941; WH's form letter ("Dear Publisher"), 2 December 1941.
29. "Remarks of Colonel E. R. Householder, Officer in Charge of Miscellaneous Division, Adjutant General's Department, War Department, at the Conference of Negro Newspaper Representatives, War Department, Munitions Building, Washington, D.C., December 8, 1941."
30. "Conference of Negro Newspaper Representatives."
31. "This Week in Black History," *Jet*, 18 September 1980, p. 18; "Army Plans New Division," Washington *Afro-American*, 13 December 1941; McGuire, "Black Civilian Aides," pp. 154-55.
32. "Conference of Negro Newspaper Representatives"; Lee, *The Employment of Negro Troops*, p. 143.
33. Lee, *The Employment of Negro Troops*, pp. 145-46.
34. Ibid., pp. 146-47; "Colored People Want Jim Crow—Army Chief," Washington *Afro-American*, 27 April 1970.
35. "Survey and Recommendations Concerning the Integration of the Negro Soldier into the Army, Submitted to the Secretary of War by the Civilian Aide to the Secretary of War, in Memo, Civ. Aide to SW through USW," 22 September 1941, p. 24.
36. "Conference of Negro Newspaper Representatives"; Lee *The Employment of Negro Troops*, p. 146.
37. WH's memorandum to the director of the Bureau of Public Relations, 13 October 1941.
38. Lee, *The Employment of Negro Troops*, p. 146; John Oliver Killens, *And Then We Heard the Thunder* (New York: Paperback Library, 1971), pp. 46-47, 49, 96.
39. Stanley High, "How the Negro Fights for Freedom," *Reader's Digest* (July 1942), 113.
40. Ibid., pp. 113-14.
41. "We Won't Fight Unless, Hastie Tells Morgan"; "Declares Stopping of Hitler Important, but Democracy Needs Be More Realistic Here," *Louisiana Weekly*, 15 November 1944; WH to Chandler Owen, 31 January 1942.

42. "Negro Groups Find an Apathy to War," New York Times, 11 January 1942.

43. Wilma L. Shannon to WH, 4 June 1942, transmitting "Survey of Intelligence Materials No. 25," 27 May 1942, Office of Facts and Figures, Bureau of Intelligence, War Department.

44. "White Attitudes Toward Negroes," 5 August 1942, Bureau of Intelligence, Office of War Information.

45. Chester Himes, Black on Black (Garden City, N.Y.: Doubleday, 1973), p. 82; WH's memorandum to the secretary of the general staff, 1 July 1942, and Colonel John R. Deane's reply, 14 July 1942.

46. WH's memorandum to the under secretary of war, 17 August 1942; Walter White, "The Right to Fight for Democracy," Survey Graphic (November 1942), 474.

47. WH's memorandum to the under secretary of war, 17 August 1942; Henry L. Stimson to Walter White, 22 May 1942.

48. Etling E. Morison, Turmoil and Tradition: A Study of the Life and Times of Henry L. Stimson (Boston: Houghton Mifflin, 1960), p. 555.

49. Editorial, The Crisis (March 1942), p. 79; William Allen White, quoted in Great Quotations, ed. George Seldes (New York: Lyle Stuart, 1960), p. 736.

50. Lee, The Employment of Negro Troops, pp. 331–32; WH to the under secretary of war, 28 January 1942; "Blast New Jim Crow Policy of Red Cross," Washington Afro-American, 31 January 1942; McGuire, "Black Civilian Aides," pp. 162–64; Albert Deutsch, "Blood Policy," PM, 23 June 1942; Lieutenant Colonel B. N. Carter to Isabelle Myrick, 18 December 1942.

51. McGuire, "Black Civilian Aides," pp. 162–67; WH to Harold A. Lett, 12 April 1943.

52. WH to Harold A. Lett, 12 April 1943.

53. Lee, The Employment of Negro Troops, p. 332; "Blood Policy of Red Cross Ill Founded," Washington Afro-American, 14 February 1942; editorial in Washington Afro-American, 8 April 1944; "Drew Gets Spingarn Medal for Blood Plasma Work," Washington Afro-American, 22 July 1944; Deutsch, "Blood Policy"; W. Augustus Low and Virgil A. Clift, eds., Encyclopedia of Black America (New York: McGraw-Hill, 1981). p. 325.

54. "Who's Lying?" Washington Afro-American, 24 January 1942; "Army, Navy Agreed on Blood Ban," Washington Afro-American, 29 November 1941; Albert Deutsch and Tom O'Connor, "Red Blood and the Red Cross," PM, 6 January 1942; Truman K. Gibson, Jr., to Lieutenant Colonel H. A. Gerhardt, 12 November 1943; WH to the secretary of war, 8 January 1942; "Army, Navy Ban on Colored Blood Downright Un-American and Stupid," Philadelphia Tribune, 3 January 1943.

55. WH to the secretary of war, 8 January 1942. Stimson Diaries, 25 September 1942.

56. Langston Hughes, Fight for Freedom: The Story of the NAACP (New York: Norton/A Berkley Medallion Book, 1962), p. 13.

CHAPTER 10. DOMESTIC ENEMIES

1. WH, "Negro Officers in Two World Wars," Journal of Negro Education 12 (Summer 1943), 316–22.

2. Ibid., pp. 322-23; WH's memorandum to the assistant chief of staff, G-1, 5 September 1942.

3. WH, "Negro Officers," pp. 320-21; "We're Doing Better Than in 1918—Hastie," Washington *Afro-American*, 19 December 1942; Ulysses Lee, *The Employment of Negro Troops* (Washington, D.C.: U.S. Government Printing Office, 1969), pp. 202-4, 211-12; Phillip McGuire, "Black Civilian Aides and the Problems of Racism and Segregation in the United States Armed Forces: 1940-1950" (Ph.D. dissertation, Howard University, 1975), p. 107.

4. Lee, *The Employment of Negro Troops*, pp. 215-16; interview with Ruth Adams, professor of social work, Howard University, Washington, D.C., 20 September 1976.

5. William E. Raynor, "Crow Car," *The New Republic* (14 June 1943), 792.

6. Major General Richard Donovan's memorandum to the adjutant general, 13 November 1942.

7. Major General Richard Donovan to Governor Sam Houston Jones, 13 November 1942.

8. Ibid.

9. Major General Richard Donovan to the Adjutant general, 13 November 1942; Henry L. Stimson and McGeorge Bundy, *On Active Service in Peace and War* (New York: Harper & Brothers, 1948), pp. 462-63; Truman K. Gibson, Jr., to the adjutant general, 21 November 1942; Henry L. Stimson to Francis Biddle, 26 November 1942; and Biddle's reply, 12 December 1942.

10. WH to L. B. Schwartz, 14 December 1942; Francis Biddle to the secretary of war, 12 December 1942.

11. WH to Major General Richard Donovan, 5 December 1942; WH to the secretary of war, through the adjutant general, 15 October 1941.

12. William Bryden's memorandum for the under secretary of war, 15 October 1941; WH's memorandum to the under secretary of war, 16 October 1941.

13. Ibid.

14. WH's memorandum to the under secretary of war, 16 October 1941; WH's handwritten notes based on military authorities' investigation of the incidents at Gurdon.

15. WH's handwritten notes . . .; WH's memorandum to the under secretary of war, 16 October 1941.

16. WH's memorandum to the under secretary of war, 16 October 1941; William Bryden's memorandum for the under secretary of war, 15 October 1941.

17. WH's memorandum to the under secretary of war, 16 October 1941; Wendell Berge to Robert P. Patterson, 12 February 1942; "Summary of Findings Concerning the Circumstances Attending the Death of Private Felix Hall, Company E, 24th Infantry," n.d.

18. WH's memoranda to the secretary of war, 17 November 1942 and 16 December 1942; WH to John P. Davis, 16 December 1942.

19. WH's memorandum to the secretary of war, 16 December 1942; "Along the N.A.A.C.P. Battlefront," *The Crisis* (November 1942), 361; WH's untitled, handwritten, and typewritten commentary (1943) on discrimination against black military personnel in public transportation.

20. WH's memorandum to the secretary of war, 16 December 1942.

21. WH's untitled, handwritten, and typewritten commentary . . .

22. Lee, *The Employment of Negro Troops*, p. 318.

23. "'Jim Crow' Practices by Public Carriers Against Negro Members of the United States Armed Forces and Defense Workers," 8 July 1942.
24. WH's memorandum to the administrative assistant to the secretary of war, 18 August 1941.
25. Lee, *The Employment of Negro Troops*, p. 319.
26. Ibid., p. 320; WH to Carl Murphy, 1 August 1942.
27. WH's untitled, handwritten, and typewritten commentary . . . ; WH to J. W. Everett, 11 June 1943.
28. WH to J. W. Everett, 11 June 1943; Walter White to Joseph B. Eastman, 8 July 1942; "'Jim Crow' Practices . . ."
29. Joseph B. Eastman to Walter White, 15 August 1942.
30. Walter White to Robert P. Patterson, 8 July 1942, and Patterson's reply, 17 August 1942; "'Jim Crow' Practices . . ."
31. WH's untitled, handwritten, and typewritten commentary . . .
32. Ibid.
33. Lee, *The Employment of Negro Troops*, p. 366.
34. Ibid., p. 357; "Hastie Sees Wider Use of Colored MP," Washington *Afro-American*, 27 December 1941.
35. WH's memorandum to Brigadier General Donald Wilson, assistant chief of staff, G-1, 27 July 1942, and Wilson's reply, 3 August 1942.
36. The 366th Infantry, 372nd Infantry, and 184th Field Artillery were stateside; the 369th Coast Artillery was overseas. The officers included: colonels (1), majors (3), captains (8), first lieutenants (2), and second lieutenants (10). WH's 1st Indorsement, 15 August 1942, to Assistant Chief of Staff Donald Wilson's memorandum to him, 3 August 1942.
37. Ibid.
38. Lee, *The Employment of Negro Troops*, pp. 359, 361-63, 623-25.
39. Ibid., p. 364.
40. Ibid.
41. WH to John J. McCloy, 30 June 1942.
42. John J. McCloy to WH, 2 July 1942.
43. WH to John J. McCloy, 30 June 1942.
44. John J. McCloy to WH, 20 July 1942.
45. Quoted in *The Voice of Black America*, ed. Philip S. Foner (New York: Capricorn Books, 1975), Vol. 1, p. 222.
46. Special Services Division's memorandum to Cornelius DuBois, 22 June 1942, in U.S. War Department, Special Services Division of Bureau of Intelligence, "Special Services Report for Survey #30 on the Negro Problem," 23 June 1942; McGuire, "Black Civilian Aides," pp. 41-42.
47. Lee, *The Employment of Negro Troops*, p. 355; WH's memorandum to the director of the Bureau of Public Relations, 17 October 1941.
48. Lee, *The Employment of Negro Troops*, pp. 384-85.
49. "Hastie assails 'patience appeal'" (Moorland-Spingarn Research Center [MSRC]).
50. McGuire, "Black Civilian Aides," p. 98.
51. Brigadier General H. B. Lewis to Roy Garvin, 12 October 1942.
52. Ibid.; McGuire, "Black Civilian Aides," pp. 98-99.
53. McGuire, "Black Civilian Aides," pp. 101-2; Lee, *The Employment of Negro Troops*, pp. 387-89.

54. "Judge Hastie," article of unspecified source and date; "Hastie assails 'patience appeal'" (MSRC).
55. WH's interview (by Jerry N. Hess) for the Harry S. Truman Library, 5 January 1972.
56. Ibid.

CHAPTER 11. INVINCIBLE MAN

1. Ulysses Lee, *The Employment of Negro Troops* (Washington, D.C.: U.S. Government Printing Office, 1966), pp. 54–56.
2. Ibid., pp. 51–56, 58–62; Walter White, *A Man Called White* (New York: Viking, 1948), p. 186.
3. Lee, *The Employment of Negro Troops*, pp. 64–65; Robert A. Rose, *Lonely Eagles* (Los Angeles: Tuskegee Airmen, 1976), pp. 12, 14; "The Enemy Within," one of four shows in the series *The Black Eagles*, by Tony Brown Productions, Inc., on WHYY-TV (PBS), Philadelphia, Pa., during February 1983.
4. WH, *On Clipped Wings: The Story of Jim Crow in the Army Air Corps* (New York: NAACP, 1943), pp. 10–11.
5. Ibid., p. 12; Alexander B. Siegel to Walter White, 7 April 1943.
6. WH, *On Clipped Wings*, pp. 3–4.
7. Ibid., pp. 8–9.
8. WH's memorandum to Lt. Gen. H. H. Arnold, 18 July 1942; Lt. Gen. H. H. Arnold's memorandum to William H. Hastie, 10 August 1942; Lee, *The Employment of Negro Troops*, pp. 163–64; Maj. Gen. George E. Stratemeyer, 2nd Indorsement, to William H. Hastie, 5 December 1942.
9. WH, 1st Indorsement, to the commanding general, Army Air Forces, 26 October 1942; WH, *On Clipped Wings*, p. 13.
10. WH, *On Clipped Wings*, pp. 13–20; WH's memorandum to the secretary of war, 5 January 1943.
11. WH's memorandum to the secretary of war, 8 January 1942; WH's address at the University of Rochester, Rochester, N.Y., 13 April 1967.
12. WH's memorandum to the assistant secretary for air, 30 June 1942; Lee, *The Employment of Negro Troops*, pp. 164–65; Ollie Stewart, "400 'Flyers' Leaving Chanute Field," Washington *Afro-American*, 25 October 1941; WH's memorandum for the secretary of war, 5 January 1943.
13. WH's memorandum to the secretary of war, 5 January 1943; "Statement of William H. Hastie, Recently Civilian Aide to the Secretary of War," 1 February 1943; WH's interview (by Jerry N. Hess) for the Harry S. Truman Library, 5 January 1972.
14. Henry L. Stimson and McGeorge Bundy, *On Active Service in Peace and War* (New York: Harper & Brothers, 1948), p. 461; The Diaries of Henry Lewis Stimson, Manuscript and Archives, Yale University Library, New Haven, Conn., 19 October 1942.
15. Alan M. Osur, *Black in the Army Air Forces During World War II* (Washington, D.C.: U.S. Government Printing Office, 1977), pp. 63–65; Felix Frankfurter to WH, 28 October 1940; Benjamin E. Haller to WH, 5 February 1943.

16. WH's address to the NAACP Emergency Conference, Detroit, Mich., 6 June 1943; "Statement of William H. Hastie . . ."; WH's interview (by Hess) for the Truman Library; WH's memorandum to the secretary of war, through the under secretary of war, 5 January 1943; WH's letter to the editor of *Time* magazine, 5 February 1942; WH to Major Alan M. Osur, 13 November 1974, enclosing a completed questionnaire.

17. WH to the secretary of war, 8 January 1942, 5 January 1943, and 6 January 1943; "Statement of William H. Hastie . . . ; "Hastie Fought Army Jim Crow to the End," Washington *Afro-American*, 13 March 1943; P. Bernard Young, Jr., "Hastie Gives Cause for Resignation," Norfolk (Va.) *Journal and Guide*, 6 February 1943; "Air Force Discrimination Reason for Resignation, Says Prepared Statement," *Louisiana Weekly*, 6 February 1943; Roy Wilkins, "The Watchtower," unspecified date and source; WH's interview (by Hess) for the Truman Library.

18. WH to the secretary of war, 5 January 1943; Major General George E. Stratemeyer's memorandum for the assistant secretary of war, 12 January 1943; WH's memorandum to the assistant secretary of war, 19 January 1943; Henry L. Stimson to WH, 29 January 1943; Phillip McGuire, "Black Civilian Aides and the Problems of Racism and Segregation in the United States Armed Forces: 1940-1950" (Ph.D. dissertation, Howard University, 1975), p. 127.

19. Notes on Patterson's press conference, n.d.; "Patterson Regrets Hastie's Resignation; Lauds Work," Chicago *Bee*, 31 January 1943; WH's interview (by Hess) for the Truman Library; Young, "Hastie Gives Cause for Resignation."

20. Louis Lautier, "Spotlight," Washington *Afro-American*, 6 February 1943.

21. W. E. B. Du Bois, "As the Crow Flies," date and source unspecified.

22. Lieutenant Monroe Dowling to WH, 27 January 1943.

23. WH to Lieutenant Monroe Dowling, 27 January 1943.

24. "100 Lawyers Pay Homage to Hastie," Baltimore *Afro-American*, 4 May 1946; "Patterson Lauds Negroes' War Role," New York *Times*, 30 April 1946.

25. Osur, *Blacks in the Army Air Forces*, pp. 71-73; "Hastie's Box Score," Pittsburgh *Courier*, 6 March 1943; "Activate Race Bomber Pilots at Selfridge," Pittsburgh *Courier*, 5 February 1944; "Army to Expand Air Training—Stimson," Washington *Afro-American*, 30 January 1943; Lee, *The Employment of Negro Troops*, pp. 174-78; McGuire, "Black Civilian Aides," pp. 130-31, 233.

26. The following account of the Tuskegee Airmen is based on these sources: "World War II's Pilots Salute 'Pinup,'" *Tony Brown's Journal* (January/March 1983) p. 11; "Black History Month TV Special," *Tony Brown's Journal* (January/March 1983), pp. 3, 10; *The Black Eagles* (see note 3); Rose, *Lonely Eagles*, pp. 28, 65-70, 156; Frank I. Weler, "V.I. Governor Views Post as Honor to His Race" (MSRC); Lee Nichols, *Breakthrough on the Color Front* (New York: Random House, 1954), pp. 223-26; WH to Lee Nichols, 4 February 1954.

27. Interview with Secretary of Transportation William T. Coleman, Jr., Washington, D.C., 26 August 1976; John F. Kennedy, *Profiles in Courage* (New York: Harper & Row/Perennial Library, 1964), p. 1; Ralph Ellison, *Invisible Man* (New York: Vintage Books, 1972).

28. "Along the N.A.A.C.P. Battlefront," *The Crisis* (April 1943), 116.

29. Adam Clayton Powell, Jr., quoted in *The People's Voice*, 6 February 1946.

30. WH to Mrs. Edward P. Lovett, 25 November 1953, enclosing a statement concerning the dedication of the Joel Spingarn High School in Washington, D.C.

31. Interview with Archibald T. LeCesne, attorney, Chicago, Ill., 4 August 1977.

32. WH, "Legal Aspects of Racial Discrimination in the Armed Forces," *National Bar Journal* 2 (June 1944), 22; "Will 1917 Hate Be Repeated?" Washington *Afro-American*, 4 January 1941; "Marine Corps Is Height of Lilly Whiteism," Washington *Afro-American*, 4 January 1941; L. D. Reddick, "The Negro in the United States Navy During World War II," *Journal of Negro History* 32 (April 1947), 202, 204; Lucille B. Milner, "Jim Crow in the Army," *The New Republic* (13 March 1944), 341.

33. "Coast Guard Defense of a Lily-White Navy," *Afro-American*, national edition (Baltimore), 12 April 1941.

34. Reddick, "The Negro in the United States Navy," p. 202; "FDR tapes: Off the Record on Japan, Blacks, Willkie," Philadelphia *Inquirer*, 31 January 1982; Daniel Jacobson, "FDR Tapes Reveal Bias Against Blacks," New York *Amsterdam News*, 23 January 1982.

35. WH to Captain L. Spencer, 13 January 1944; WH to Alexander P. Haley, 13 January 1944.

36. Alexander P. Haley to WH, 26 November 1943; Gordon Parks, *A Choice of Weapons* (New York: Harper & Row/Perennial Library 1973), pp. 209-10. This account of the Fisher-Loury case is based on *A United States Army "Scottsboro Case"* (New York: NAACP Legal Defense and Educational Fund, n.d.).

37. Vito Marcantonio and WH to President Franklin D. Roosevelt, 10 June 1944, which was printed in the undated International Labor Defense publication *For Equality of Military Justice* (HULS).

38. "Used Third Degree in Army 'Scottsboro Case,'" Pittsburgh *Courier*, 5 February 1944.

39. Brigadier General Edward S. Greenbaum to WH, 1 April 1944, enclosing Under Secretary of War Robert P. Patterson's opinion, "Clemency Applications of General Prisoners Frank Fisher, Jr., and Edward P. [sic] Lowry [sic]," 31 March 1944 (HULS).

40. Paterson's opinion, "Clemency Applications of . . . Fisher . . . and . . . Lowry [sic]"; WH to Vito Marcantonio, 13 April 1945, enclosing a draft, "Petition of Clemency" in the matter of Frank Fisher, Jr., and Edward P. Loury (HULS).

41. WH's untitled and undated remarks (HULS).

CHAPTER 12. CALL DUPONT 6100

1. WH to Dr. Jake Billikoff, 15 February 1943.

2. Interviews with Pauli Murray, attorney and author, Alexandria, Va., 7 April 1977, and Judge Billy Jones, Circuit Court of Illinois, held at Houston, Texas, 3 August 1976.

3. Spottswood W. Robinson III, "No Tea for the Feeble: Two Perspectives on Charles Hamilton Houston," 20 *Howard Law Journal*, 3 (1977); WH, Foreword,

In Any Fight Some Fall, by Geraldine Segal (Rockville, Md.: Mercury Press, 1975), p. 5.

4. WH, Foreword, *In Any Fight Some Fall*, p. 6.

5. Ibid.; Robinson, "No Tea for the Feeble," p. 3; McNeil, "Charles Hamilton Houston," p. 123.

6. WH, Foreword, *In Any Fight Some Fall*, p. 6.

7. Ibid., pp. 6-7.

8. Ibid., pp. 7-8.

9. McNeil, "Charles Hamilton Houston," pp. 124-25.

10. Robinson, "No Tea for the Feeble," p. 4.

11. Ossie Davis, "We Laughed with Godfrey, Laughing till We Cried . . ." Washington *Post*, 5 December 1976.

12. Joseph D. Whitaker, "Amos 'n' Andy Set an Image," Washington *Post*, 11 April 1976.

13. Charles Hamilton Houston, "Findings on the Black Lawyer," 3 May 1928, p. 15 (Rockefeller Archive Center, Hillcrest, Pocantico Hills, North Tarrytown, N.Y.).

14. WH's address to the National Bar Association, Atlanta, Ga., 5 August 1971.

15. Gilbert Ware, "A Word to and About Black Lawyers," *The Crisis* (August-September 1978), 247-48.

16. Interview with Dr. J. Clay Smith, Equal Employment Opportunity Commission, Washington, D.C., 22 January 1980.

17. Jerold S. Auerbach, *Unequal Justice* (New York: Oxford University Press, 1976), pp. 65-66, 216.

18. Raymond Pace Alexander, "The National Bar Association—Its Aims and Purposes," *National Bar Journal*, 1 (1941), 4.

19. Edward Brathwaite, *Rights of Passage* (London: Oxford University Press, 1967), p. 80.

20. Eugue L. Meyer, "'The Only Colored Man . . .' at Justice," Washington *Post*, 11 April 1976; Robinson, "No Tea for the Feeble," p. 2.

21. Robinson, "No Tea for the Feeble," pp. 2-3; Meyer, "'The Only Colored Man . . .' at Justice"; Eugene L. Meyer and Joseph D. Whitaker, "Blacks Moving into Key Legal Posts," Washington *Post*, 11 April 1976.

22. Joseph D. Whitaker, "Judge Bryant Struggled to Reach His High Post," Washington *Post*, 15 April 1976.

23. WH, "Toward an Equalitarian Legal Order, 1930-1950," *Annals of the American Academy of Political and Social Science* 407 (May 1973), 30-31.

24. Meyer and Whitaker, "Blacks Moving into Key Legal Posts."

25. WH, "Persons and Affairs," *New Negro Opinion*, 10 February 1934.

26. "Ban Against Lawyers Is Abolished," Washington *Afro-American*, 15 February 1941; "Lawyers to Oppose Official Appointments," Washington *Afro-American*, 1 March 1941; "From Press of the Nation," *The Crisis* (April 1941), 135; Alexander, "The National Bar Association," pp. 2-3.

27. "Hastie Attempt to Have Bar Associate with Negro Congress Creates Row" and "District's Rival Bar Associations Both Functioning" (Beck Cultural Exchange Center [BCEC]).

28. "13 Lawyers in Offshoot from D.C. Bar Group," "Bolting Bar Association Members Form New Group," and "Hastie Attempt to Have Bar Associate with Negro Congress Creates Row" (all in BCEC).

29. Meyer and Whitaker, "D.C. Firm a 'School' for Black Jurists," Washington *Post*, 12 April 1976; W. E. B. Du Bois, *The Gift of Black Folk* (New York: Washington Square Press, 1980), p. 65.

30. Interview with Judge Billy Jones; "Hastie Returns as Dean of Howard Law School," Washington *Afro-American*, 4 September 1943; School of Law, Howard University, *Annual Report 1943-44* (Washington, D.C., 1944), pp. 1-2, 4, 8-12; School of Law, Howard University, *Annual Report 1944-45* (Washington, D.C., 1945) p. 2 (Howard University Law School [HULS]).

31. Interview with Pauli Murray.

32. Ibid.; T. R. Powell to Pauli Murray, 13 June 1944; Pauli Murray to A. Calvert Smith, n.d.; Pauli Murray to Registrar, School of Law, Harvard University, 4 January 1944; M. M. (initials only) to Pauli Murray, 5 January 1944; L. K. Garrison to Pauli Murray, 23 June 1944; Pauli Murray to Mr. Justice Felix Frankfurter, 13 May 1944; Grace G. Tully to Pauli Murray, 9 June 1944; George H. Chase to the President [Franklin D. Roosevelt], 5 June 1944; Dorothy K. Clark to Pauli Murray, 21 June 1944 (all in HULS).

33. Pauli Murray to Mr. Justice Felix Frankfurter, 13 May 1944; Pauli Murray to T. R. Powell, 8 June 1944; Pauli Murray to A. Calvert smith, n.d. (all in HULS); interview with Pauli Murray.

34. Pauli Murray to A. Calvert Smith, n.d.; Pauli Murray to Eleanor Roosevelt, 24 June 1944; Pauli Murray to Thomas Reed Powell, 24 June 1944 (all in HULS).

35. Erwin Griswold to WH, 3 August 1944 (HULS).

36. Ibid.

37. Ibid.

38. Ibid.; WH to Erwin Griswold, 8 August 1944; WH to May T. Peacock, 21 November 1944; WH to James E. Brenner, 23 May 1940; WH to Will Shafroth, 3 January 1940 (all in HULS).

39. T. R. Powell to WH, 27 June 1944; Pauli Murray to T. R. Powell, 19 June 1944 (both in HULS).

40. WH to T. R. Powell, 1 July 1944; T. R. Powell to WH, 27 June 1944 (both in HULS).

41. Ibid.

42. Interview with Pauli Murray; Pauli Murray, "Harvard Law School Turns Down Howard Law Honor Graduate," dated 16 and 17 June 1944—4:30 A.M.; Pauli Murray to A. Calvert Smith, n.d. (HULS).

43. This section about Judge Carter of the United States District Court is based on my interview with him in New York City, 9 January 1978; WH to Edwin W. Patterson, 22 July 1940; WH to the Committee on Admissions, International House, 19 November 1940; Richard Kluger, *Simple Justice* (New York: Alfred A. Knopf, 1976), passim; WH to Walter White, 21 July 1945; WH to Robert L. Carter, 26 July 1945; Robert L. Carter to WH, 15 June 1945.

44. Interviews: Judge Billy Jones and attorney Eugene H. Clarke, Jr., Philadelphia, Pa., 13 December 1977.

45. WH's handwritten remarks of an unspecified date and occasion.

46. Interview with Mr. Justice Thurgood Marshall, Supreme Court of the United States, Washington, D.C., 17 February 1977.

47. Interview with Oliver W. Hill, attorney, Richmond, Va., 7 April 1977.

48. Confidential interview.

49. Interview with Oliver W. Hill.

50. Interview with Eugene H. Clarke, Jr.

51. Interview with Ruth Harvey Charity, attorney, held at New Orleans, La., 4 August 1977.

52. Interview with Vincent M. Townsend, Jr., attorney, held at Houston, Texas, 4 August 1976.

53. Interviews with Ruth Harvey Charity and Archibald T. LeCesne held at New Orleans, La., 4 August 1976.

54. Interview with Vincent M. Townsend, Jr.

55. Interview with Eugene H. Clarke, Jr.

56. Interview with Pauli Murray.

57. Interview with Eugene H. Clarke, Jr.; Edith Hamilton quoted in "Manchester's Book Pays J.F.K. Library $750,000," Baltimore *Evening Sun*, 21 June 1968.

58. Roger N. Baldwin's typescript in the Oral History Collection at Columbia University, p. 151 (copyright by The Trustees of Columbia University in the City of New York and used with permission).

59. Genna Rae McNeil, "Charles Hamilton Houston (1895–1950) and the Struggle for Civil Rights" (Ph.D. dissertation, University of Chicago, 1975), p. 253.

60. WH's remarks on an unspecified occasion and date.

61. WH to Dr. Lewis H. Fenderson, 14 July 1967, enclosing his manuscript "No Royal Road."

62. WH's address at Colby College, Waterville, Me., 2 March 1967.

63. Interview with Conrad O. Pearson, attorney, Durham, N.C., 8 April 1977.

64. Howard University School of Law, "Report of the Acting Dean for the School Year Ending June 30, 1941" (mimeographed), pp. 3–4, 6, 17, 25.

65. Pauli Murray, "A Blueprint for First Class Citizenship," *The Crisis* (November 1944), 358–59; interviews: Pauli Murray and Ruth Harvey Charity; Patricia Roberts Harris, secretary of Housing and Urban Development, keynote address at OPEN's First Judge William H. Hastie Award Symposium, Philadelphia, Pa., 11 November 1978.

66. This account of Odell Waller's ordeal is based on the following sources: interview with Pauli Murray; "The Battle to Save Sharecropper Waller," *The Crisis* (January 1941), 23; "N.A.A.C.P. Aids in Defense of Sharecropper," *The Crisis* (December 1940), 390; Ted LeBerthon, "Pauli Murray: Modern Joan of Arc," *War Worker*, Part One (August 1944), in Pauli Murray's files; "Waller Case in Brief," Washington *Afro-American*, 11 July 1942; "Waller Still in Death Row After Fifth Reprieve," Washington *Afro-American*, 27 June 1942; Morris Milgram, Introduction, *Dark Testament and Other Poems*, by Pauli Murray (Comstock Hill, Conn.: Silvermine, 1970).

67. "Legal Lynching Is Deplored at D.C. Scottsboro Meet," Washington *Afro-American*, 18 April 1936.

68. Pauli Murray, "A Blueprint for First Class Citizenship," *The Crisis* (November 1944), 358.

69. I. F. Stone, *The Truman Era* (New York: Vintage Books, 1973), pp. 132–33; Martin Kaplan, "After 38 Years, Stone Comes Back to the Club," Washington *Star*, 6 June 1980.

70. Howard Whitman, "Washington—Disgrace to the Nation," *Woman's*

NOTES

275

Home Companion (February 1950), 46; Joseph D. Lohman and Edwin R. Embree, "The Nation's Capital," *Survey Geographic* (7 January 1947), 37.

71. WH, "Persons and Affairs," *New Negro Opinion*, 1 September 1934, 24 November 1934, and 1 December 1934.

72. President's Committee on Civil Rights, *To Secure These Rights* (Washington, D.C.: U.S. Government Printing Office, 1947), pp. 90–91.

73. WH to William A. H. Birnie, 16 March 1944; Mary Linda Helfant, "Springfield Schools Have Become Laboratory in Education for Democracy," Springfield (Mass.) *Union and Republican*, 5 September 1943; WH to Mrs. Henry Grattan Doyle, 21 November 1944; "Logan Raps School Board Teacher Jim-Crow Ruling," Washington *Afro-American*, 18 November 1944 (all in HULS).

74. "Resolution Adopted at Membership Meeting of the District of Columbia Branch of the National Association for the Advancement of Colored People," 19 November 1944, enclosed in letter from WH to Mrs. Henry Grattan Doyle, 21 November 1944 (HULS).

75. Citizens' Committee Against Segregation in Recreation, Report of Initial Meeting, 20 July 1945; Venice T. Spraggs, "Inter-Race Group Fights D.C. Jim Crow Parks," Chicago *Defender*, n.d.; United States Court of Appeals for the District of Columbia, *Willie Farrall et al., Appellants,* v. *District of Columbia Amateur Athletic Union, a Corporation, et al., Appellees*, No. 9084 (1946); Minutes of the August 4th and 5th [1945] Meetings of the Steering Committee of the Citizens' Committee Against Segregation in Recreation; WH's General Statement Opposing Exclusion From Public Recreational Facilities in the District of Columbia Because of Race (included in the document, Statements or Excerpts of [citizens'] Committee Against Segregation in Recreation before the Board [of Recreation], 17 July 1945; all in HULS).

76. "Play Area Segregation Protests to Be Heard by Board Tuesday"; Abe Fortas to Harry S. Wender, 10 July 1945; "Play Area Segregation Legality Defended by Recreation Chief"; WH to the Editor, Washington *Star*, 30 October 1945 (all in HULS).

77. WH to President Franklin D. Roosevelt, 17 November 1944; Frances H. Kenin to the author, 17 April 1980, with enclosures; "Hastie Letter Puts Democrats on Spot," New York *Amsterdam News*, 26 August 1944; Temporary Committee for Improved Public Transportation, "A Manual of Facts on the Employment of Negroes in the Local Transit Industry," (Washington, D.C.: TCIPT, March 1945), pp. 1–3 (HULS).

78. WH to the Public Utilities Commission, 11 September 1944 and 22 March 1945; Isadore A. Letcher's affidavit, 9 October 1944; press release of unspecified source and date; "Outline and Background of the Capital Transit Controversy," press release of unspecified source, 15 January 1945 (all these items in HULS).

79. TCIPT's memorandum to Washington Organizations (1945); Henry Lee Moon to C. B. Baldwin, 5 January 1945; WH to the Public Utilities Commission, 22 March 1945; Public Utilities Commission of the District of Columbia, Order No. 2912, 20 April 1945; WH to Kathleen R. Clift, 12 February 1945, and to the Public Utilities Commission, 22 March 1945 (all in HULS); "Outline and Background of the Capital Transit Controversy"; "Transit Co. Capitulates," Washington *Afro-American*, 19 December 1942; TCIPT, "A Manual of Facts on the

Employment of Negroes in the Local Transit Industry," pp. 6–7, 11–12; President's Committee on Fair Employment Practices, "Statement of Charges Case No. 70," 30 December 1944 (HULS).

80. Michael Carter, "FEPC Not Dead or Even Dying, Says Lawrence Cramer," Washington *Afro-American*, 27 February 1943; Margaret Lewis, "3 New FEPC Members Are from Dixie," Washington *Afro-American*, 10 July 1943; "Along the N.A.A.C.P. Battlefront," *The Crisis* (February 1943),1; NAACP's telegram to Hon. Franklin D. Roosevelt, 5 November 1943 (HULS).

81. "Hastie Makes Plea for FEPC; Praises Work," Washington *Afro-American*, 13 March 1943; WH, "Full Employment and the Negro Worker," outline of an address to the CIO Conference on Full Employment, site unspecified, 15 January 1944, and a mimeographed announcement (HULS); "Should Congress Pass a Law Prohibiting Employment Discrimination?" *Congressional Digest* 24 (June 1945), 188, 190; "Statement of William H. Hastie on the Pending Employment Practice Bills Before a Sub-Committee of the Committee on Education and Labor," 14 March 1945; "Houston Quits FEPC," Washington *Afro-American*, 8 December 1945; editorial in *The Crisis* (January 1946), 9.

82. WH's comments were untitled and undated but were evidently addressed to the National Lawyers Guild (HULS).

83. WH, "Toward an Equalitarian Legal Order, 1930–1950," *Annals of the American Academy of Political and Social Science* 407 (May 1973), 23–24; Agnes E. Meyer, "Negro Housing: Our Dismal Record," Washington *Post*, 6 February 1944; United States Court of Appeals, District of Columbia, *Clara I. Mays et al. v. William T. Burgess et al.*, No. 8831 (1945), p. 11n.

84. "Statement of the Emergency Committee on Housing in Metropolitan Washington Before the Special Sub-Committee of the Senate Committee on the District of Columbia Concerning the Program of the National Capital Housing Authority," n.d.; interview with Judge Spottswood W. Robinson III, United States Court of Appeals, Washington, D.C., 27 January 1977; Robert C. Weaver, "Race Restrictive Housing Covenants," *Journal of Land and Public Utility Economics* 20 (August 1944), 183–93.

85. Walter White to President Franklin D. Roosevelt, 26 October 1944 (HULS); "Statement of William H. Hastie on Behalf of the National Association for the Advancement of Colored People in Support of General Housing Bill—S. 1592 (1945)"; "U.S. Housing Jim Crow Scored by Judge Hastie," Pittsburgh *Courier*, 22 December 1945; Walter White to President Harry S. Truman, 19 May 1945 (HULS); *To Secure These Rights: The Report of the President's Committee On Civil Rights* (Washington, D.C.: U.S. Government Printing Office, 1947), p. 92; Statement prepared for Lowell Lomax [of the] *Afro-American*, by John Ihlder, executive officer, National Capital Housing Authority, 7 August 1944, and National Capital Housing Authority's letter to Delegates, Federation of Citizens' Associations, 15 December 1943 (HULS).

86. Louis Lautier to WH, 18 November 1943.

87. WH to National Capital Housing Authority, 25 January 1944, and a single sheet entitled "Housing," n.d. (HULS).

88. WH to Federal Public Housing Authority, 11 February 1944, and a document entitled "Original Complaint," n.d. (HULS); "U.S. Housing Jim Crow Scored by Judge Hastie," Pittsburgh *Courier*, 22 December 1945; *To Secure These Rights*, pp. 91–92.

89. Loren Miller, "Race Restrictions on the Use or Sale of Real Property," *National Bar Journal* 2 (June 1944), 24–28; WH to Pauli Murray, 11 April 1945 (Pauli Murray's files).

90. "Miss Mays Must Vacate Property," Washington *Afro-American*, 29 September 1945; *Transcript of Record*, Supreme Court of the United States, October Term, 1944, *Clara I. Mays et al., Petitioners v. William T. Burgess et al.*, pp. 3–4, 23–24, 28, 35; "Notes and Comments," *National Bar Journal* 3 (December 1945), 364.

91. United States Court of Appeals, District of Columbia, No. 8831, *Clara I. Mays et al. v. William T. Burgess et al.* (1945), pp. 2–6.

92. United States Court of Appeals, *Clara I. Mays v. William T. Burgess*, pp. 6–7.

93. Ibid., pp. 7–10, 12.

94. "Neighborhood Covenants 'Criminal Conspiracy,'" Washington *Afro-American*, 2 December 1944; United States Court of Appeals, *Clara I. Mays v. William T. Burgess*, p. 8.

95. In the Supreme Court of the United States, October Term, 1944, No. 1208, *Clara I. Mays, Petitioner, v. William T. Burgess, Frances E. Burgess, H. P. Gumbrecht, Agnes B. Mularkey, Mary C. Carleton, Respondents*, Petition for Writ of Certiorari to the United States Court of Appeals for the District of Columbia and Brief in Support Thereof, pp. 4–5.

96. WH to Pauli Murray, 11 May 1945 (Pauli Murray's files); interview with Judge Spottswood W. Robinson III.

97. WH to Pauli Murray, 11 May 1945 (Pauli Murray's files); "Miss Mays Must Vacate Property," Washington *Afro-American*, 29 September 1945; Sidney A. Jones, Jr. "Legality of Race Restrictive Covenants" *National Bar Journal* 4 (1946), 22–23.

98. WH to Ralph Winstead, 2 January 1945 (HULS): Lohman and Embree, "The Nation's Capital," p. 36; Bruce Bliven, "Black Skin & White Marble," *The New Republic* (20 December 1948), 15.

99. Interview with Elwood H. Chisholm, associate general counsel, Howard University, Washington, D.C., 20 September 1976.

100. Thurgood Marshall to WH, 20 March 1947.

101. Interview with Oliver W. Hill.

102. Interview with Judge Spottswood W. Robinson III; "NAACP Lawyer Slugged in Tennessee Courthouse," Washington *Afro-American*, 7 March 1942; "X Marks the Spot," Washington *Afro-American*, 7 March 1942; "Ransom Sues for $10,000," Washington *Afro-American*, 18 April 1942.

103. "Ransom Sues for $10,000"; "Nashville Mayor Reports Arrest in Attack on Ransom," Washington *Afro-American*, 7 March 1942.

104. Interview with Judge Billy Jones.

105. Interview with Dean Charles T. Duncan, Howard University School of Law, Washington, D.C., 10 September 1975.

106. "Thurgood Marshall Intimidated After Columbia Trial," *Pittsburgh Courier*, 30 November 1946.

107. Interview with Brenda S. Spears, attorney, New York City, 6 September 1982.

108. WH's untitled and undated remarks, apparently to the National Lawyers Guild (HULS).

CHAPTER 13. A FIGHTER FOR US

1. Langston Hughes, *Fight for Freedom: The Story of the NAACP* (New York: Norton/A Berkley Medallion Book, 1962), p. 37.
2. Victor W. Rotnem, "The Federal Civil Right 'Not to Be Lynched,'" 28 *Washington University Law Quarterly* 57-58 (1943); "Lawyers Insist U.S. Halt Racial Abuses," Washington *Afro-American*, 5 December 1942.
3. WH, "Persons and Affairs," *New Negro Opinion*, 15 February 1934.
4. Mary Lu Nuckols, "The NAACP and the Dyer Anti-Lynching Bill: A Barometer of Emerging Negro Political Power" (M.A. thesis, University of North Carolina, 1963), pp. 14-16; Milton Mayer, "The Issue Is Miscegenation," in *White Racism*, ed. Barry N. Schwartz and Robert Disch (New York: Dell, 1970), p. 208; Walter White, *Rope and Faggot* (New York: Arno Press, 1969), pp. 58, 76.
5. Helen R. Bryan to WH, 20 December 1935, and WH's reply, 31 December 1935; WH, "Persons and Affairs," *New Negro Opinion*, 15 February 1934; U.S. Congress, Senate, Committee on the Judiciary, *Crime of Lynching*, Hearings before a subcommittee of the Senate Committee on the Judiciary on H.R. 801, 76th Cong., 34th Sess., 1940, pp. 85, 91-92.
6. WH, "People's Power Is Ballot, Hastie Says," Philadelphia *Inquirer*, 22 October 1963; James M. Nabrit, Jr., "Disabilities Affecting Suffrage Among Negroes," *Journal of Negro Education* 8 (July 1939), 387-88; interview with James M. Nabrit, Jr., president emeritus of Howard University, Washington, D.C., 30 September 1976; "Judge Hastie Cites Plight of U.S. Negro," New York *Times*, 8 April 1945; Sidney A. Jones, Jr., "The White Primary and the Supreme Court," *National Bar Journal* 3 (1945), 20-21; Ralph J. Bunche, *The Political Status of the Negro in the Age of FDR* (Chicago: The University of Chicago Press, 1973), p. 30; Darlene Clark Hine, *Black Victory* (Millwood, New York: Kto Press, 1970), p. 201.
7. Nabrit, "Disabilities," pp. 388-89; *Newberry v. United States*, 256 U.S. 232 (1921); *Nixon v. Herndon*, 23 U.S. 536 (1927); Thurgood Marshall, "The Rise and Collapse of the 'White Democratic Primary,'" in *The Making of Black America*, ed. August Meier and Elliott Rudwick (New York: Atheneum, 1971), II, pp. 274-79.
8. *Nixon v. Condon*, 286 U.S. 73 (1932).
9. *Grovey v. Townsend*, 295 U.S. 45 (1935); "Connally's Nemesis," Washington *Afro-American* (17 February 1940); Marshall, "Rise and Collapse," pp. 277-78; *United States v. Classic*, 313 U.S. 299 (1941).
10. WH, "An Appraisal of Smith v. Allwright," *Lawyers Guild Review* 5 (1945), 68-69; WH to Milton R. Konvitz, 13 July 1943 (NAACP Papers); Hine, *Black Victory*, pp. 180-81, 187n.
11. Hine, *Black-Victory*, pp. 201, 206-7.
12. Ibid.; NAACP *Annual Report* (1941), p. 25; "Supreme Court Gets Vote Case," Norfolk (Va.) *Journal and Guide*, 20 November 1943; interview with Karen Hastie Williams, attorney, Washington, D.C., 4 March 1978.
13. *Transcript of Record*, Supreme Court of the United States, October Term 1943, No. 51, *Lonnie E. Smith, Petitioner v. S. E. Allwright, Election Judge, and James E. Liuzza, Associate Election Judge; 48th Precinct of Harris County, Texas*, pp. 81, 85, 110.

14. "Supreme Court Gets Vote Case"; "High Court Hears White Primary Case," Washington *Afro-American*, 13 November 1943; Hine, *Black Victory*, p. 218.

15. WH to Arthur D. Shores, 8 December 1943; Carter Wesley to Thurgood Marshall, 11 December 1943.

16. WH to Carter Wesley, 16 December 1943.

17. Steven F. Lawson, *Black Ballots* (New York: Columbia University Press, 1976), p. 44.

18. Interview with Mr. Justice Thurgood Marshall, Supreme Court of the United States, Washington, D.C., 17 February 1977.

19. Ibid.

20. Ibid.

21. WH's Holmes Lecture at Washington University, St. Louis, Mo., 18 November 1964; *Smith v. Allwright*, 321 U.S. 649 (1944).

22. WH "An Appraisal of Smith v. Allwright," pp. 68–70.

23. Ibid., pp. 70–71.

24. Joseph P. Lash, *From the Diaries of Felix Frankfurter* (New York: Norton, 1975), pp. 73–74; Hine, *Black Victory*, p. 219.

25. "Supreme Court Reverses Self," Washington *Afro-American*, 8 April 1944; WH, "An Appraisal of Smith v. Allwright," p. 71; WH to Erwin N. Griswold, 10 April 1944; WH to Thurgood Marshall, 5 May 1944, and Charles Elmore Cromley to Thurgood Marshall, 9 May 1944 (both in NAACP Papers); Hine, *Black Victory*, p. 223.

26. Erwin N. Griswold to WH, 6 April 1944.

27. Darlene Clark Hine to the author, 30 December 1980; William H. Huff to WH, 6 April 1944.

28. August Meier and Elliott Rudwick, *Along the Color Line* (Urbana, Ill.: University of Illinois Press, 1976), p. 132; Lawson, *Black Ballots*, pp. 26, 52.

29. James Marshall to the author, 26 August 1976; James Marshall to WH, 7 April 1944.

30. WH to James Marshall, 15 April 1944.

31. WH's handwritten notes contained in an envelope addressed to him by Carter G. Wesley, 31 January 1940, but labeled "Franchise Misc notes at Swarthmore Institute," n.d.; WH to Channing H. Tobias, 8 April 1944.

32. Channing H. Tobias to Thurgood Marshall, 4 April 1944.

33. Lawson, *Black Ballots*, pp. 42–48, 52–53.

34. WH, "The American Negro and the Franchise," 1945 (HULS).

35. "FEPC Scores McNutt in Letter to F.D.R." Washington *Afro-American*, 13 March 1943.

36. Interview with Mr. Justice Thurgood Marshall.

37. "No Time to Scream," Washington *Afro-American*, 15 April 1944; WH to Judge Francis E. Rivers, 17 April 1944; Hastie, "An Appraisal of Smith v. Allwright," p. 66.

38. WH to Channing H. Tobias, 8 April 1944.

39. Interview with Judge Robert L. Carter, United States District Court, New York City, 9 January 1978; Jack Bass and Walter De Vries, *The Transformation of Southern Politics* (New York: Basic Books, 1976), p. 306.

40. Interview with Judge Constance Baker Motley, United States District Court, New York City, 12 September 1978.

41. Ibid.; Judge Constance Baker Motley to the author, 3 October 1978.

42. WH's untitled, handwritten and typed statement (1943) about discrimination against black military personnel in transportation; WH to J. W. Everett, 11 June 1943; WH, "Persons and Affairs," New Negro Opinion, 11 August 1934.

43. WH, "Persons and Affairs," New Negro Opinion, 11 August 1934.

44. "Morgan vs. State of Virginia," an article of unspecified source and date (Pauli Murray's Papers); Irene Morgan v. Commonwealth of Virginia, Record No. 2974, In the Supreme Court of Appeals of Virginia at Richmond, Petition for Writ of Error, and "Irene Morgan v. Commonwealth of Virginia," Record No. 2974, Opinion by Justice Herbert B. Gregory, Wytheville, Virginia, 6 June 1945 (mimeographed; both in HULS); interview with Judge Spottswood W. Robinson III, United States Court of Appeals, Washington, D.C., 27 January 1977.

45. "Morgan vs. State of Virginia."

46. Ibid.; Irene Morgan v. Commonwealth of Virginia, Record No. 2974 (mimeographed).

47. Irene Morgan v. Commonwealth of Virginia, Record No. 2974.

48. In the Supreme Court of Appeals of Virginia, Record No. 2974, Irene Morgan v. Commonwealth of Virginia, Reply Brief for Plaintiff in Error, pp. 41–42, 45–46; Raymond Pace Alexander, "Recent Trends in the Law of Racial Segregation on Public Carriers," National Bar Journal 4 (December 1947), 403–8; Sidney A. Jones, Jr., "The Supreme Court's Role in Jim Crow Transportation," National Bar Journal 3 (June 1945), 115–16, 120; interview with Judge Spottswood W. Robinson III.

49. Interview with Judge Spottswood W. Robinson III.

50. Interview with Judge Spottswood W. Robinson III.

51. Ibid.; In the Supreme Court of Appeals of Virginia, Record No. 2974, Irene Morgan v. Commonwealth of Virginia, Brief on Behalf of the Commonwealth, n.d., pp. 9–10, 12–15, 28–29.

52. Irene Morgan v. Commonwealth of Virginia, Record No. 2974, Brief on Behalf of the Commonwealth, pp. 34–39.

53. "Co-eds' Bus Jim-Crow Heads for High Court," Washington Afro-American, 16 December 1944; WH to Pauli Murray, 11 April 1945 (Pauli Murray's files); interview with Pauli Murray, attorney and author, Alexandria, Va., 7 April 1977.

54. In the Supreme Court of Appeals of Virginia, Record No. 2974, Irene Morgan v. Commonwealth of Virginia, Petition for Rehearing, pp. 3–11, 26–27, 33–34.

55. Ibid., pp. 6–8, 14, 20–21, 26–27.

56. Interview with Judge Spottswood W. Robinson III; Murray I. Gurfein, "Appellate Advocacy, Modern Style," Litigation 4 (Winter 1978), 9; Spottswood W. Robinson III, "William Henry Hastie—The Lawyer," 125 University of Pennsylvania Law Review (1976), 10–11.

57. Interviews: Judge Spottswood W. Robinson III and Judge William B. Bryant, United States District Court, Washington, D.C., 27 January 1977.

58. Robinson, "William Henry Hastie—The Lawyer," pp. 9–10.

59. "Monthly Report of [NAACP] Legal Department, March 1946"; "Virginia Jim Crow," The Crisis (May 1946), 151; Gurfein, "Appellate Advocacy, Modern Style," p. 9; interview with Judge Spottswood W. Robinson III.

60. Interview with Judge Spottswood W. Robinson III.

61. President John William Ward's address at commencement exercises held at Amherst College, Amherst, Mass., 6 June 1976; interviews: John William Ward, president, Amherst College, 18 August 1977; Judge Spottswood W. Robinson III; and Judge William B. Bryant.

62. Interviews: John William Ward and Hon. William T. Coleman, Jr., secretary of transportation, Washington, D.C., 26 August 1976.

63. Louis H. Pollak's address at the dinner honoring Chief Judge William H. Hastie, Philadelphia, Pa., 15 May 1971; interviews: Hon. William T. Coleman, Jr., and Dean Louis H. Pollak, University of Pennsylvania Law School, Philadelphia, Pa., 10 November 1976.

64. Pollak's address at the dinner honoring Chief Judge Hastie.

65. Robinson, "William Henry Hastie—The Lawyer," pp. 9–10.

66. Ibid., p. 8n; Judge Constance Baker Motley to the author, 3 October 1978, referring to the cases listed in Randall W. Bland, *Private Pressure on Public Law: The Legal Career of Justice Thurgood Marshall* (Port Washington, N.Y.: Kennikat Press), 1973, pp. 183–84; interview with Mr. Justice Thurgood Marshall.

67. Robinson, "William Henry Hastie—The Lawyer," pp. 8–9.

68. Aeschylus quoted in Judy Zimmermann to Mrs. Robert F. Kennedy, 10 June 1968, in *An Honorable Profession*, ed. Pierre Salinger et al. (Garden City, N.Y.: Doubleday, 1968), p. 75.

CHAPTER 14. CARIBBEAN OUTPOST

1. WH's interview (by Jerry N. Hess) for the Harry S. Truman Library, 5 January 1972.

2. Ibid.

3. Harold L. Ickes, *The Secret Diary of Harold L. Ickes*, Vol. 3 (New York: Simon and Schuster, 1954), pp. 404–5, 443.

4. Harold L. Ickes, "My Twelve Years with F.D.R." *Saturday Evening Post*, 26 June 1948, p. 81.

5. Ibid.; WH's interview (by Hess) for the Truman Library; *The Paper of Harold L. Ickes: Diaries*, 15 December 1945 (on microfilm in the Library of Congress).

6. Oliver Pilat, "Senate Fight Due on Negro for Isle Post," New York *Post*, n.d.; U.S. Congress, Senate, Committee on Territorial and Insular Affairs, "Nomination of William H. Hastie for Appointment as Governor of the Virgin Islands," Hearings before a subcommittee of the Senate Committee on Territorial and Insular Affairs, 1946, Vol. 1, pp. 1–7, 10–27, 36–37, and Vol. 4, p. 187 (hereafter "Nomination of William Hastie for Governor"); Abe Murdock to Pauli Murray, 22 April 1946; "Hastie Grilled for 3 Hours Comes Out on Top," St. Thomas *West End News*, 26 March 1946.

7. Louis Lautier, "Capital Spotlight," Washington *Afro-American*, 6 April 1946.

8. "Nomination of William Hastie for Governor," Vol. 1, p. 37, and Vol. 2, p. 97.

9. Ibid., Vol. 2, pp. 97–98; Virginia Gardner, "Meet Governor Hastie," *New*

Masses, 12 February 1946, p. 10; "Hastie Answers Eastland," Baltimore *Afro-American*, 30 March 1946.

10. "Nomination of William Hastie for Governor," Vol. 2, pp. 98-101, 105; "Secretary of Interior Krug Endorses Hastie at Hearing," Washington *Afro-American*, 30 March 1946.

11. "Nomination of William Hastie for Governor," Vol. 2, pp. 101-4; fragment of a newspaper article.

12. "Nomination of William Hastie for Governor," Vol. 2, pp. 104-6; "Secretary of Interior Krug Endorses Hastie at Hearing."

13. Harry McAlpin, "Army Sticks to Policy of Segregating Blood Plasma," Washington *Afro-American*, 8 April 1944; Joe Shepherd, "22 Cops Halt Protest Against Blood Jim Crow," Washington *Afro-American*, 9 September 1944.

14. "Nomination of William Hastie for Governor," Vol. 2, pp. 39, 106.

15. "Secretary of Interior Krug Endorses Hastie at Hearing."

16. "Nomination of William Hastie for Governor," Vol. 2, pp. 40-42.

17. Ibid., pp. 41-45; Venice T. Spraggs, "Hastie Tilts with Dixie Demagogues at Hearing on Virgin Islands Post," *Chicago Defender*, 3 March 1946; "Senator Eastland Tried to Shake Hastie at Hearing," Pittsburgh *Courier*, 30 March 1946.

18. "Nomination of William Hastie for Governor," Vol. 2, pp. 45-53.

19. Ibid., Vol. 3, pp. 76-84, 87-89.

20. Ibid., p. 116.

21. "Rule Out Testimony of Witness Hostile to Hastie," Washington *Afro-American*, 6 April 1946.

22. "Nomination of William Hastie for Governor," Vol. 3, p. 136.

23. Ibid., pp. 160, 172.

24. Ibid., pp. 190-97, 209-17; fragment of a newspaper article.

25. "Nomination of William Hastie for Governor," Vol. 4, pp. 217-18.

26. Ibid., pp. 225-26, 229; "Rule Out Testimony of Witness Hostile to Hastie."

27. "Nomination of William Hastie for Governor," Vol. 6, pp. 302-15, and Vol. 8, pp. 437-39.

28. Ibid., Vol. 8, pp. 446-83.

29. Ibid., pp. 481-82.

30. WH's interview (by Hess) for the Truman Library; "Hastie to Take Virgin Islands Helm May 17," *Chicago Defender*, 11 May 1946; "A Hasty Interview with Governor Hastie," *PM* (New York City) 5 May 1946; Ralph Matthews, "Virgin Islands Ready for Hastie," Washington *Afro-American*, 11 May 1946.

31. Matthews, "Virgin Islands Ready for Hastie"; "Hasties Bid Farewell to U.S. (newspaper photograph in the Hastie Papers); "Hastie Installed Governor," New York *Amsterdam News*, 25 May 1946; "Governor Hastie and Party Arrive in St. Thomas; Given Grand Welcome," St. Croix *Avis*, 17 May 1946.

32. "Huge Crowd Sees Hastie Take Oath," St. Thomas *Photo News*, 18 May 1946; "Governor Hastie Takes Oath in Presence of Huge Crowd in Emancipation Garden," St. Thomas *Daily News*, 18 May 1946; Inaugural address of William H. Hastie, governor of the Virgin Islands, Charlotte Amalie, 17 May 1946.

33. "Governor Hastie Takes Oath in Presence of Huge Crowd in Emancipation Garden"; "Hastie, First V.I. Colored Governor, Will Take Oath Here Friday Noon"; Ralph Matthews, "Saga of the Lockharts Rich in Lore and Romance of the Virgin Islands," Baltimore *Afro-American*, 25 May 1946; interviews with

Dr. Roy A. and Vivian Anduze, physician and wife, St. Thomas, V.I., 21 July 1977.

34. "Elaborate Island-Wide Program for Governor, Interior Secretary Reception," St. Croix *Avis*, 17 May 1946; Matthews, "Saga of the Lockharts Rich in Lore and Romance of the Virgin Islands"; interviews with Dr. Roy A. and Vivian Anduze.

35. Ralph Matthews, "Native Named Acting Governor of V.I." Baltimore *Afro-American*, 2 February 1946.

36. Ralph Matthews, "Job as Governor of Virgin Islands No Bed of Roses," Washington *Afro-American*, 12 January 1946.

37. J. Antonio Jarvis, "A Few Thoughts at a Historical Moment," St. Thomas *Daily News*, 18 May 1946; James A. O'Bryan, "Mr. Hastie Faces the Future," St. Thomas *Daily News*, 18 May 1946.

38. Editorial, St. Thomas *Photo News*, 9 April 1946; Peter Edson, "U.S. Has Chance to Do a Big Thing in Tiny Virgin Isles" (MSRC).

39. "Nomination of William Hastie for Governor," Vol. 2, pp. 76-90, 95-96; "A hasty interview with Governor Hastie."

40. "A hasty interview with Governor Hastie"; "fredi says"; "The Virgin Islands," *House and Garden* (September 1948), 108; "Charlotte Amalie, I Love You . . ." *Glamour* (December 1948), 102; Jeanne Perkins Harman, *The Love Junk* (London: Hammond, Hammond 1959), p. 120.

41. B. M. Phillips, "Caribbean Coverage," Washington *Afro-American*, 16 March 1946.

42. *Annual Report of the Governor of the Virgin Islands to the Secretary of the Interior* (Washington, D.C.: U.S. Government Printing Office, 1947), pp. 1-2; WH's interview (by Hess) for the Truman Library.

43. Walter White to Oscar Chapman, 6 February 1950; Walter White to Z. Alexander Looby, 9 February 1950 and 2 May 1950 (all in the Collection of American Literature, Beinecke and Rare Book and Manuscript Library, Yale University).

44. Donald L. Hibbard to the Honorable [Harold L.] Ickes, 4 February 1946, and the Hibbard Report on the Virgin Islands, 1 February 1946 (National Archives).

45. *Annual Report of the Governor of the Virgin Islands to the Secretary of the Interior* (1947), pp. 2-5, 13; Darwin D. Creque, *The U.S. Virgins and the Eastern Caribbean* (Philadelphia: Whitmore, 1968), pp. 123-24; interview with Dr. Roy A. Anduze.

46. Interview with Dr. George D. Cannon, physician, New York City, 10 January 1978.

47. Harman, *The Love Junk*, pp. 59-63; interview with Louis Shulterbrandt, businessman, St. Thomas, V.I., 23 July 1977; interview (by Dr. Harriet F. Berger) with Louis Shulterbrandt, St. Thomas, V.I., 27 October 1978; interview with Aubrey C. Ottley, postmaster, St. Thomas, V.I., 16 July 1977; Creque, *The U.S. Virgins and the Eastern Caribbean*, pp. 126-37.

48. *Annual Report of the Governor of the Virgin Islands to the Secretary of the Interior* (Washington, D.C.: U.S. Government Printing Office, 1949), pp. 3, 13-15.

49. Ibid., p. 15; interview with Hon. Edward R. Dudley, administrative

judge, Supreme Court of the State of New York, held at New York City, 10 January 1978.

50. Interviews with Judge Edward R. Dudley and Dr. G. James Fleming, political scientist, Baltimore, Md., 31 July 1979; Patricia Coffin, "The Virgin Islands: Reno's Newest Rival" and an article by John O'Donnell.

51. WH's interview (by Hess) for the Truman Library; WH to William W. Boyer, Jr., 2 June 1949 (Boyer's files); interviews with Professor William W. Boyer, Jr., department of political science, University of Delaware, held at St. Thomas, V.I., 21 July 1977, and Isidor Paiewonsky, businessman, St. Thomas, V.I., 20 July 1977; "Pictorial Review: Visit of President Harry S. Truman to the Virgin Islands, February 22 and 23, 1948" (MSRC); P. Bernard Young, Jr., "Warm Welcome Greets Truman on Arrival in Virgin Islands," Norfolk *Journal and Guide*, 28 February 1948; President Harry S. Truman, "Remarks in St. Thomas on a Visit to the Virgin Islands," *Public Papers on the Presidents of the United States* (Washington, D.C.: U.S. Government Printing Office, 1964), p. 155.

52. WH's interview (by Hess) for the Truman Library; P.Bernard Young, Jr., "Islanders Greet Truman," Norfolk (Va.) *Journal and Guide*, 28 February 1948.

53. "President: Pleasure and Business," *Newsweek* (8 March 1948); Ward Canaday to WH, 1 March 1948.

54. WH's interview (by Hess) for the Truman Library; "President: Pleasure and Business"; Young, "Islanders Greet Truman"; "Pictorial Review: Visit of President Harry S. Truman to the Virgin Islands, February 22 and 23, 1948" (MSRC).

55. WH's interview (by Hess) for the Truman Library; "Truman Goes South and Enjoys It"; "President: Pleasure and Business."

56. Earle B. Ottley, *Trials and Triumphs* (privately printed, 1982), pp. 67–74; 87; editorials, St. Thomas *Daily News*, 30 July and 4 November 1946; "American Veterans Committee Column," St. Thomas *Daily News*, 14 October 1946.

57. "Tell The Truth, Earle," St. Thomas *Daily News*, 28 October 1946; "Square Dealers Lecture at Garden," St. Thomas *Daily News*, 31 October 1946.

58. "V.I.P.G. Column," St. Thomas *Daily News*, 29 October 1946; Downing's letter, St. Thomas *Daily News*, 31 October 1946.

59. WH's address at commencement exercises held at the College of the Virgin Islands, St. Thomas, V.I., 12 June 1966; "Nomination of William Hastie for Governor," Vol. 3, pp. 142–44, and Vol. 8, p. 444; Gordon K. Lewis *The Virgin Islands* (Evanston, Ill.: Northwestern University Press, 1972), p. 19.

60. Lewis, *The Virgin Islands*, pp. 95–118.

61. Editorial, St. Thomas *Daily News*, 3 July 1946.

62. Editorials, St. Thomas *Daily News*, 3 July 1946, and St. Thomas *Photo News*, 12 August 1946.

63. Editorial, St. Thomas *Daily News*, 3 September 1946.

64. Inaugural address of William H. Hastie; "Governor Issues Interpretation of Hatch Act," St. Thomas *Daily News*, 5 October 1948; "Governor Expresses Regret on Result of Referendum," St. Thomas *Daily News*, 12 November 1948.

65. "Brown Arrested; Charged with Fraud," St. Thomas *Daily News*, 11 February 1947; "Charged with Fraud, Brown Goes on Trial Tomorrow," St. Thomas *Daily News*, 27 March 1947; "Gov. Comments on Brown's Case," St.

Thomas *Daily News*, 1 April 1947; Ottley, *Trials and Triumphs*, pp. 79-80; interviews: Judge Edward R. Dudley and Omar Brown, businessman, St. Thomas, V.I., 21 July 1977.

66. Interview with Judge Edward R. Dudley; Lewis, *The Virgin Islands*, pp. 155-56.

67. Allan Morrison, "Top Judge." *Ebony* (March 1964), 118; interviews: Senator Frits Lawetz, Legislature of the Virgin Islands, St. Thomas, V.I., 19 July 1977, and Alphonso A. Christian, Sr., attorney, St. Thomas, V.I., 21 July 1977.

68. Editorial, St. Thomas *Photo News*, 10 April 1947; *Annual Report of the Governor of the Virgin Islands to the Secretary of the Interior* (1947), p. 13; "Governor Proposes Plan Streamlining Govt. Functions," St. Thomas *Daily News*, 12 April 1947; editorial, St. Thomas *Daily News*, 5 May 1947.

69. Editorial, St. Thomas *Daily News*, 14 April 1947; "Congressman Jensen Compliments Hastie on Reorganization of Govt. Here," St. Thomas *Daily News*, 23 April 1947; "Hastie Accepts Reorganization Amendment," St. Thomas *Daily News*, 27 May 1947; "Gov. to Complete Reorganization upon His Return"; "Ten Commissions, Agencies Abolished by Gov. Hastie," St. Thomas *Photo News*, 20 May 1947; "Council Overrides Hastie's Veto."

70. Ottley, *Trials and Triumphs*, p. 84; editorial, St. Thomas *Photo News*, 26 November 1946; "Gov Stresses Need for Taxes at Opening Session of Council," St. Thomas *Daily News*, 10 January 1947; editorial, St. Thomas *Daily News*, 14 May 1947; interview with Judge Edward R. Dudley.

71. "Hastie Files Suit in District Court Against Harris, Gordon," St. Thomas *Daily News*, 13 May 1947; "Governor Hastie Files Suit Against Gordon, Harris," St. Thomas *Photo News*, 13 May 1947; editorials, St. Thomas *Photo News*, 14 May 1947, and St. Thomas *Photo News*, 14 May 1947 and 28 August 1947; interviews: Dr. Roy A. and Vivian Anduze; Judge Edward R. Dudley; Aubrey C. Ottley; and Omar Brown.

72. Interview with Judge Edward R. Dudley.

73. *Saturday Evening Post* press release.

74. "Hastie Denounces Action of 'Willful Minority' in Assembly," St. Thomas *Daily News*, 16 May 1947; Ottley, *Trials and Triumphs*, p. 85.

75. "V.I. Leg. Assembly Quits in Disgust" and "Excerpts from Statements Made on Assembly Floor on Tuesday Evening, 3 P.M."

76. Ibid; "V.I. Leg. Assembly Adjourns in Disgust," St. Thomas *Daily News*, 21 May 1947.

77. Editorials, St. Thomas *Daily News*, 22 May 1947, and St. Thomas *Photo News*, 22 May 1947; "Hastie Back from St. Croix," St. Thomas *Daily News*, 22 May 1947; "Raps Council for 'Political Billingsgate.'"

78. Ottley, *Trials and Triumphs*, p. 92; Governor Lawrence W. Cramer to WH, 27 September 1940, and WH's rely, 5 October 1940 (HULS).

79. Ottley, *Trials and Triumphs*, pp. 78-82, 88-90.

80. Editorial, St. Thomas *Daily News*, 11 December 1948.

81. Interview with Raymond Plaskett, WH's butler, St. Thomas, V.I., 20 July 1977.

82. "Stork Visits Hastie," St. Thomas *Daily News*, 5 March 1947; interview with Dr. W. Mercer Cook, ambassador and educator, Washington, D.C., 26 August 1976, and James A. Bough, attorney, St. Thomas, V.I., 21 July 1977.

83. Interviews: Raymond Plaskett; Dr. Roy A. and Vivian Anduze.

84. Interviews: Dr. Roy A. Anduze and James A. Bough.

85. Ottley, *Trials and Triumphs*, p. 87; WH's interview (by Hess) for the Truman Library.

CHAPTER 15. TRUMAN'S RESCUER

1. Lucille B. Milner, "Jim Crow in the Army," *The New Republic* (13 March 1944), 342; WH, "Only U.S. Action Can Protect Negroes in Army, Says Leader," *PM* (New York City), 14 June 1943.

2. Robert J. Donovan, *Conflict and Crisis* (New York: Norton, 1977), pp. 30–31, 390–91.

3. WH's interview (by Jerry N. Hess) for the Harry S. Truman Library, 5 January 1972; WH to William W. Boyer, Jr., 2 June 1949 (Boyer's files); "Truman Goes South and Enjoys It"; "Hastie Brands O'Donnell's Report 'Utterly False'," Norfolk (Va.) *Journal and Guide*, 13 March 1948.

4. Joseph C. Goulden, *The Best Years 1945–1950* (New York: Atheneum, 1976), pp. 363–64; George H. Curtis to the author, 9 June 1982, enclosing a copy of the memorandum (Harry S. Truman Library).

5. "Civil Rights Issue in Election," *U.S. News & World Report*, 20 February 1948, 22; Donald R. McCoy and Richard Ruetten, *Quest and Response* (Lawrence, Kan.: University of Kansas Press, 1973), pp. 69, 73–74, 79–96, 99–100.

6. Donovan, *Conflict and Crisis*, pp. 390–91.

7. Ibid., p. 391.

8. Ibid., p. 411.

9. Barton J. Bernstein, "The Ambiguous Legacy: Civil Rights," *Politics and Policies of the Truman Administration*, ed. Barton J. Bernstein (Chicago: Quadrangle Books, 1970), pp. 285–89; McCoy and Ruetten, *Quest and Response*, pp. 101–12; Irwin Ross, *The Loneliest Campaign* (New York: New American Library, 1968), pp. 120–26; Margaret Truman, *Harry S. Truman* (New York: Morrow, 1973), p. 392; James R. Roebuck, Jr., "Virginia in the Election of 1948" (M.A. thesis, University of Virginia, 1969), p. 60.

10. Ross, *The Loneliest Campaign*, p. 23; Philip H. Vaughn, *The Truman Administration's Legacy for Black America* (Reseda, Calif.: Mojave Books, 1976), p. 34.

11. Richard M. Dalfiume, "The 'Forgotten Years' of the Negro Revolution," *The Black Americans*, ed. Seth M. Scheiner and Tilden G. Edelstein (New York: Holt, Rinehart and Winston, 1971), pp. 421–36; Henry Lee Moon, *Balance of Power* (Garden City, N.Y.: Doubleday, 1948), pp. 10, 31, 197–200.

12. John L. Clark, "Dixie Would Put Truman on Spot on Civil Rights," Pittsburgh *Courier*, 2 October 1948.

13. Interview with Dr. G. James Fleming, political scientist, Baltimore, Md., 20 February 1979.

14. McCoy and Ruetten, *Quest and Response*, p. 138.

15. Interview with Dr. G. James Fleming.

16. Interviews with Henry E. and Jeanne Perkins Harman, Hastie's friends, Valdosta, Ga., 21 August 1977.

17. WH's interview (by Hess) for Truman Library.

18. Ibid.; interview with George L-P Weaver, consultant, Organization for Rehabilitation and Change, Washington, D.C., 13 July 1978.

19. McCoy and Ruetten, *Quest and Response*, pp. 139–41; William L. Dawson's cable to WH, 28 July 1948; WH's cable to William L. Dawson, 4 August 1948; George H. Curtis to the author, 9 June 1982, enclosing a copy of Stephen J. Spingarn's memorandum, "Judge William H. Hastie," n.d. (Harry S. Truman Library).

20. Spingarn's memorandum (see note 19).

21. "Itinerary of Governor William H. Hastie, Campaign Tour for the Re-Election of President Truman."

22. Drew Pearson, "Thurmond Invites a House Guest," Washington *Post*, 13 October 1948; Ted Poston, "Dixiecrat Thurmond Invites Negro Gov. Hastie to Visit Him;" "Invitation to Negro! Thurmond Didn't Do It," New York *Post Home News*, 14 October 1948.

23. Poston, "Dixiecrat Thurmond Invites Negro Gov. Hastie to Visit Him."

24. Roebuck, "Virginia in the Election of 1948," p. 60; interview with Dr. James R. Roebuck, Jr., assistant professor of history, Drexel University, Philadelphia, Pa., 24 July 1981.

25. "Virgin Is. Head Scores Wallace," New York *Daily News*, 14 October 1948; John P. Davis's memorandum to WH, 13 October 1948; "Itinerary of Governor William H. Hastie, Campaign Tour for the Re-election of President Truman"; "Governor of Virgin Islands Commends Truman Stand," Cleveland *Plain Dealer*, 18 October 1948.

26. "Governor of Virgin Islands Commends Truman Stand"; WH's interview (by Hess), for Truman Library.

27. Harvard Sitkoff, *A New Deal for Blacks* (New York: Oxford University Press, 1978), pp. 52–54; Raymold Wolters, *Negroes and the Great Depression* (Westport, Conn.: Greenwood Press, 1970), pp. 3–79.

28. Cited in *In Their Own Words*, ed. Milton Meltzer (New York: Crowell, 1967), pp. 127, 129.

29. "Governor of Virgin Islands Commends Truman Stand"; Sitokff, *A New Deal for Blacks*, p. 44; cited in *In Their Own Words*, p. 127.

30. "Itinerary of Governor William H. Hastie . . ."; "Negroes' Support Urged for Truman"; interview with Harry E. Harman.

31. "Hastie Impressed on Arrival Here by Truman Rally," Harrisburg (Pa.) *Evening News*, 22 October 1948.

32. Interview with George L-P Weaver.

33. Truman, *Harry S. Truman*, p. 392.

34. President Harry S. Truman's address to the NAACP, Washington, D.C., 29 June 1947.

35. WH's handwritten notes; "Governor of Virgin Islands Commends Truman Stand"; "GOP Declared Using Victory Propaganda," Buffalo *Courier-Express*, 1 November 1948; Theodore M. Berry to WH, 9 November 1948.

36. Interview with Judge Howard E. Bell, Civil Court of New York City, held at New Orleans, La., 3 August 1977.

37. McCoy and Ruetten, *Quest and Response*, pp. 140–41, 144–47.

38. Paul Shearer, "Facts About Ohio's Role in Elections," *Ohio State News*, 13 November 1948.

39. Margaret Truman Daniel to the author, 20 August 1979; WH to Jaime Benitez, 5 November 1948.

40. Interview with Mr. Justice Thurgood Marshall, Supreme Court of the United States, Washington, D.C., 17 February 1977.

41. Genevieve Blatt to WH, 15 November 1948; William B. Mahoney to WH, 29 November 1948; Congressman Roy J. Madden to WH, 23 November 1948; Theodore M. Berry to WH, 9 November 1948.

42. Arthur Krock, "President Can Claim Big National Triumph," New York Times, 7 November 1948; William C. Berman, *The Politics of Civil Rights in the Truman Administration* (Columbus, Ohio: Ohio State University Press, 1970), p. 129; Henry Lee Moon, "What Chance for Civil Rights?" *The Crisis* (February 1949), 42; McCoy and Ruetten, *Quest and Response*, p. 144; Shearer, "Facts About Ohio's Role in Elections"; Edward T. Clayton, *The Negro Politician* (Chicago: Johnson, 1964), pp. 75–76.

43. John O. Holly to WH, 3 December 1948; Governor Frank J. Lausche to WH, 29 December 1948.

44. WH's interview (by Hess) for the Truman Library.

45. President Harry S. Truman to WH, 9 and 13 November 1948; WH to President Harry S. Truman, 10 November 1948.

46. WH's interview (by Hess) for the Truman Library.

47. Interviews: Harry E. and Jeanne P. Harman; Alphonso A. Christian, Sr., attorney, St. Thomas, V.I., 23 July 1977; Dr. Roy A. and Vivian Anduze, physician and wife, St. Thomas, V.I., 21 July 1977; James A. O'Bryan, educator and activist, St. Thomas, V.I., 4 February 1978; and Aubrey C. Ottley, postmaster, St. Thomas, V.I., 16 July 1977.

48. Interview with Ariel A. Melchior, Sr., newspaper publisher and editor, St. Thomas, V.I., 22 July 1977.

49. Interview with Alphonso A. Christian, Sr.

50. Ibid.; interview with Dr. Roy A. and Vivian Anduze.

51. Interview with Dr. Roy A. Anduze.

52. Interview with Isidor Paiewonsky, businessman, St. Thomas, V.I., 20 July 1977.

53. Interview with Alphonso A. Christian, Sr.

54. Interviews: Dr. Roy A. and Vivian Anduze; Aubrey C. Ottley; Hon. Edward R. Dudley, administrative judge, Supreme Court of the State of New York, New York City, 10 January 1978; Eldra Shulterbrandt, educator, St. Thomas, V.I., 23 July 1977; Darwin D. Creque, *The U.S. Virgins and the Eastern Caribbean* (Philadelphia: Whitmore, 1968), pp. 126–29; WH's interview (by Hess) for the Truman Library.

55. Frank I. Weler, "V.I. Governor Views Post as Honor to His Race" (MSRC); Spottswood W. Robinson III, "William Henry Hastie—The Lawyer," 125 *University of Pennsylvania Law Review* 8n, (1976); interview with Eldra Shulterbrandt.

56. Interview with Isidor Paiewonsky.

CHAPTER 16. THE GREATEST NEED

1. WH's interview (by Jerry N. Hess) for the Harry S. Truman Library, 5 January 1972.

2. Interviews with Judge A. Leon Higginbotham, Jr., and Chief Judge Collins J. Seitz, both of the United States Court of Appeals for the Third Circuit,

Philadelphia, 30 April 1980 and 1 September 1976, respectively; "Patronage Kings Ponder Colored Judge Prospects," Norfolk *Journal and Guide*, 1 October 1949; "Truman Names Hastie to Circuit Bench," Philadelphia *Inquirer*, 16 October 1949; John Francis William to Walter White, 18 October 1945 (NAACP Papers); Mark Hyman, "Philadelphia's Bonfire of Opposition Against Hastie Gains Momentum," n.d., and Claude A. Barnett, "Hastie Gets Second Judgeship Appointment . . . ," 17 October 1949 (mimeographed press release).

3. Interviews with Carlyle M. Tucker, attorney, Philadelphia, Pa., 9 February 1978, and Judge Harvey N. Schmidt, Common Pleas Court, Philadelphia, Pa., 7 November 1977; Hyman, "Philadelphia Bonfire of Opposition Against Hastie Gains Momentum"; Joseph V. Baker, "Myers' Backing of Hastie for Court Stirs Row"; Barnett, "Hastie Gets Second Judgeship Appointment . . ."; "Hastie's Name Up for Appeals Bench," Washington *Afro-American*, 11 August 1945; "Hastie's Failure to Get U.S. Court Post Analyzed," Washington *Afro-American*, 22 September 1945; "Dawson Emerges as Political Leader," Pittsburgh *Courier*, 10 November 1945.

4. Editorial, *The Crisis* (October 1945), 281; Judge Henry W. Edgerton to WH, 25 October 1949; Herbert R. Cain, Jr., to WH, 25 August 1949; Arthur C. Thomas to WH, 6 October 1949.

5. Baker "Meyers' Backing of Hastie for Court Stirs Row"; "Myers Denies he Is Backing Hastie," Philadelphia *Afro-American*, 13 September 1949; editorial, Philadelphia *Afro-American*, 13 September 1949; "Hastie Opposed for Judgeship," St. Thomas *Daily News*, 7 October 1949.

6. Lem Graves, Jr., "Washington Notebook . . ." Pittsburgh *Courier*, 29 October 1949; Louis R. Lautier to the Editor, Pittsburgh *Courier*, 24 September 1949.

7. WH to Louis R. Lautier, 10 October 1949; WH to Herbert R. Cain, Jr., 30 September 1949; "Truman Names First Negro to Appeals Court," Chicago *Tribune*, 16 October 1949.

8. John A. Davis to WH, 19 October 1949; Harold L. Ickes to WH, 28 October 1949, and Hastie's reply, 25 November 1949; Governor Ernest Gruening to WH, 21 October 1949; George Slaff to WH, 26 October 1949; William T. McKnight to WH, 19 October 1949; Pauli Murray to WH, 24 October 1949, and WH's reply, 28 October 1949.

9. Russell Porter, "11 Communists Convicted of Plot . . ." New York *Times*, 15 October 1949; Benjamin J. Davis, Jr., *Communist Councilman from Harlem* (New York: International Publishers, 1969), p. 181.

10. "The Judiciary" *Time* (24 October 1949), 24–25; WH's address to the National Bar Association, Atlanta, Ga., 5 August 1971.

11. Correspondence in Howard University Law School (HULS) collection; Porter, "11 Communists Convicted of Plot . . ."; Davis, *Communist Councilman from Harlem*, p. 181.

12. "Truman Blasts Wallace Gang," Los Angeles *Times*, 15 July 1949; Dorothy B. Gilliam, *Paul Robeson* (Washington, D.C.: New Republic, 1976), p. 126.

13. "NAACP Secretary Puts Nation First," Los Angeles *Times*, 13 July 1949; "Truman Blasts Wallace Gang," Los Angeles *Times*, 13 July 1949.

14. Gilliam, *Paul Robeson*, pp. 140–43.

15. Ibid., pp. 137–38, 141; Wilson Record, *The Negro and the Communist Party* (New York: Atheneum, 1971), p. 313; "Need of FEPC Law Stressed at Rally," Los Angeles *Times*, 16 July 1949; "Civil Rights Law Pushed by NAACP."

16. Alan Barth, *The Loyalty of Free Men* (New York: Viking, 1951), pp. 8–11,

51–55; David Caute, *The Great Fear* (New York: Simon and Schuster, 1978), pp. 491–93, 505.

17. Joseph C. Goulden, *The Best Years 1945–1950* (New York: Atheneum, 1976), pp. 296–97.

18. Lem Graves, Jr., "Leaders Question Cause of Loyalty Probe Within Race," Pittsburgh *Courier*, 23 July 1949; Gilliam, *Paul Robeson*, p. 140.

19. Benjamin Quarles, *The Negro In the Making of America* (New York: Colliers, 1966), pp. 206–7; Barth, *The Loyalty of Free Men*, p. 7; Caute, *The Great Fear*, pp. 15, 18, 282–83; Alger Hiss, "How McCarthyism Silenced America," *Barrister* (Fall 1980), 53; I. F. Stone, *The Truman Era* (New York: Random House, 1973), p. ix.

20. Robert J. Donovan, *Conflict and Crisis: The Presidency of Harry S. Truman, 1945–1948* (New York: Norton, 1977), pp. 292–95.

21. Ibid., pp. 295–98; Athan Theoharis, "The Escalation of the Loyalty Program," *Politics and Policies of the Truman Administration*, ed. Barton J. Bernstein (New York: New Viewpoints, 1974), pp. 244–46.

22. "Items Concerning the Position of William H. Hastie with Reference to the Communist Party and Communist Ideology," n.d.

23. WH's address at the NAACP Convention, Los Angeles, Calif., 13 July 1949 (NAACP Papers).

24. George Streator, "Negro Group Hints White May Resign," New York *Times*, 17 July 1949.

25. Ibid.; WH's commencement address at Central State College, Wilberforce, Ohio, 7 June 1959.

26. Herbert F. Goodrich to WH, 16 October 1949.

27. WH to Constance Daniel, 27 July 1950.

28. "Truman Names First Negro to Appeals Court"; "Senate Body Defers Court Nominations," New York *Times*, 18 October 1949; "Re: Nomination of Honorable William Henry Hastie to be Judge of the United States Court of Appeals for the Third Circuit," n.d. (National Archives).

29. Harold L. Ickes to WH, 24 July 1950.

30. WH to Senator Pat McCarran, 7 April 1950; Senator William Langer to WH, 10 April 1950. Chief Judge John Biggs, Jr., to the following: WH, 8 June 1950, enclosing a copy of his letter to E. J. Fox, Jr., 8 June 1950; Attorney General Peyton Ford, 10 June 1950; and WH, 27 April 1950. Francis Biddle to Honorable John Biggs, Jr., 25 April 1950; interview with Bernard G. Segal, attorney, Philadelphia, Pa., 30 November 1977.

31. Judge Herbert F. Goodrich to Honorable Forrest C. Donnell, 11 April 1950.

32. Judge Albert B. Maris to Honorable Forrest C. Donnell, 10 April 1950.

33. Francis Biddle to Chief Judge John Biggs, Jr., and to Honorable Arthur T. Vanderbilt, both on 25 April 1950; Chief Judge John Biggs, Jr., to Honorable Pat McCarran, 8 April 1950; WH to Honorable Owen J. Roberts, 26 July 1950.

34. Judge Francis E. Rivers to WH, 9 May 1950, enclosing a copy of his letter to Senator Forrest C. Donnell, 21 April 1950; WH to Judge Francis E. Rivers, 16 May 1950.

35. Judge John J. Parker to Senator Pat McCarran, 16 May 1950, enclosed in Parker's letter to WH, 4 November 1950; Leslie Perry's memorandum to WH, 24 May 1950.

36. Leslie Perry's memorandum to WH, 26 May 1950.

37. Clarence Mitchell's memorandum to WH, 23 May 1950, enclosing a copy of Roy Wilkins's letter to William O. Walker, 22 May 1950.

38. WH to Theodore Poston, 1 June 1950; WH to Walter White, 9 June 1950 (NAACP Papers); Walter White to WH, 14 June 1950 (National Archives); WH to Alfred Baker Lewis, 30 June 1950.

39. U.S. Congress, Senate, Committee on the Judiciary, *Hearings Before the Subcommittee of the Committee on the Judiciary, On Confirmation of the Nomination of Honorable William Henry Hastie of the Virgin Islands, To Be Judge of the United States Court of Appeals for the Third Circuit*, Vol. 1 (1950), pp. 5–6, 9, 12.

40. "New York Jurist Accused to Senate," New York *Times*, 2 April 1950.

41. *Hearings*, Vol. 1, p. 16.

42. Ibid., Vol. 2, pp. 16–17.

43. Ibid., Vol. 2, pp. 55–58, 141–43.

44. Ibid., Vol. 4, p. 167.

45. Ibid., Vol. 3, pp. 83–87, 107–8.

46. Ibid., Vol. 3, pp. 114–15.

47. Ibid., Vol. 3, p. 133.

48. Ibid., Vol. 3, p. 134.

49. Ibid., Vol. 4, pp. 179–84.

50. Ibid., Vol. 2, pp. 18–19.

51. Ibid., Vol. 2, pp. 17, 22; Vol. 4, pp. 188, 191, 228, 237.

52. Louis Lautier, "Hastie Denies 'Smear' Charges," Kansas City (Mo.) *Call*, 23 September 1949; editorial, Kansas City (Mo.) *Call*; "Hastie May Go on Bench" and "Item Concerning the Position of William H. Hastie with Reference to the Communist Party and Communist Ideology."

53. Florence Murray, "National Lawyers Guild, in Liberal Move, Opens its Doors to Negro Barristers," Washington (D.C.) *Tribune*, 27 February 1937; interview with Morris L. Forer, attorney, Philadelphia, Pa., 20 May 1980.

54. Interview wth Mr. Justice Thurgood Marshall, Supreme Court of the United States, Washington, D.C., 17 February 1977.

55. Roy Wilkins to WH, 28 June 1950; Roy Wilkins to Leon Washington, 2 August 1949; Walter White to WH, 28 June 1950.

56. WH, "Persons and Affairs," *New Negro Opinion*, 28 April 1934.

57. WH to Walter White, 9 June 1950.

58. Walter White, "Judge Hastie's Confirmation by Congress Long Overdue," a release from Graphic Syndicate, New York City, to subscribing newspapers, on 1 June 1950, enclosed in WH's letter to Walter White, 9 June 1950.

59. Joseph V. Baker, "Delay on Judge Hastie's Approval Protested to Truman," Philadelphia *Inquirer*, 19 April 1950; Marquis Childs, "Hastie Files Gather Dust," New York *Post*, 10 May 1950.

60. Childs, "Hastie Files Gather Dust"; "McCarran Denies Action on Judgeships Is Held Up," Washington *Evening Star*, 23 June 1950; Jack Steele, "Hastie Queried by Senators on Loyalty"; Doris May Harris to Senator Patrick McCarran, 14 July 1950; interview with Judge Doris May Harris, Common Pleas Court, Philadelphia, Pa., 21 November 1981; George L-P Weaver to All Civil Rights Committee and Staff Members, 5 July 1950; *Hearings*, Vol. 4, p. 184.

61. Philip Murray to Majority Leader Scott Lucas, 30 June 1950, and Lucas's reply, 4 July 1950.

62. "CIO Asks Senate Judiciary Committee Recommend Confirmation of William Hastie," CIO press release, 13 July 1950.

63. "Senators Vote 9-1 to Confirm Hastie as Judge"; H. Thomas Austern to the author, 19 September 1980; "Negro Judge Confirmed," New York *Times*, 20 July 1950; Childs, "Hastie Files Gather Dust."

64. E. Lewis Ferrell to WH, 12 June 1950.

65. George M. Johnson to WH, 16 February 1951.

66. WH to Chief Judge John Biggs, Jr., 24 October 1949.

67. WH to Honorable Earl Chudoff, n.d.

68. Langston Hughes, "The Need for Heroes," *The Crisis* (June 1941), 185.

69. Stephen Spender, "I Think Continually of Those Who Were Truly Great," *Rights* (February 1965), 2.

EPILOGUE

1. WH remarks at Independence Hall, Philadelphia, 5 April 1976.

2. Interviews: Judge Spottswood W. Robinson III, United States Court of Appeals, Washington, D.C., 27 January 1977; Wesley S. Williams, Jr., and Karen Hastie Williams, attorneys, Washington, D.C., 4 March 1978; Dr. Cyril A. Riley, physician, Philadelphia, Pa., 26 September 1976.

3. Interview with Judge Charles E. Wyzanski, Jr., United States District Court, Cambridge, Mass., 23 May 1977.

4. WH to the Editor, *Color Magazine*, 14 October 1953, enclosing corrected pages of Dorothy Anderson's manuscript.

BIBLIOGRAPHICAL ESSAY

My primary source of information was the collection of Judge William H. Hastie's papers now housed at Harvard Law School and cited with permission. A modest but instructive collection is at the Beck Cultural Exchange Center in Knoxville, Tennessee. I use no symbols to refer to the collection at Harvard, but the collection at Beck I designate by the letters BCEC. As identified in the notes, other helpful sources include papers that I researched at the Howard University Law School (HULS), which have since been deposited at Howard University Archives; the Moorland-Spingarn Research Center (MSRC) at Howard University; Enid BAA Public Library in Charlotte Amalie; Yale University; Amherst College; the National Archives; and the NAACP Papers at the Library of Congress.

I see no point in listing in a bibliography books and other sources that are included in the notes. There are, of course, exceptions, and in this instance they include: Barton J. Bernstein, ed., *Politics and Policies of the Truman Administration* (New York: New Viewpoints, 1974); Ulysses Lee, *The Employment of Negro Troops* (Washington, D.C.: U.S. Government Printing Office, 1966); Gordon K. Lewis, *The Virgin Islands* (Evanston, Ill.: Northwestern University Press, 1972); Alan M. Osur, *Blacks in the Army Air Forces During World War II* (Washington, D.C.: U.S. Government Printing Office, 1977); Benjamin Quarles, *The Negro in the Making of America* (New York: Colliers Books, 1966); Harvard Sitkoff, *A New Deal for Blacks* (New York: Oxford University Press, 1978); and Walter White, *A Man Called White* (New York: Viking, 1948).

Articles that merit mention are William H. Hastie, "Toward an Equalitarian Legal Order 1930–1950," *Annals of the American Academy of Political and Social Science* (May 1973), 18–31; Beverly Smith, "The

First Negro Governor," *Saturday Evening Post*, 17 April 1948, pp. 15–17, 151, 153–57; and Robert C. Weaver, "William Henry Hastie, 1904–1976," *The Crisis* (October 1976), 267–70. Weaver's is one among many articles in *The Crisis* that were invaluable in researching American life and law in the thirties and forties. Similarly indispensable was the coverage provided by black newspapers, particularly the Pittsburgh *Courier*, Washington *Afro-American*, Baltimore *Afro-American*, New York *Amsterdam News*, Washington *Tribune*, Philadelphia *Tribune*, Norfolk (Va.) *Journal and Guide*, and in the Virgin Islands, the *Daily News*, *St Crois Avis*, and *Photo News*.

My faith in oral history as an exceptionally rewarding research technique was confirmed by the enrichment of my research by numerous interviewees, including Alton A. Adams, Sr.; Judge Arlin M. Adams; Lee Adams; Ruth Adams; Gustave Amsterdam; Dr. Roy A. Anduze; Vivian Anduze; Dorothy Berkowitz; Thomas H. Austern; Enid Baa; Ben Bayne; Irene Bayne; Howard Beagles; Judge Howard E. Bell; Dr. Harriet F. Berger; Theodore M. Berry; Chief Judge John Biggs, Jr.; Dr. Alfred W. Blatter; Professor William W. Boyer, Jr.; Edith Bornn; Roy Bornn; James A. Bough; Wiley A. Branton; Mr. Justice William J. Brennan, Jr.; E. Leonard Brewer; John H. Britton, Jr.; Omar Brown; Sterling A. Brown; Chief Judge William B. Bryant; George O. Butler; Judge Herbert E. Cain, Jr.; Kathryn S. Campbell; Edgar C. Campbell; Dr. Thomas L. Canavan; Dr. Philip V. Cannistraro; Dr. George D. Cannon; Margaret Carson; Judge Robert L. Carter; Julius LeVonne Chambers; Ruth Harvey Charity; Bertha Childs; Elwood H. Chisholm; Judge Alphonso A. Christian, Sr.; Dr. Kenneth B. Clark; Eugene H. Clarke, Jr.; Dr. W. Montague Cobb; Hon William T. Coleman, Jr.; Dr. W. Mercer Cook; Carl A. Cowan; Professor Archibald Cox; Darwin D. Creque; Dr. John Aubrey Davis; Dr. John W. Davis; Mrs. Herman J. Daves; Judge Edward R. Dudley; George H. T. Dudley, Jr.; Charles T. Duncan; Mr. and Mrs. Boyd Eatmon; James C. Evans; Professor Clarence C. Ferguson; Dutton Ferguson; H. Naylor Fitzhugh; Dr. G. James Fleming; Hazel H. Fleming; Judge Phillip Forman; Professor Paul A. Freund; Margaret A. Gaiter; Lloyd K. Garrison; Judge Leonard I. Garth; Professor Walter Gellhorn; Adolph A. Gereau; Dr. Millard Gladfelter; Frances O. Grant; Jack Greenberg; Geraldo Guirty; James V. Hackney; Louise Hall; Harry E. and Jeanne P. Harman; Conrad K. Harper; Judge Doris M. Harris; William H. Hastie, Jr.; Fritz Henle; Judge A. Leon Higginbotham, Jr.; Senator Roger C. Hill; Oliver W. Hill; Mrs. Thomas R. Hocutt; Judge Louis Hoffman; Joseph Hopkins; Clotille Houston; Judge Joseph C. Howard, Sr.; Mary G. Hundley; John J. Johnson, Jr.; Judge Billy Jones; Carolyn G. Jones; James D. Jones; Judge Nathaniel R. Jones; Judge Harry E. Kalodner; Paul E. Konney; Lee H.

Kozol, Jr.; Senator Frits Laewetz; Archibald T. LeCesne; Bettie C. Lee; Judge George N. Leighton; June A. V. Lindqvist; H. T. Lockard; Herbert Lockhart, Jr.; Professor Louis Loss; Bernice Loss; Edna Love; Dorothy Lutz; Conard J. Lynn; Judge Albert B. Maris; Burke Marshall; Mr. Justice Thurgood Marshall; Dr. Benjamin E. Mays; Robert Grayson McGuire, Jr.; Eleanor McGuire; Georgianna McGuire; Robert J. McKean, Jr.; Dr. Genna Rae McNeil; Ariel Melchior, Sr.; Ariel Melchior, Jr.; Oliver B. Merrill; Lorna Milgram; Morris Milgram; Clarence M. Mitchell, Jr.; Juanita Jackson Mitchell; Judge Constance B. Motley; Dr. Pauli Murray; Dr. James M. Nabrit, Jr.; Leon S. Nance; Phileo Nash; Chief Judge Theodore R. Newman, Jr.; James A. O'Bryan; Aubrey C. Ottley; Michele F. Pacifico; Isidor Paiewonsky; Judge James B. Parsons; Conrad O. Pearson; Dorothy Fegg; Judge Eileen Petersen; Basil Phillips; Raymond Plaskett; Forrest C. Pogue; Judge Louis H. Pollak; Judge Lawrence W. Prattis; Dr. Benjamin Quarles; Ruth Ramsey; Charles B. Ray; Louis L. Redding; Helen Reid; Professor Herbert O. Reid; Mrs. J. E. Richard; Eugene Richardson, Jr.; Dr. Cyril A. Riley; A. A. Roberts; Lenerte Roberts; Chief Judge Spottswood W. Robinson III; Dr. James R. Roebuck; Dr. James F. Rogers; Dean Albert Sachs; Judge Harvey N. Schmidt; Professor Louis B. Schwartz; Professor Austin W. Scott; John P. Scott; Sarah S. Scott; Alma Scurlock; Chief Judge Collins J. Seitz; Bernard G. Segal; Judge W. Eugene Sharpe; Louis and Eldra L. M. Shulterbrandt; Ruby Simmonds; George Slaff; Alfred Edgar Smith; Lula Smith; Dr. J. Clay Smith, Jr.; Margaret Smith; Professor Ralph R. Smith; Brenda S. Spears; John F. Stewart; Judge Juanita Kidd Stout; Dr. and Mrs. Wilbur Strickland; Judge and Mrs. William S. Thompson; William J. Trent, Jr.; Carlyle M. Tucker; Charles Tyson; Judge Francis L. Van Dusen; Bertha S. Vasquez; Dr. Marvin Wachman; Etta Childs Walker; Dr. Caroline F. Ware; Dr. Bernard C. Watson; Dedmond Watson; George L-P Weaver; Dr. Robert C. Weaver; Judge Joseph F. Weiss, Jr.; Roy Wilkins; Jeane B. Williams; Dr. Michael R. Winston; Wendell Wray; Judge Charles Wright; Giles R. Wright; Judge Charles E. Wyzanski, Jr.; Marian W. P. Yankauer.

Index

Ackiss, Thelma D., 75–76
Adams, E. S., 102
Agricultural Adjustment Administration (AAA), 26–27: policies, effects on blacks, 218
Air Corps: blacks in, 124–32; segregation in, 126; integration, 129–30
Alexander, Raymond Pace, 225
Alexander, William, 218
Allen, A. T., 57–58
Allen, Macon B., 29
Alston, Melvin O., 62–63
Alston case, 65
American Bar Association (ABA): exclusion of blacks, 29, 144; rejection of Hastie, 145; endorsement of Hastie, for Third Circuit Court, 233, 236
American Civil Liberties Union (ACLU), 43
American Federation of Labor (AFL), discriminatory practices, 38
American Fund for Public Service (ASPS), 43
American Missionary Association (AMA), 3
Americans for Democratic Action (ADA), 238
American Tennis Association (ATA), 26
Amherst College: first black graduate, 12; fraternities, 14–15; black students' entertainments at, 15–16; blacks on track team, 17–18; Hastie's academic excellence in, 18–20; treatment of black students, 12–13
Amherst Writing, racist articles, 13–14
Anduze, Roy A., 91, 210, 211, 222

Anduze, Vivian, 23, 222
Arens, Richard, 236, 238
Army: black officers, 110–11; racism in, 152. *See also* Military
Army Scottsboro Case, 135–41
Arnold, Henry H., 126, 130
Associated Negro Press, 101
Association of American Railroads, 117
Austin, Louis E., 46

Baa, Enid, 91
Baldwin, Roger N., 43, 45, 69, 157–58
Ballard, Lulu, 27
Barnett, Claude A., 101
Barnhill, M. W., 49–53
Barristers Club of Philadelphia, 240
Bass, Jack, 184
Bell, Derrick A., Jr., 32
Bell, George, 6
Bell, Howard E., 220
Bethune, Mary McLeod, 23, 103, 162
Biddle, Francis, 112–13, 116, 175, 183–84, 231, 235
Biggs, John, 234–35, 239
Black, Aline Elizabeth, 62
Black, Hugo, 143
Black militants, Hastie's observations about, 11
Black MPs, 119–20
Black officers, racism encountered by, 110–12
Black press, 122: and military racism, 100–101, 120–23; sabotage of, by military, 121

297

Blacks: in military, 95–106; loyalty to war effort, in WW II, 105; recruitment into the Army, 121–22; attitude toward Truman, 213
Black soldiers: murders of, 112–14, 115–16; assaults on, 116; forced segregation, 117–19; disorders, 119; violence against, Hastie on, 141
Black suffrage, 176–78
Blair, Lawrence K., 16
Blood bank: military, segregation in, 107–9; segregation, NAACP protest against, 194
Bok, Derek A., 30
Bordentown Manual Training School (BMTS). See Old Ironsides
Bornn, Roy, 84, 90
Bough, James A., 87, 211
Boyer, James W., 102
Branson, Lewis, 116
Brennan, William J., Jr., 30
Brewer, Earl, 159
Brewster, Ralph O., 193–95, 235–36
Brotherhood of Sleeping Car Porters (BSCP), 38
Brown, Omar, 206, 210, 223; attack on Hastie, 209
Brown, Sterling A., 16, 19, 26
Brown, v. Board of Education, 53–54, 189–90
Brummitt, Dennis G., 49, 50–51, 57
Bryant, William B., 146–47
Bryden, William, 114–15
Bullard, Robert Lee, 95
Bunche, Ralph J., 34, 67–68, 177, 225, 227, 229
Bureau of Intelligence, 121
Burns, Howard F., 234, 236

Cain, Herbert R., Jr., 226
Cambridge, Godfrey, 144
Canaday, Ward M., 202–3
Caniero, Francisco, 151
Cannon, George D., 202, 228
Capital Press Club, 163
Carmody, John J., 236
Carr, Raymond, murder of, 112
Carter, Hodding, 130
Carter, Robert L., 151–53, 184
Chapman, Oscar L., 165–66
Charity, Ruth Harvey, 156
Chestnut, W. Calvin, 61
Childs, Samuel, 3
Chisholm, Elwood H., 170–71

Christian, Alphonso A., Sr., 86, 91, 93, 207, 222–23
Citizens' Committee Against Segregation in Recreation (CCASR), 165
Civil Aeronautics Authority (CAA), training of black pilots, 124–25
Civilian Conservation Corps (CCC), Hastie's challenge of, 89–90
Civilian Pilot Training (CPT), 124–25
Clark, Tom, 231
Clarke, Eugene H., Jr., 153–54, 155, 157
Clifford, Clark M., 214–15
Cobb, James A., 169
Cobb, W. Montague, 7–8, 10, 12–13, 15–19
Coe, Rose Marcus, 37
Coleman, William T., Jr., 33, 133, 189
Communist National Textile Workers, 38
Communists, 38, 56, 67, 70, 73, 196–97: in 1948 election, 214; and blacks, 230–31; subversion trial, 1949, 228; and NAACP, 232–33, 239; Wallace's party and, 217; Hastie's views on, 239
Conference on Civil Liberties (1934), 69
Congress of Industrial Organizations (CIO), 38, 166, 240
Cook, John F., 7
Cook, Mercer, 8, 13, 15–16, 18–19, 20, 210–11
Cook, Will Mercer, 13
Coolidge, John, 15
Cramer, Lawrence W., 85
Creque, Darwin D., 87
Crisis, The, 70–72, 107, 233; ban on, 72–73
Cummings, Homer S., 85–86
Cummings, Thomas L., 172
Curry, Donald A., 114–15

Daniel, Margaret Truman, 220
Darden, Colgate W., 161–62
Daves, Mrs. J. Herman, 5
Davis, Benjamin J., Jr., 13, 15–17, 30, 196, 228
Davis, Benjamin O., Jr., 100, 132
Davis, Benjamin O., Sr., 97, 101
Davis, John Aubrey, 66–68, 78–80, 227
Davis, John Preston, 34, 36–38, 40, 218
Davis, Ossie, 144
Dawson, William L., 216, 219, 226
DeCastro, Morris F., 198–99
Department of Interior: Hastie appointed as assistant solicitor, 81; racism at, 81–82
Department of Justice. See Justice Department
Dewey, Thomas E., 214, 221; Hastie's criticism of, 218–20

Dickerson, Earl B., 166
District of Columbia. *See* Washington, D.C.
District of Columbia Bar Association library, closed to blacks, 147
Dixiecrats, 220
Dodson, Thurmond L., 78
Donnell, Forrest C., 234-37, 240
Donovan, Richard, 112-14
Donovan, Robert J., 213, 231
Dorsch, Frederick, 196-97
Double V protest, 104-5
Dowling, Monroe, 131
Dowling, Carlos A., 87, 205; attack on Hastie, 209
Drew, Charles R., 8, 10, 13, 15, 17-18, 34, 108, 227
Dubinsky, David, 38
Du Bois, W.E.B., 23-25, 131; on fight against segregation, 70-72
Dudley, Edward R., 206-8
Dunbar High School, 6-12, 19-20
Duncan, Charles T., 53, 173
Durham, W. J., 179, 180

Early, Stephen, 97
Eastland, James O., 236, 238, 240; opposition to Hastie, 193-94
Edgerton, Henry W., 226
Education, for blacks, differing views, 24-25
Edwards, Judy, 135
Einstein, Albert, 23
Eisenhower, Dwight D., 106, 120, 132
Ellender, Allen J., 193-96
Engels, Robert L., 136-40
Ernst, Morris L., 43
Evans, James C., 132

Fahy Charles, 81, 183
Fairchild, Arthur S., 202
Fair Employment Practices Commission, 215
Fair Employment Practices Committee (FEPC), 166
Federal Public Housing Authority, racism in, 168
Ferguson, Dutton, 74-75
Ferguson, Homer, 235
Finnerty, John, 161
Fisher, Frank, Jr., 135-41
Fisk University, 3
Fitzhugh, H. Naylor, 68, 74-75, 80
Fleming, G. James, 216
Foreman, Clark, 81

Fortas, Abe, 165-66
Fort Bragg, 119
Foster, Thomas B., murder of, 115-16
Francis, Rothschild, 83
Frankfurter, Felix, 28, 31, 44, 48, 81, 98, 129, 149, 182
Franklin, Nicholas, 6
Federick, J. D., 112-14
Freund, Paul A., 227
Fuller, Solomon, 23

Gannett, Lewis S., 43
Garland, Charles, 43
Garland, Judy, 230
Garland, Fund, 43-45
Garnet Elementary School, 7
Garrison, Lloyd K., 95, 149, 150-51, 237
Gay, Walter E., 225
Gellhorn, Walter, 14
Georgetown, black education in, 6-7
Germans, in Virgin Islands, 91-92
Gibbs, William B., Jr., 60
Gibson, Truman K., Jr., 98, 131
Goodrich, Herbert F., 233-35
Gordon, Roy P., 87, 90, 196, 198, 205, 208; attack on Hastie, 209
Graham, Frank P., 52
Granger, Lester B., 23
Grant, Frances O., 23, 27
Graves, Lemuel, 204, 226-27, 230
Green, Arnold B., 167-68
Greene, Norma, 116
Gregory, James M., 21
Griswold, Erwin N., 149-50, 182, 227-28
Grovey, R. R., 177-78
Grovey v. Townsend, 177-82
Gruening, Ernest, 227
Guirty, Geraldo, 91
Gulliou, Allen W., 119
Gurfein, Murray I., 188

Haley, Alexander P., 134-35; defense of naval policy, 134
Hall, Felix, 115
Hall, Louise Russell, 4
Hamburger Grill, 67
Hanst, K. D., 112
Harlen-Terrell Lawyers' Association, 148
Harman, Harry E., 216
Harris, Abram, 39
Harris, Oswald E., 87, 208; attack on Hastie, 209
Harris, Patricia Roberts, 160

Harry, George Winston, 12
Harvard Club, 33
Harvard Law Review, 30–31, 143
Harvard Law School: Hastie at, 27–34; achievement ethic, 31; Hastie's academic excellence at, 31; anti-Semitism in, 33; racism in, 33; camaraderie of blacks at, 34; exclusion of women, 149–51; open to women, 227
Harwood, Charles, 192
Hasgett v. Werner, 179
Hastie, Alma Syphax, 86, 92
Hastie, Beryl (Lockhart), 194, 199, 222–23
Hastie, Karen, 197–98, 210
Hastie, Roberta (Childs), 3, 10–11, 86, 194: racial pride, 4; devotion to son, 4–6; character, 6; education, 6–8; tutoring of son, 7, 9; in New Negro Alliance, 68; inspirational character, 211
Hastie, William Henry, 31–32: birth, 4; relationship with parents, 4–6; resistance to racism, 4; white playmates, 4; athleticism, 10, 16–18; observation about militancy, 11; president of college Phi Beta Kappa, 18; elected to Phi Beta Kappa, 19; faculty appointment at Old Ironsides, 21; personality, 27, 170–71; as tennis player, 27; on editorial board of *Harvard Law Review*, 30–31; plan to become an engineer, 34; as NAACP's representative, 36; on racism in labor unions, 38; doctorate in juridicial science, 48; participation in *Hocutt*, 48–53; on issue of teachers' pay, 55–56; participation in *Mills*, 61–62; participation in *Alston*, 62–65; U.S. District Court judge for Virgin Islands, 85–94; work for black combat pilots, 124–32; resignation from War Department, 130–32; involvement in New Caledonia rape case, 136–41; attitude toward law students, 148–49, 154, 156; attitude toward law, 154; on blacks' battle for racial equality, 158–59; compared to Joe Louis, 170–71; in fight for black voting rights, 179–85; importance to civil rights cases, 185, 190; racial encounter in train dining car, 185; skill at logical argument, 188–90; appointment as Governor of the Virgin Islands, confirmation hearing, 192–97; political affiliations, 193, 197; description of his family and its influence on him, 194–95; accusations against, of communism, 196–97; inauguration as governor of Virgin Islands, 197–99; relationship with Turman, 204, 213–14; relationship with Virgin Islands legislators, 207–10, 222–24; support of Truman in 1948 election, 216–24; on United States Court of Appeals, 225; opinion of red-baiting, 231–32; confirmation to United States Court of Appeals, 233; characterization of "subversive" government officials, 232, 237–38; views about communism, 239
Hastie, William Henry, Jr., 11, 210–11
Hastie, William Henry (father), 3–4
Hawkins, Annette, 7
Hay, A. J., 115–16
Hayes, Arthur P., 153
Hayes, George E. C., 165, 169
Henderson, E. B., 9, 165
Hewin, J. Thomas, Jr., 63, 171–72
Hibbard Report, 201
Hill, Arnold, 97
Hill, Oliver W., 45, 63–64, 154–55, 171–72, 187
Hill, Valdemar A., Sr., 87, 90
Himes, Chester, 106
Hine, Darlene Clark, 182
Hiss, Alger, 32, 231
Hocutt case, 47–53, 65
Hocutt, Thomas R., 46–53
Holly, John O., 221
Holmes, Talley, 27
Hoover, Herbert C., 83
Hoover, J. Edgar, 236
Hopkins, Harry, 162
Horne, Lena, 100, 132
Horst Wessel, 91–93
Householder, E. R., 101–2
House Un-American Activities Committee (HUAC), 196, 229–30, 238, 239
Housing: restrictive covenants, 167, 169; segregated, Hastie's fight against, 167–70
Houston, Charles Hamilton, 5, 12, 29–30, 32, 34, 41, 44–45, 48, 69, 81, 85, 145–46, 148, 153–54, 157–58, 188: relationship with Hastie, 142; character, 143; education, 143; endorsement of Hugo Black, 143; hatred for racism, 143; military experience, 143
Houston, William L., 69, 81, 86: "Cousin Plug," 5; law firm, 148
Howard, Perry W., 146, 148
Howard University, 45, 91: Hastie's teaching at, 48; atmosphere, 49; commitment to public service, 160

INDEX

Howard University Law School: Hastie as dean, 93–95, 142, 144; Hastie at, 133–34; Houston's views on, 144; faculty-student relations, 155; effect on students, 156–57; faculty, 171–172, Hastie in civil rights alliance, 190–91
Huff, William Henry, 182
Hughes, Langston, 109, 241
Hughes, W. A. C., Jr., 62
Huntt, Leslie F., 196–97

Ickes, Harold L., 81, 85–86, 192, 217, 227, 234
Interior Department, practice of segregation, 165–66
International Labor Defense (ILD), 37, 56, 58, 136; and Scottsboro case, 70
International Ladies Garment Workers Union (ILGWU), 38

Jackson, Hamilton, 83
Jackson, Robert H., 147, 182
James Walker Tennis Club, 26
Jamison, Jackie, 26
Jarvis, J. Antonio, 199
Jefferson Barracks, 129, 132
Johnson, Campbell C., 97
Johnson, James Weldon, 23, 43–44
Johnson, John J., 6
Johnson, Mordecai W., 48–49
Joint Committee on National Recovery, 36, 218
Jones, Billy, 142, 153, 160, 173
Jones, Sam Houston, 112–13
Julian, Percy, 227
Justice Department: black lawyers in, 146; reluctance to encorce black voting rights, 183–84

Kempton, Murray, 161
Kilgore, Harlan M., 233–34
Killens, John Oliver, 103
King, Albert, murder of, 113–14
King, William H., 86
Kluger, Richard, 49, 55
Knox, Frank, 97, 134
Krug, Julius A., 193, 195, 197, 216

Labor unions, racism by, 38–39
Langer, William, 233–34, 238
Langston Law Club, 226
Lausche, Frank J., 221
Lautier, Louis, 131, 193, 226
Lawetz, Frits, 207

Lawson, Belford V., Jr., 66, 68, 74–76, 78
Lawyers: black, 29–30, 45; white, Houston's attitude toward, 45; black, in Judge Advocate General's Department, 111; black, Houston's views on, 143–44; black, disregarded in works about American life, 144; black, difficulties faced by, 145–47; white, use urged by Howard University professor, 147; black, southern practice of "sponsorship" by white lawyers, 173; black, opposition to Hastie's appointment to Third Circuit Court of Appeals, 225–27
League for Non-Violent Disobedience against Military Segregation, 215
LeCesne, Archibald T., 134, 156
Letts, F. Dickerson, 75–76
Lewis, Gordon K., 83, 205, 207
Lewis, John P., 151
Lincoln Normal Institute, 3
Lindqvist, Elsa A., 92–93
Lockhart, Beryl, 90–91
Lockhart, Herbert E., Sr., 90–91
Logan, Rayford W., 164
Looby, Z. Alexander, 172, 173–74
Loury, Edward R., Jr., 135–41
Lovett, Edward P., 74, 86
Lovett, Robert A., 126, 128–29, 130
Loyalty investigations, 231–32, 237
Loyalty Review Board, 231
Lucas, Scott, 240
Lynchings, 47, 175–76, 177, 228–29

MacLeish, Archibald, 105, 106
Magee, James C., 107
Magens Bay, acquisition of, 202, 223
Marcantonio, Vito, 136–41
March, Frederick, 230
Margold, Nathan R., 44, 81, 178; nomination to appellate bench, 234
Maris, Albert B., 235
Marshall, George C., 99–101, 103
Marshall, James, 182–83, 195
Marshall, Louis, 182–83
Marshall, Thurgood, 32, 45–46, 54–55, 60, 62–65, 86, 97, 122, 130, 154, 157, 169–71, 175, 177–80, 182–184, 190, 220, 239; risks life in defending blacks, 173–74
Maryland, teachers' salary litigation, 60–62
Mass, David E., 208
Matheny-Zorn, Otto, 18
Matthews, Ralph, 199
Mays, Clara I., 168–70
Mays v. Burgess, 226

McCarran, Patrick A., 233-40
McCarthy, Joseph R., 233
McCloy, John J., 120-21
McCoy, Cecil A., 46-53, 57
McDuffy, Clyde, 8
McGohey, John F. X., 228
McGrath, J. Howard, 216
McKean, Robert J., Jr., 20
McKeon, Newton F., 16, 18
McKinney, Ernest, 8
McLaughry, D. O. "Tuss," 17
McReynolds, James C., 188
Medalie, George Z., 189
Medina, Harold R., 228
Mehlinger, James Edward, 145
Mehlinger, Louis R., 145-46
Meiklejohn, Alexander, 14, 18
Melchior, Ariel A., Sr., 199, 222
Menard, Willis, 10
Merrill, Oliver B., 13, 16
Meyer, Agnes E., 167
Military: racism in, 120, 213, 215; integration, 132-33
Millard, Stephen H., 15
Miller, Kelly, 147
Mills, Walter, 61
Mills case, 65
Miner, Myrtilla, 7
Ming, William R., 155
Mollison, Irvin C., 226
Moon, Henry Lee, 24, 215
Moore, Herman E., 198, 208
Moorhead, John S., 89
Morgan, Irene, 185-90
Morgan v. Virginia, 186-90
Morris, Robert, 144
Mother, role of, in inspiring great men, 10-11
Motley, Constance Baker, 184-85
Murdock, Abe, 193, 196
Murray, Pauli, 142, 149-50, 153, 157, 160, 170, 187, 227; defense of Odell Waller, 160-62
Murray, Philip, 38, 240
Myers, Francis J., 225-26

Nabrit, James M., Jr., 155
National Association for the Advancement of Colored People (NAACP), 165, 194, 226: all-out war against racism, 35; Amenia Conference, 35, 38; and Joint Committee on National Recovery, 35; economic program, 36, 38-39; Advisory Committee on Economic Activities, 39; Committee on Future Plan and Program, 39; and NNC, 41; campaign against school segregation, 42-43; Hastie's work for, 42; Garland Fund grant, 43-44; and Scottsboro case, 70; views on discrimination in public transportation, 117-18; handling court-martial convictions, 153; Washington Branch, Hastie's work for, 164-65; fight for voting rights, 178; litigation, Hastie's indispensability in, 190; importance to civil rights, 191; Truman's address to, 219; and communists, 233; leftists in, 233
National Bar Association (NBA), 29, 144-45
National Capital Housing Authority (NCHA), discriminatory practices, 167-68
National Conference on the Economic Status of the Negro (1935), 39-40
National Housing Agency, racism in, 168
National Lawyers Guild, 193, 236-37, 238-39, 240
National Negro Congress (NNC), 40-41, 193-94; Hastie's affiliation with, 238
National Press Club, Hastie refused service at, 163
National Recovery Administration (NRA), blacks' monitoring, 36-37
National Urban League, 9
Navy, in Virgin Islands, 83-84. See also Military
Negro Industrial League, 36
Negro Soldier, The, 122
Newberry v. United States, 177
Newbold, Nathan C., 47, 57-58
New Deal: for blacks, 37, 39-40, 218; Hastie on, 41-42
New Negro Alliance, 66-80, 194: Hastie's roles in, 68-70; Hastie as administrator for, 72-74; Hastie as attorney, 74; negotiations with A&P, 74; vs. Sanitary Grocery Company, 74, 77-80; vs. High Ice Cream Company, 75; vs. Kaufman's Department Store, 75-76
New Negro Opinion, 68, 75, 77
Nice, Harry W., 61
1925 Olio, The, 15; appraisal of Hastie, 19
94th Engineer Battalion, incidents with, 114-15
99th Fighter Squadron, 125, 129
Nixon, L. A., 177
Norris-La Guardia Act of 1932, 75-76, 78-80

INDEX

North Carolina: racial attitudes in, 49–50; salary discrimination in, 55
North Carolina College, 47
North Carolina Negro Teachers' Association, 58–59

Oberlin College, 13
O'Bryan, James A., 200
Office of Defense Transportation, 118
Officer Candidate Schools (OCS), blacks in, 110–11, 129
O'Grady, John, 236
Old Ironsides, 21–28: students, 22–23; teachers, 23; Biscuits (tennis players), 26–27; Muffins (tennis players), 26
Opportunity, ban on, 73
Organic Act of 1936, 83–87, 208
Osur, Alan M., 132
Ottley, Aubrey C., 87–88
Ottley, Earle B., 209; Progressive Guide's attack on, 204–5

Paiewonsky, Isidor, 223, 224
Parker, John J., 63–65, 235
Parks, Gordon, 100, 135
Patterson, Robert P., 97–98, 103, 115, 118, 122, 130–31, 139, 195
Pearson, Conrad O., 46–53, 57–60, 159–60
Pearson, Drew, 217
Pearson, Paul, 90
Perry, Leslie, 235–36
Pinkett, A. S., 73
Plaskett, Raymond, 210–11
Plessy v. Ferguson, 42–44, 54, 117, 189
Pollak, Louis H., 190
Poole, Rufus G., 86
Poston, Ted, 100
Potter, Henry, 6
Pound, Roscoe, 31
Powell, Adam Clayton, Jr., 104, 134
Powell, Ruth, 187
Powell, T. R., 149–50
Price, James H., 161
Primaries: denial of black participation in, 177–83
Proudfoot, Maud, 92–93
Prout, John, 7

Racism: at Amherst College, 12–13; in military, 110–23; fighting, Hastie on, 158–59; fighting, blacks' hindrance of, 159–60; fighting, whites' aid with, 159
Railroads, discrimination in employment, 166

Randolph, A. Philip, 38, 40, 97. 162, 166, 215
Ransom, Leon A., 62–63, 86, 155, 157, 165, 169, 172–173, 187
Rape charges, against blacks, 175–76
Ray, Charles, 23
Reagan, Ronald, 230
Redding, Louis L., 30, 34
Red-Tailed Angels, 133
Reed, Stanley F., 182, 237
Reid, Helen, 27
Resolute Beneficial Society, 6
Rice, Reverend Walter A., 21, 25
Richards, Henry V., 87
Rivers, Francis E., 235
Roberts, Owen J., 179, 181
Robeson, Paul, 23, 229–30
Robinson, Jackie, 229
Robinson, Spottswood W., III, 154, 169–70, 185–86, 188, 189, 190
Roosevelt, Eleanor, 23, 162
Roosevelt, Franklin Delano, 37, 93–94, 97, 105, 134–35, 178, 183; "do nothing" policies toward blacks, 166
Roper, Bill, 17
Rowe, James, 98, 214–15
Ruffin, George Lewis, 30
Rutledge, Wiley E., 189

Sanitary Grocery Company, 74, 77–80
Saxon, John F., 137
Scott, Stuart N., 227
Scottsboro case, 37, 56, 70, 194
Scurlock, Alma, 91
Seams, C. O. "Mother," 27
Seavey, Warren A., 96
Segal, Bernard G., 239–40
Segregation: in education, NAACP attack on, 42–43; Hastie on, 42, 71; in 1933, 42; Hastie's dispute with Du Bois about, 70–72; in military, 95–97; in railroads, 117; in D.C. schools, 164; in interstate commerce, fight against, 186–90
Selective Service Act of 1940, 96–97, 125
Separate-but-equal doctrine, 42–44, 54, 189
Sharecropper Union, 38
Shepard, James E., 47–48
Shulterbrandt, Louis, 199
Sinatra, Frank, 230
Slaves: Hastie's ancestors, 3; in Virgin Islands, 93
Smith, A. Calvert, 151
Smith, Lonnie E., 177; voting rights case, 179–84
Smith, Walter Beddell, 130

Smith v. Allwright, 190
Smothers, Henry, 7
Soper, Morris, 64
Southern Conference for Human Welfare, 194
Spaulding, C. C., 47–48, 58
Spears, Brenda S., 174
St. Thomas *Daily News*, 90–91
Staples, Abram P., 187
Stassen, Harold, 237
Stevens, Hope R., 196
Stimson, Henry L., 96–98, 113, 116, 129, 130, 136; and military racism, 122–23
Stone, Harlan F., 161, 182
Stone, I. F., 163
Storey, Moorfield, 182–83
Stratemeyer, George E., 127, 130
Sumner, Charles, 144
Supreme Court, as classroom for Howard law students, 157
Surles, Alexander D., 100, 103, 122
"Sweetheart of All My Dreams," 13
Syphax, William, 7

Taft, Robert A., 220
Talented Tenth, 24
Talladega College, 3
Teachers' pay, NAACP campaign for equality in, 55–65
Temporary Committee for Improved Public Transportation (TCIPT), 166
Texas, white primaries, fight against, 177–83
Thomas, Arthur C., 226
Thomas, Nevall H., 9, 20
Thompson, Ted, 27
Thompson, William S., 147
Thorne, M. Franklin, 66
Thurmond, Strom, 215, 217, 230
Tinsley, J. M., 52
Tobias, Channing, 183
Toll, Maynard J., 31–32
Totten, Ashley L., 196
Townsend, Vincent M., Jr., 156–57
Trent, William, 89–90
Trotman, James, 27
Truman, Harry S., 83–84, 192, 225, 240: attitude toward civil rights, 167; visit to Virgin Islands, 202–4; in 1948 election, courting of blacks, 213–16; trouble with blacks, 213; ordered end to discrimination in civil service and armed forces, 215; and delegation against lynching, 229; and Paul Robeson, 229

Tucker, Carlyle, 225
Tugwell, Rexford Guy, 193
Turner, Mary, 176
Tuskegee Airmen, 132–33; combat record, 133
Tuskegee Institute, 24; pilot training program, 125–27, 132

Ulio, James A., 120
Union Seminary, 7
United Citrus Workers, 38
United Mine Workers (UMW), 38
United States Court of Appeals: Third Circuit, Hastie appointed to, 225–26; Third Circuit, opposition to Hastie's appointment, 225–27, 233–34; for District of Columbia, Hastie not appointed to, 226
United States v. *Classic*, 178–83
Universal Military Training (UMT), 215
University of North Carolina: desegregation suit against, 46–53

Valentine, William R., 22–23
Vanderbilt, Arthur T., 235
Virginia, teachers' salary case, 60, 62–65
Virgin Islands, 83, 93: Colonial Law of 1906, 84; Legislative Assembly, 85; salary discrimination, 87–88; alien problem, 88; racial discrimination, 90; slave rebellions, 93; Hastie as Governor, 192–212; tourism, 195–96, 203; economic situation encountered by Governor Hastie, 198, 200, 201; rum tax, return to islands, Hastie's proposal for, 198, 200, 201; political divisiveness, 199–200, 204–5; political situation encountered by Governor Hastie, 199–200; beauty, 200; poverty, 200–201; bigotry in, 201–2; native legislators, political immaturity, 201–2; economic development, under Governor Hastie, 202–3; Truman's visit to, 202–4; colonial status, 205; self-government, Hastie's efforts for, 206, 209–10; Legislative Assembly, relationship with Governor Hastie, 207–10, 222–24; Police Commission appointments, Hastie's objection, 208; politics, Hastie's views on, 211–12; progress, under Governor Hastie, 223
Virgin Islands Progressive Guide, 87–90, 204–5: Hastie's disputes with, 204, 208–10; political corruption, 206; showdown with Hastie, 223
von Scholton, Peter, 93

INDEX

Voter registration, black efforts toward, 172, 176–77

Waddy, Joseph C., 146
Walker, Etta Childs, 4
Wallace, Henry A., 213–14, 229; Hastie's opinion of, 217–18
Waller, Odell, 160–62
Ward, James, 74
War Department: Hastie's work in, 95–109; opposition to Hastie in, 103–7; Bureau of Public Relations, and military racism, 120–22; Hastie's resignation, cited in Senate confirmation hearings, 195
Washington, Booker T., 23–25
Washington, D.C.: black education in, 6–7; 1919 race riot, 9; blacks in, in 1933, 66; racism in, 163–66, 170; board of education, racist policy, challenged by Hastie, 164–65; Board of Recreation, segregation of recreational facilities, 165; segregated housing, 167
Washington, Ora, 27
Washington Bar Association, 145; support of National Negro Congress, 148
Washington Committee for the Southern Conference on Human Welfare, 193
Washington Council for Community Planning, 193
Washington Real Estate Board: racism in, 168

Weaver, George L-P, 219
Weaver, Robert C., 5–6, 9, 12, 19, 34, 36, 81, 97
Weaver, Tony, 135
Wesley, Carter W., 179
Wesley, Charles, 179–80
White, Walter, 36–39, 41, 44–45, 47–48, 52–53, 55–60, 64, 69, 95, 97, 118, 122, 126, 143, 178, 201, 216, 225, 232–33, 236, 237, 239, 240; assessment of Hastie, 62
Wiley, Alexander, 235, 236
Wilkins, Roy, 36, 38, 39, 101, 229, 232–33, 236, 239
Wilkinson, Garnet, 9, 72–73
Williams, Croxton, 206–7
Williams, Wesley, 147
Williams College, 16
Williston, Samuel, 30
Winston, Dr. Michael R., 9
Wolters, Raymond, 39
Women: in defense program, 103; banned from Harvard Law School, 149–51
Women's Army Auxiliary Corps, blacks in, 106, 110–11
Women's suffrage, in Virgin Islands, 85–87
Work, Fred, 23
Workers Defense League, 161
Wright, Cleo, 175–76
Wyzanski, Charles E., Jr., 33, 242

Young, P. Bernard, Jr., 204